Springer Series in Optical Sciences   Volume 37

Edited by Theodor Tamir

Springer Series in Optical Sciences

Editorial Board: J.M. Enoch D.L. MacAdam A.L. Schawlow K. Shimoda T. Tamir

Volume 42 **Principles of Phase Conjugation**
By B. Ya. Zel'dovich, N. F. Pilipetsky, and V. V. Shkunov

Volume 43 **X-Ray Microscopy**
Editors: G. Schmahl and D. Rudolph

Volume 44 **Introduction to Laser Physics**
By K. Shimoda

Volume 45 **Scanning Electron Microscopy**
Physics of Image Formation and Microanalysis
By L. Reimer

Volume 46 **Holography and Deformation Analysis**
By W. Schumann, J.-P. Zürcher, and D. Cuche

Volume 47 **Tunable Solid State Lasers**
Editors: P. Hammerling, A. B. Budgor, and A. Pinto

Volume 48 **Integrated Optics**
Editors: H.-P. Nolting and R. Ulrich

Volume 49 **Laser Spectroscopy VII**
Editors: T. W. Hänsch and Y. R. Shen

Volume 50 **Laser-Induced Dynamic Gratings**
By H. J. Eichler, P. Günter, and D. W. Pohl

Volume 51 **Tunable Solid State Lasers for Remote Sensing**
Editors: R. L. Byer, E. K. Gustafson, and R. Trebino

Volumes 1–41 are listed on the back inside cover

V. P. Zharov   V. S. Letokhov

Laser Optoacoustic Spectroscopy

With 95 Figures

Springer-Verlag Berlin Heidelberg GmbH

Dr. VLADIMIR P. ZHAROV

Moscow High Technical School, Department of Optoelectronics
SU-107005 Moscow, USSR

Professor Dr. VLADILEN S. LETOKHOV

Institute of Spectroscopy, Academy of Sciences of the USSR
SU-142092 Moscow Academgorodok, USSR

ISBN 978-3-662-14479-4   ISBN 978-3-540-39492-1 (eBook)
DOI 10.1007/978-3-540-39492-1

Library of Congress Cataloging in Publication Data. Zharov, V.P. Laser optoacoustic spectroscopy. (Springer series in optical sciences ; v. 37) Bibliography: p. Includes index. 1. Optoacoustic spectroscopy. 2. Laser spectroscopy. I. Letokhov, V. S. (Vladilen S.) II. Title. III. Series. QD96.O6L48 1986   535.8'4   86-1768

2153/3130-543210

Preface

The optoacoustic method has by now an almost one-century-long history of application in spectroscopy, but it was only with the advent of the laser that it became a convenient and effective method among the vast family of spectroscopy techniques. The great variety of these techniques is capable of tackling most diversified tasks, such as the achievement of a high sensitivity and a high spectral or temporal resolution. The optoacoustic method is one of the simplest and most versatile ways to attain a high sensitivity for both gaseous and condensed media. It is precisely for this reason that the method has found wide use, and that we have decided to publish a monograph reviewing the information on this method available in the literature and gathered by us at the Institute of Spectroscopy during the past few years. We hope that such a systematic exposition of the material scattered throughout numerous scientific journals will be of use to many potential readers. The reader will undoubtedly notice the absence in our monograph of references to some recent works, but unfortunately, this is inevitable when the translation and publication of a book in a foreign language takes several years. Nevertheless, we tried our best to cover the entire field from the material available to us, but unfortunately, some recent publications might be missing due to the time lag for the translation and publication in a language foreign to us.

We would like to express our gratitude to Springer-Verlag and to Dr. H.K.V. Lotsch for the cooperation in publishing this book in English and thank Mrs. I. Blass for preparing the camera-ready manuscript. We shall look forward to any comments the reader might care to make regarding this monograph.

December 1985 *V.P. Zharov V.S. Letokhov*

Contents

Part 1   Optoacoustic Techniques

1. Introduction 2
   1.1 Historical Review 2
       1.1.1 Optoacoustic Effects in Gases 2
       1.1.2 Optoacoustic Effects in Condensed Media 4
       1.1.3 Other Literature 5
   1.2 Optoacoustic Methods in Laser Spectroscopy 6
       1.2.1 Basic Problems in Laser Spectroscopy 6
       1.2.2 Classification of Laser Spectroscopy Methods 7
       1.2.3 Comparison of Various Methods 9
   1.3 Classification of Calorimetric Methods 11
       1.3.1 Types of Calorimetric Methods. Time of Relaxations ... 11
       1.3.2 Optothermal Method 13
       1.3.3 Refraction Methods 14
       1.3.4 Optoacoustic Method 15

2. Optoacoustic Gas Spectroscopy with Laser Sources 16
   2.1 Laser Optoacoustic Devices 16
   2.2 Formation of Optoacoustic Signal 18
       2.2.1 Calculation of IR-Radiation Absorption at Vibrational-
             Rotational Transitions 19
       2.2.2 Calculation of Optoacoustic Signals in a Nonresonant
             Spectrophone 24
   2.3 Nonresonant Spectrophones 28
       2.3.1 Acoustic Pickups.(Microphones) 29
       2.3.2 Optimization of Spectrophone Parameters 39

3. Optoacoustic Spectroscopy of Condensed Media 45
   3.1 Optoacoustic Spectroscopy of Liquids with Direct Detection ... 45
       3.1.1 Formation of Optoacoustic Signals 46
       3.1.2 Liquid Measuring Cells 51
   3.2 Optoacoustic Spectroscopy of Solids 56
       3.2.1 Formation of Optoacoustic Signal 56
       3.2.2 Characteristics of Solid Measuring Cells 58
   3.3 Optoacoustic Spectroscopy of Condensed Media with Signal
       Detection in a Gas Medium (Photoacoustic Spectroscopy) 61
       3.3.1 Fundamentals 62
       3.3.2 Measuring Cells 67
   3.4 Optoacoustic Spectroscopy of Solids in Liquids 70
   3.5 Comparison of Various Methods 72

4. Optothermal Spectroscopy 74
   4.1 Optothermal Spectroscopy of Gases 74

    4.1.1 The Optothermal Effect 74
    4.1.2 Types of Thermal Pickups 75
    4.1.3 Pulsed Regime 77
    4.1.4 Continuous Regime 80
  4.2 Optothermal Spectroscopy of Condensed Media 81
    4.2.1 Laser Measurement of Weakly Absorbing Media 82
    4.2.2 Analysis of Highly Absorbing Media 83

5. Principles of Laser Optoacoustic Instruments 84
  5.1 Basic Sources of Background Signals 84
    5.1.1 Background Signals from Windows 85
    5.1.2 Background Signal from Scattered Radiation 86
    5.1.3 Background Signal in the Sample 87
    5.1.4 Comparison of Background Signals 88
  5.2 Elimination of Background Signal 89
    5.2.1 Modifications of Cell Design 89
    5.2.2 Special Measurement Conditions 91
    5.2.3 Compensating Measuring Schemes 93
  5.3 Resonant Cells 96
    5.3.1 Resonant Spectrophone 96
    5.3.2 Resonant Cells for Optoacoustic Spectroscopy with
           Indirect Detection 102
  5.4 Spectrophones with Spatial Resolution 104
  5.5 Cells with an Enlarged Temperature Range 107
  5.6 Other Cell Designs 110
    5.6.1 Cells with Electric and Magnetic Fields 110
    5.6.2 Gas-Flow Spectrophones 111
    5.6.3 Optoacoustic Cells for Aggressive Gas and in Vacuum ... 113
    5.6.4 Combined OA Cells 115
  5.7 Measuring Techniques of Optoacoustic Instruments 115
    5.7.1 Generalized Scheme 115
    5.7.2 Optical Schemes 116
    5.7.3 Electronic Schemes 119

6. Analytical Characteristics of Optoacoustic Instruments 122
  6.1 Definition of Characteristics 122
    6.1.1 Sensitivity 122
    6.1.2 Sensitivity Threshold or Detection Limit 123
    6.1.3 Accuracy of Measurement 125
    6.1.4 Resolution 125
    6.1.5 Information Capacity 125
    6.1.6 Dynamic Range 126
  6.2 Sensitivity Threshold 126
    6.2.1 Ultimate Sensitivity 126
    6.2.2 Analysis of Gaseous Media 127
    6.2.3 Analysis of Condensed Media 130
  6.3 Measurement Accuracy 131
    6.3.1 Static Errors 131
    6.3.2 Dynamic Errors 136
  6.4 Graduation and Calibration 137
    6.4.1 Graduation by Concentration 137
    6.4.2 Graduation by Absorption 138
    6.4.3 Graduation by Pressure 141
    6.4.4 Calibration 143

Part 2 Applications

7. Laser Optoacoustic Spectroscopy of Gas Media 148
   7.1 High-Resolution Spectroscopy of Weakly Absorbing Media 148
       7.1.1 Laser Optoacoustic Spectrometers 148
       7.1.2 Linear Spectroscopy 154
       7.1.3 Spectroscopy in Electric and Magnetic Fields 162
       7.1.4 Doppler-Free Spectroscopy 163
   7.2 Spectroscopy of Excited Molecular States 166
       7.2.1 Thermal Excitation of Vibrational States 167
       7.2.2 Laser Excitation of Vibrational States 168
   7.3 Nonlinear Absorption in Molecules 171
       7.3.1 Saturation Effect 171
       7.3.2 Multiphoton Absorption 173
   7.4 Optoacoustic Raman Spectroscopy 179
   7.5 Measurement of Relaxation Times 182
       7.5.1 Pulsed Method. Heating and Cooling of Gas 183
       7.5.2 Continuous Method with a Nonresonant Spectrophone 186
       7.5.3 Continuous Method with Resonant Spectrophones 188
   7.6 Chemical Reactions 190
       7.6.1 Photochemical Reactions 190
       7.6.2 Analysis of Reaction Products. Catalytic Reactions ... 193

8. Laser Optoacoustic Spectroscopy of Condensed Media 197
   8.1 Spectroscopy of Weakly Absorbing Media 197
       8.1.1 Liquids 197
       8.1.2 Cryogenic Solutions 200
       8.1.3 Solids 201
       8.1.4 Multiphoton Absorption 205
       8.1.5 Raman Scattering 207
   8.2 Surfaces 208
       8.2.1 Highly Absorbing Substances 208
       8.2.2 Analysis of Coatings 210
       8.2.3 Analysis of Metals 215
   8.3 Circular Dichroism 218
       8.3.1 Experimental Schemes 218
       8.3.2 Sensitivity of CD Method 220
       8.3.3 Comparison of Different Methods 221
   8.4 Photoactive Media 222
       8.4.1 Quantum Yield of Luminescence 223
       8.4.2 Photochemical Reactions 226
       8.4.3 Measurement of Photovoltaic Energy 227

9. Laser Optoacoustic Analytical Spectroscopy 229
   9.1 Analysis of Molecular Traces in Gases 229
       9.1.1 Laser Optoacoustic Gas Analyzers 229
       9.1.2 Concentration Sensitivity in Various Experiments 233
   9.2 Analysis of Impurities in Condensed Media 237
       9.2.1 Impurities in Liquids (Including Cryogenic Case) 237
       9.2.2 Impurities in Solids 239
       9.2.3 Impurities on Surfaces 240
   9.3 Isotopic Analysis of Molecules 241
       9.3.1 Isotopically Selective Detection of Molecules 242
       9.3.2 Measurement of Isotopic Ratios and Their Variations .. 243
   9.4 Analysis of Aerosols 246
       9.4.1 In the Gas Phase 246
       9.4.2 In Solid-Phase Sedimentations 249
       9.4.3 In Liquids 251

9.5 Selectivity of Optoacoustic Analysis 251
    9.5.1 Enhancement of Selectivity for Binary Mixtures 252
    9.5.2 Analysis of Multicomponent Mixtures with Etalon Spectra 256
    9.5.3 Optoacoustic Spectroscopy Combined with Chromatography 258

10. Laser Optoacoustic Microspectroscopy 265
10.1 Principles of Optoacoustic Microspectroscopy 265
    10.1.1 Optically Nonuniform Samples. "Optical" OA Microscopy . 266
    10.1.2 Thermally Nonuniform Samples. "Thermal" OA Microscopy . 268
    10.1.3 Acoustically Nonuniform Samples. "Acoustic" OA
         Microscopy 269
    10.1.4 Operation of Laser OA Microscopes 270
10.2 Two Dimensional Optoacoustic Microscopy 271
    10.2.1 Optoacoustic Microscopes with Indirect Detection 272
    10.2.2 Optoacoustic Microscope with Direct (Piezoelectric)
         Detection 275
10.3 Three-Dimensional Optoacoustic Microscopy 279
10.4 Comparison with Other Methods of Microscopy 281

11. Nonspectroscopic Applications of Optoacoustic Spectroscopy 285
11.1 Control and Stabilization of Laser Parameters 285
    11.1.1 Measurement and Control of Power 286
    11.1.2 Frequency Control and Stabilization 287
    11.1.3 Search for New Laser Lines 288
11.2 Measurement of Thermodynamic Parameters 289
    11.2.1 Gaseous Media 290
    11.2.2 Condensed Media 292
11.3 Thermal Properties of Solid Surfaces 293
    11.3.1 Surface Crystallinity 294
    11.3.2 Laser-Light Interaction with Sample 296
    11.3.3 Observation of Phase Transitions 297

12. Conclusion 299

References 301

Main Notation 321

Subject Index 325

Part 1 Optoacoustic Techniques

1. Introduction

Optoacoustic (OA) methods of measurements currently find growing applications in spectroscopy and in studying the interaction of light with substance. The OA method is based on the effect, discovered and described by BELL [1.1,2], TYNDALL [1.3], and ROENTGEN [1.4] around 1880, which consists in the formation of acoustic waves in a sample under irradiation by light modulated at an audio frequency. The origin of this effect is explained by nonradiative transitions transforming part of the absorbed radiation energy to the thermal energy of the medium, which under certain conditions gives rise to acoustic vibrations.

Bell called the device designed to study the OA effect photophone, since he developed it to create a new type of telephone communication. Later, when the OA effect combined with monochromatic light was used to study absorption spectra of gases and vapor, the measuring OA cell was termed a spectrophone. When detecting small concentrations of impurities in different media using powerful laser radiation, the OA cell has sometimes been called an OA detector.

1.1 Historical Review

1.1.1 Optoacoustic Effects in Gases

Soon after its discovery, the OA effect was almost forgotten until 1938, when VIENGEROV encountered it again in his experimental work [1.5]. He applied the OA effect to quantitative and qualitative analysis of gas mixtures. Subsequently, PFUND [1.6] and LUFT [1.7] developed this approach further. Such studies culminated in designing rather simple, sensitive, selective and commercially available devices called dispersionless infrared gas analyzers [1.8]. Later, further research demonstrated broad applications to solve other tasks, particularly to determine the lifetimes of excited vibrational molecular states [1.9,10], to measure the intensity of weak IR radiation [1.11], to study photochemical reactions [1.12], etc. The applicability of the OA method was demonstrated in the microwave [1.13], the ultraviolet [1.14] and in the visible [1.15] ranges. In 1956 the negative OA effect was revealed

2

[1.16] and used in the far-infrared experiments [1.17]. Basic results in physical-chemical measurements with noncoherent radiation sources were comprehensively studied in [1.18].

With the advent of lasers with their unique properties, the OA method acquired new development potential. Due to the high power of the monochromatic radiation it has become possible to increase the sensitivity of the method by several orders. The first use of the laser as a radiation source in an OA method dates back to 1968, when KERR and ATWOOD [1.19] used a pulsed tunable ruby laser to measure the absorption band contour of water vapor in room air with a good spectral resolution ($\Delta\lambda = 0.2$ Å) and absorption sensitivity ($\alpha_{min} \approx 3 \times 10^{-7} \text{cm}^{-1}$). Using a CW CO_2 laser, they measured an absorption factor of $\alpha_{min} = 1.2 \times 10^{-7} \text{cm}^{-1}$ in the $CO_2 + N_2$ mixture.

In 1971 KREUZER used the OA method to detect molecular impurities in gases [1.20]. With a He-Ne laser operating at $\lambda = 3.39$ μm and tuned within a small range by changing the axial magnetic field, he detected methane in nitrogen with a concentration up to 0.01 ppm. Later on, KREUZER and his colleagues demonstrated the high sensitivity of the method by detecting NO molecules in air using a "spin-flip" laser [1.21] and indicated the potential of the method or the analysis of impurities by employing CO and CO_2 lasers [1.22,23]. The detection of NO molecules in the stratosphere with a concentration of 10^8 mol / cm^3 by a spin-flip spectrometer in a stratostat was a brilliant demonstration of the practicabilities of optoacoustics [1.24]. The potentialities of OA for gas analysis and spectroscopic research have been demonstrated with pulsed HF and CO_2 lasers [1.25], a pressure- and current-tuned diode GaAs laser [1.26] and dye lasers [1.27].

At the same time the OA method was subjected to important modifications. In [1.28] a resonant-type absorption cell was used for the first time to detect low-concentration impurities in gases. The merit of this cell is that the effect of background noise due to absorption in its end windows can be reduced. To increase the sensitivity of resonant cells an optical scheme with multiple passage of the beam through the cell has been realized in [1.29]. A combination of an OA cell with a device for inducing magnetic or electric fields has enabled the Zeeman [1.30,31] and Stark [1.32] OA spectra of low-absorption gas media to be studied. Promising results for increasing selectivity of the OA analysis of complex mixtures have been obtained by KREUZER using a laser spectrophone combined with a gas chromatograph [1.33].

From about 1970, the acoustic method to detect weak absorption in gases has been developed by KRUPNOV et al. using coherent sources of millimeter and submillimeter radiation [1.34,35]. For this spectral range they have obtained maximum results on sensitivity and spectral resolution.

3

Lasers combined with OA methods have solved or opened ways for solving
quite new tasks of spectroscopy. High peak power of pulsed radiation, for example, together with a spectrophone with high spatial resolution has made it
possible to develop OA spectroscopy (OAS) of nonlinear and multiphoton absorption in focused laser beams [1.36,37]. The measuring technique of a spectrophone with variable gas temperature in its chamber [1.38] as well as the use
of two-frequency excitation [1.31,39] enable studies of excited vibrational
molecular states. The high concentration sensitivity of OA methods combined
with tunable IR lasers broadens the potentials of isotopic analysis of molecules, particularly for selective detection of low-abundance isotopes and
measurement of small variations of isotopic ratios for some elements [1.40].
Although suggested already in 1971 [1.41], nonlinear OA doppler-free spectroscopy has now been realized in [1.42]. Finally, a method for studying Raman
scattering based on OA detection of the energy deposited in the molecular
gas has been experimentally realized [1.43]. Thus, within the last decade
the use of lasers has widened essentially both the parameters and the areas
of application of OA methods to the study of gases.

1.1.2 Optoacoustic Effects in Condensed Media

From about 1973, the OA effect has been successfully used to investigate
condensed media. ROSENCWAIG developed the technique for detecting the energy
absorbed in a sample by measuring the acoustic vibrations in the gas embedding
the sample [1.44-47]. This approach in OA analysis has been called photoacoustic spectroscopy (PAS) by Rosencwaig. The prefix "photo" distinguishes this
approach of OAS with indirect detection from other optoacoustical methods.

Using noncoherent-radiation sources, Rosencwaig demonstrated the significant
advantages of such a technique for the spectral analysis of nongaseous substances, including such media as various powders, biological samples, etc.,
previously considered inconvenient for classical spectroscopy. Several types
of commercial OA spectrometers have been designed and are effectively used in
chemistry, biology, medicine and other fields of science and engineering.

Concurrently, the laser was used in developing OAS with indirect detection.
KERR [1.48] and PARKER [1.49] demonstrated its applicability for measuring surface absorption in crystals at the level of 10^{-2}%. In [1.50-52]
bulk and surface absorption were measured in transparent dielectric media
based on frequency and phase techniques of signal selection. Optoacoustic
spectroscopy was also used to analyze highly reflecting samples and to measure
oxide layers [1.53]. Focusing radiation onto the sample with simultaneous spa-

tial scanning (laser OA microscope) allowed the study of surface and subsurface structure of high-absorption samples [1.54].

Alongwith indirect detection of absorption in a sample using OA signals from the gas surrounding the sample, it is also possible to measure directly the acoustic vibrations in solids using a piezoelement in contact with the sample [1.55]. This method has been used to measure absorption at the level $10^{-4} - 10^{-5}$ cm^{-1}, of some optical materials at the HF and DF laser lines [1.56]. By additional measurement of the acoustic signal phase one can determine the thermodynamic characteristics of transparent samples [1.57].

It should be noted that in 1958 GROSS et al. could observe directly the OA effect both in solids [1.58] and in liquids [1.59], using noncoherent radiation sources. The sensitivity of the OA method in this case, however, was too low for analytical applications. The use of powerful pulsed dye lasers allowed the threshold sensitivity for liquids to be increased up to 10^{-6} cm^{-1} [1.60,61]. Because of this high sensitivity, the OA method could be used in liquids to study multiphoton absorption [1.62], measure quantum yields [1.63], investigate quenching processes [1.64], analyze weakly absorbing media [1.61], detect low-concentration impurities in liquids [1.65], study stimulated Raman scattering [1.66], etc. In some cases, when investigating absorption in solids, it has been found useful to simultaneously detect the acoustic waves in the liquid being in contact with the sample [1.67].

1.1.3 Other Literature

Even though the period of research is long enough and the volume of information obtained is great, the results of the OA method applied to laser spectroscopy have seemingly not been reviewed in the literature. DEWEY briefly considered the results of only the very early OA experiments in molecular laser spectroscopy [1.68]. ROSENGREN touched upon some technical points of laser spectrophone design [1.69]. A collection of papers edited by PAO [1.70] generalized the early works on OA using lasers. The progress achieved with coherent sources in the microwave range were reviewed in [1.35]. In his book [1.71], ROSENCWAIG mainly generalized the results of OAS with indirect detection of acoustic vibrations and with the use of noncoherent-radiation sources. The results of some original studies of OA with IR lasers were considered in [1.72]. Some interesting applications of the OA method in laser spectroscopy were described in [1.73], but the number of problems considered is somewhat limited.

1.2 Optoacoustic Methods in Laser Spectroscopy

The use of lasers as radiation sources in spectroscopic studies has resulted in essential modifications of conventional methods and in the development of new techniques.

1.2.1 Basic Problems in Laser Spectroscopy

Such unique properties of laser radiation as high power, directivity, monochromaticity together with recently developed means of frequency tuning have made it possible to solve or proceed to solve some important problems not solvable by classical spectroscopy. Among these are:

1) Reaching the resolution limit determined not by the instrumental width of the spectral device but by spectral line broadening of the substance. This is especially essential for gas spectroscopy in the infrared region where only with unique setups was it possible to attain resolution of 0.01 cm^{-1}; this is one hundred times higher than the Doppler width of spectral lines [1.74].

2) Eliminating the Doppler broadening in the spectroscopy of atoms and molecules in gas phases. Highly coherent laser light forms the basis of nonlinear laser spectroscopy enabling the study of the spectral line structure usually screened by Doppler broadening due to thermal motion of particles [1.75,76].

3) Reaching the sensitivity limit in a spectral analysis of atoms and, in the future, of molecules. With laser radiation it is now possible to detect single atoms and trace amounts of molecules [1.77,78].

4) Studying spectra and relaxation of atoms and molecules from excited states, i.e., kinetic spectroscopy of excited states. Laser radiation allows selective excitation of a considerable part of any atom and molecule to any quantum state, then their relaxation to the ground state can be observed [1.74,79].

5) Remote spectral analysis. With the use of collimated laser radiation it is possible to study Raman scattering, resonant absorption and emission of atoms and molecules as well as to excite fluorescence of atoms and molecules distant from the laser. In this way information on atomic and molecular composition of, say, the atmosphere can be obtained. These methods are particularly important at present in studies of the atmosphere and the control of environmental pollution [1.80,81].

6) Local spectral analysis. By focusing laser radiation it is possible to realize spectroscopy of trace amounts of substances localized in a small volume, in principle up to λ^3 (λ being the laser wavelength). This possibility

has been successfully realized in local emission spectral analysis [1.82]. Yet the potentialities of local spectral analysis extend much wider to medical and biological applications.

Apart from the special treatments mentioned above, fundamental techniques of laser spectroscopy and their merits are given in [1.83-85].

1.2.2 Classification of Laser Spectroscopy Methods

The methods of linear laser spectroscopy can be generally classified by analyzing the physical phenomena occurring in the interaction of the laser radiation with the medium under study. The primary effect in a medium under resonant laser radiation is excitation of the upper level of the resonant transition on account of absorption of photons with energy $h\nu = E_2 - E_1$ (Fig.1.1), E_2, E_1 being the energies of the upper and lower levels of the transition participating in absorption, ν the radiation frequency and h Planck's constant.

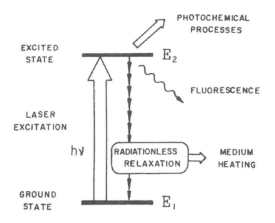

Fig. 1.1. Main channels of absorbed energy transformation in samples

The absorbed power in the sample, I_{abs}, is determined in accordance with

$$I_{abs} = I[1 - \exp(-\alpha L)] \simeq \alpha IL \quad , \tag{1.1}$$

with $\alpha L \ll 1$, where I is the radiation intensity at the sample input, and L is the sample length. The absorption coefficient per unit length, α, is determined by

$$\alpha = \sigma(N_1 - N_2) \quad , \tag{1.2}$$

where N_1 and N_2 are the densities of absorbing particles at the levels 1 and 2 of the resonant transition, and σ is the cross section of absorption. The populations N_1 and N_2 of the levels 1 and 2 can change under laser radiation on account of strong excitation from the ground state (saturation effect). The absorption cross section σ is constant for one-quantum transitions, but may essentially depend on intensity for multiquantum absorption. Here, in the introduction, we do not discuss all the various possibilities, leaving them for later chapters. We only consider the measurement of the absorption coefficient by OA methods.

Secondary phenomena in a medium are both radiative and nonradiative relaxation of excitation (Fig.1.1). Radiative relaxation causes the medium to reirradiate with an appropriate quantum yield η_r and a characteristic time τ_r (or several different times τ_r in the presence of several radiative relaxation channels). Nonradiative relaxation after all heats the medium or (and) the walls of the measuring cell. The quantum yield of nonradiative relaxation η_{nr} and the characteristic times of excitation relaxation to heat differ materially for gases, condensed media, the type of quantum levels under excitation, etc. In some cases account should be taken of the absorbed energy spent on the medium's photoactivation (photoionization, photodissociation, chemical reactions of excited particles, etc.) with an appropriate quantum yield η_{pa}.

The total quantum yield of all the channels of excitation relaxation equals unity, i.e.,

$$\eta_r + \eta_{nr} + \eta_{pa} = 1 \quad . \tag{1.3}$$

Accordingly, the intensity I_{abs} of laser radiation absorbed in a unit volume is distributed over all three possible channels

$$I_{abs} = I_r + I_{nr} + I_{pa} \quad , \tag{1.4}$$

where $I_r = \eta_r I_{abs}$, $I_{nr} = \eta_{nr} I_{abs}$, $I_{pa} = \eta_{pa} I_{abs}$ are absorbed power losses by reradition, heating of the medium or cell walls and photoactive processes, respectively.

According to the main channels of absorbed power transformation in a sample there are three basic techniques of laser spectroscopy (Fig.1.2): (a) the absorption method; (b) the radiative method; and (c) the calorimetric method. In the first technique information on sample properties and composition can be obtained from changing the parameters (intensity, polarization at a given frequency) of the laser radiation that passed through the sample or reflected from it. With radiative methods it is the light reradiated due to fluorescence or scattering that contains spectral information. And, finally, in the third technique the power absorbed in the sample is directly detected from changing the parameters (pressure, temperature, etc.) of the sample itself. In this

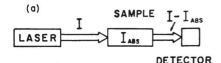

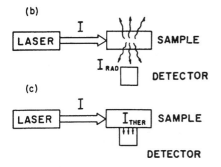

Fig. 1.2 a-c. Simplified schemes of the basic methods of linear laser spectroscopy: (**a**) absorption method; (**b**) fluorescence method; (**c**) calorimetric method

classification the OA method is calorimetric technique since the power absorbed in the sample is detected by thermal features.

1.2.3 Comparison of Various Methods

Methods based on absorption are more widely applicable since they allow analysis of substances in different aggregate states in various spectral ranges, and do not depend on the ratio between characteristic relaxation times. The method comparing the radiation intensity before and after entering a sample is the most popular one. Its ultimate sensitivity depends mainly on the optical path of light in the sample, and the ability of the detecting system to record small relative changes in intensity of the radiation passing through the sample. The laser beam has a very small divergence and therefore, to increase absorption, it can be transmitted through the sample many times. In most practical cases the threshold sensitivity of this method is limited by fluctuations of laser radiation due to technical problems which reduces the threshold of the minimum detectable power variation to levels $0.01 - 1\%$. This corresponds approximately to an absorption factor of $10^{-5} - 10^{-7} \mathrm{cm}^{-1}$ over an effective absorption length of up to $10^{3} \mathrm{cm}$ than can be realized under laboratory conditions with the use of multipass cells. Most measurements of absorption with lasers have been carried out in the IR region at molecular rotational or vibrational-rotational transitions. The absorption method is linear and so there are no particular requirements on the power level, allowing effective use of low-power diode lasers in high-resolution spectroscopy [1.74].

A modification of the absorption method is the technique of intracavity selective losses, whereby the absorption cell is placed inside the cavity of a multimode laser with a broad amplification line [1.86,87]. Due to the high sensitivity of independent modes to selective absorption losses the threshold sensitivity of the method turns out to be rather high, of the order of $10^{-8} cm^{-1}$. The theoretical threshold sensitivity determined by quantum fluctuations of the order of $10^{-11} cm^{-1}$. The application of this method is limited mainly by the visible and near-IR ranges for which appropriate lasers, dye and neodymium and color-center lasers, are available.

A fluorescence method of laser spectroscopy requires that some portion of the excitation should relax through radiative channels. This condition can be well fulfilled in detecting atoms and molecules in the UV, visible and near-IR spectral regions. In contrast to an absorption method, it is the only fluorescence signal that enters the photodetector (with an inevitable weak background). The sensitivity is extremely high, which enables one to detect single atoms in the laser beam [1.77].

The third class includes such calorimetric methods as the optoacoustic (OA), optothermal (OT), "thermolens" methods and others. Among them OAS has developed the most due to its simplicity and high sensitivity. In contrast to the fluorescence method, it is important for OAS that there should be a nonradiative channel of excitation relaxation. This can be realized if the radiative relaxation time is longer than the nonradiative relaxation time. In this case a considerable part of absorbed laser power is transformed to thermal energy of the medium. This requirement can be fulfilled well, for example, in analysis of molecular gases in the IR spectral region at gas pressures higher than $10^2 - 10^3$ Pa. At present the applicability of OAS has been demonstrated in analysis of substances both in gaseous and condensed media. The highest threshold sensitivity, from 10^{-9} to $10^{-10} cm^{-1}$, has been attained in a gas analysis [1.88].

The fluorescence and OA methods complement each other, which follows from (1.4). In principle, both methods can be applied to any type of transition. A characteristic feature of OAS is the formation of an output signal due to excited particle energy redistribution resulting from collisional relaxation over all the particles in the measuring cell. Therefore, this method is most efficient in detecting relative molecular concentrations at comparatively high total gas pressure. The optimum field of application of the fluorescence method is the detection of small absolute concentrations of particles or their relative concentrations at rather low pressures. As a rule, both methods require comparatively small volumes or samples to be analyzed. This is their obvious advantage over the absorption method, especially when the multipass cell has a big volume. Table 1.1 compares the methods under consideration and their basic characteristics.

10

Table 1.1. Comparison of linear laser-spectroscopy methods

Method Characteristics	Absorption	Intracavity absorption	Fluorescence	Optoacoustic
Necessary conditions for application	-	lasers with a wide amplifi- cation line	radiative channel of relaxation	nonradiative channel of relaxation
Spectral range	from UV to far IR	visible and near IR	UV and visible	from UV to far IR
Sensitivity (in units of absorption)[cm^{-1}]	$10^{-5} - 10^{-7}$	$10^{-8} - 10^{-9}$	up to single atoms	$10^{-7} - 10^{-10}$
Spatial resolution	bad	bad	up to λ	up to 10^{-2} - 1 mm
Time resolution (for the given sensitivity) [s]	1	$10^{-3} - 10^{-6}$	$1 - 10^{-12}$	$1 - 10^{-3}$

1.3 Classification of Calorimetric Methods

The OA method belongs to the class of calorimetric methods of laser spectroscopy. Their common characteristic feature is that the information on properties and composition of the medium under investigation can be obtained by
directly detecting the power absorbed in it from the thermal changes in
physical and thermodynamic parameters of the medium itself. The characteristic properties of calorimetric methods are: 1) zero-type measurement (no
signal without absorption); 2) increase in sensitivity with increasing radiation power (up to the absorption saturation regime); 3) fundamental limitation of the ultimate sensitivity by thermal fluctuations in the medium under
study.

Calorimetric methods together with powerful laser-radiation sources have
proved very efficient and universal because they can be applied to analysis
of substances in gaseous, liquid and solid phases over a comparatively wide
spectral range, from ultraviolet to microwave.

1.3.1 Types of Calorimetric Methods. Time of Relaxations

Calorimetric methods are based on a sequence of rather complex processes followed by an interaction of laser radiation with the medium (Fig.1.3). Their
particularities and optimum field of application are briefly considered below.

Essential for calorimetric methods is the relation between the following
relaxation times of the excited to the ground state: τ_r is the time of radi-

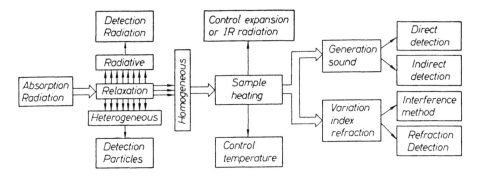

Fig. 1.3. Basic physical effects in a sample under irradiation, and methods to detect them

ative relaxation, τ_{hom} is the time of nonradiative relaxation due to colli-sions in the volume (homogeneous relaxation), τ_{het} is the time of nonradia-tive excitation relaxation to the walls of the measuring cell (heterogeneous relaxation). The relations between these times determine the dominant relax-ation channel and thereby affect the choice of an optimum thermal sensor. Table 1.2 gives typical values of relaxation time for various cases.

For molecular gases the relation between the relaxation times changes with the gas pressure P since $\tau_{hom} \propto P^{-1}$, $\tau_{het} \simeq \tau_T(\varepsilon_A) \propto P$, and the value of τ_r does not depend on pressure. Here τ_T is the time of thermal relaxation of the gas in the measuring cell, ε_A is the accomodation coefficient which determines the probability of relaxation of excited molecules during their collisions with the wall. Figure 1.4 presents the pressure variation of the resultant

Table 1.2. Typical values of the relaxation time

State of medium	Type of τ	Spectral range		
		UV and visible	IR	submilli-meter
	τ_r	$10^{-4} - 10^{-7}$	$10^{-1} - 10^{-3}$	
Gas and vapor ($P = 10^5$ Pa = 760 torr)	τ_{hom}	$10^{-5} - 10^{-7}$	$10^{-8} - 10^{-9}$	$10^{-8} - 10^{-9}$
	τ_{het}	$10^{-2} - 10^0$	$10^{-2} - 10^0$	$10^{-2} - 10^0$
Liquid	τ_r	$10^{-3} - 10^{-8}$	$10^{-7} - 10^{-9}$	
	τ_{hom}	$10^{-8} - 10^{-12}$	$10^{-10} - 10^{-13}$	$10^{-10} - 10^{-12}$
Solid	τ_r	$10^{-3} - 10^{-8}$	$10^{-7} - 10^{-9}$	
	τ_{hom}	$10^{-8} - 10^{-13}$	$10^{-10} - 10^{-13}$	$10^{-10} - 10^{-13}$

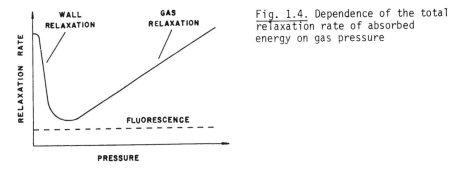

Fig. 1.4. Dependence of the total relaxation rate of absorbed energy on gas pressure

relaxation rate τ^{-1} typical of any molecules:

$$\tau = (\tau_r^{-1} + \tau_{hom}^{-1} + \tau_{het}^{-1})^{-1} \quad . \tag{1.5}$$

Relaxation of excited particles on account of collisions with other particles is usually prevalent at high gas pressures, whereas the relaxation due to cell walls prevails at low pressures. For most molecules the minimum which also depends on cell dimensions, lies approximately within the pressure range $10^2 - 10^3$ Pa when the cell diameter $D = 1$ cm. The contribution of fluorescence whose role is particularly great in the visible spectral range is evident in the neighborhood of the minimum. In the region of low pressures, starting approximately with $0.1 - 1$ Pa, the relaxation time no longer depends on pressure (the constant section of the curve in Fig.1.4). This is explained by collisionless flight of the excited molecules to the gas cell.

1.3.2 Optothermal Method

At low pressures the most effective method is direct detection of their thermal [optothermal (OT) method] or mechanical action of excited particles on the surface of an appropriate sensitive element, e.g., bolometers, pyroelectrics or piezoelements. In [1.89] attention was drawn to such an approach, and later it was experimentally tested by exciting molecules with continuous [1.90 - 91] and pulsed laser radiation [1.92]. The ultimate sensitivity of this method is limited by thermal noise of the sensor which is about $10^6 - 10^7$ mol/cm^3, calculated for the minimum number of detectable molecules. The OT method can be used in linear and nonlinear high-resolution laser spectroscopy to study multiphoton effects, to detect the products of photochemical reactions, etc.

When the role of the radiative relaxation channel is dominant, i.e., $\tau_r \ll \tau_{hom}, \tau_{het}$, the fluorescence method features high sensitivity, as already noted. In some cases, especially in the IR spectral range, calorimetric

13

detection of fluorescent radiation may prove useful. For example, in rather long absorption cells with $D/L \ll 1$, D being the cell diameter and L the cell length, almost all the fluorescent radiation can be absorbed by the cell walls with resultant heat transfer to the gas or the thermal pickup on the wall. This function can be performed by a pyroelectric pickup which simultaneously detects both the excited molecules relaxing on the surface and the fluorescent radiation.

At higher gas pressure, approximately starting with 10^3 Pa, homogeneous relaxation because of intermolecular collisions becomes prevalent which causes the translational molecular energy to increase and, in turn, the temperature. Compared to the previous case, this relaxation channel is not optimum for detecting absorbed power, since the energy of excited particles is redistributed over all the particles in the measuring cell as a result of collisions. Gas temperature variations can also be recorded by thermal pickups. Applied to gases, the theoretical values of threshold sensitivity of the OT and OA methods are comparable. But since the available thermal pickups are imperfect, the sensitivity threshold of the OT methods, achieved in practice, is smaller by one or two orders.

Such measuring techniques can be applied to a weak-absorption analysis of substances in the condensed phase, ordinary thermocouples are used as thermal pickups [1.93]. The OT methods can also be effectively used in the analysis of strong absorption in thin samples placed on thermal pickups. Heating the leads to an enhancement of thermal radiation of the media, which can be detected by highly sensitive IR photodetectors [1.94]. This technique is suitable for remote measurements and for selective gas analysis.

It is possible to detect the absorbed energy in nongaseous samples by detecting the changes in sample dimensions accompanying heating. This method is most efficient in the analysis of liquids.

1.3.3 Refraction Methods

The variation in temperature of the medium across the laser beam causes the refractive index of the medium to change. The "thermolens" and interference lend themselves to efficient means of measurement.

In the "thermolens" method small variations of the refractive index are measured by recording the defocusing of an additional probe beam through the region of excitation. This method is widely used to analyze substances in the liquid phase and more rarely to analyze gases. This is because it is highly sensitive in the case of liquids, in some cases exceeding the sensitivity of OAS. Usually the main laser for excitation of the medium, operates under pulsed or continuous conditions with intensity modulation, while the probing laser beam is continuous. The ultimate sensitivity of the "thermolens" method is limited by photodetector noise [1.95]. In practice, the threshold sensitivity of the method is usually limited by intensity fluctuations of the laser radiation. At present it is being used successfully to study intramolecular and intermolecular transitions [1.96-98], and to measure weak absorption in various media, etc.

When the probe laser-beam propagates at an angle to the exciting laser-beam, a small deviation of probe beam takes place. This phenomenon is called the mirage effect, and its detection can be utilized to determine the laser energy absorbed in the sample. This technique is promising for the study of solid samples in case of probe-beam propagation parallel to a surface [1.99,100]. Noncontact recording which allows strong electric or magnetic fields or temperature to act on the sample is an advantage of the method.

Variation in refractive index can also be measured by the interference method [1.101]. Its sensitivity is comparable with that of the previous method, but it is complicated in its practical realization. The formation of acoustic waves and the variation of refractive index occur in the sample almost simultaneously. The interference method may be useful to visualize the acoustic processes in the sample [1.102].

1.3.4 Optoacoustic Method

Periodic heating of the sample in the laser beam gives rise to acoustic vibrations. Their direct detection with appropriate sensors (condenser, electret or piezoelectric microphones) makes it possible to achieve the highest sensitivity for this class of OA methods, especially for gases (up to 10^{-10} cm^{-1} with radiation power of about 1 W). Indirect detection of acoustic vibrations in the medium contacting the sample is less sensitive, but the possibilities of such a technique are much greater for analysis of surface absorption.

OAS is one of the most sensitive methods among different calorimetric methods, although its potentialities have not yet been fully realized. Among its other merits are comparative simplicity, sufficient universality, wide dynamic range, possibility of measurement automatization and potential for increasing its selectivity.

2. Optoacoustic Gas Spectroscopy with Laser Sources

The physical processes responsible for the OA phenomenon are the same for both coherent and incoherent radiation sources. They include optical resonant excitation of the medium followed by radiation absorption, non-radiative relaxation of excited particles, heating of the medium, and the formation of acoustic vibrations. The use of coherent radiation sources has quite a number of advantages and leads to essential features both in the processes of generating OA signals and in the principles of designing proper measuring schemes.

High radiation power makes it possible to excite a considerable fraction of the molecules up to saturation, thereby realizing its ultimate sensitivity and making it appropriate for nonlinear laser spectroscopy. High monochromaticity and frequency tuning allow studying the absorption spectra of many molecules with high resolution otherwise inaccessible for existing OA devices as well as realizing selective excitation of not only molecules of a certain sort but also individual quantum states of these molecules. Spatial coherence provides, first of all, high laser-beam collimation which increases OA sensitivity through multipass cells. Secondly, it allows focusing the radiation into small volumes, which increases the spatial resolution of the method and the energy flux, and opens the way for research on nonlinear and multiphoton effects in strong laser fields.

This chapter deals with the particularities in the formation of OA signals under laser radiation. On this basis we consider the theory and operational principles for analyzing gas media with laser OA devices.

2.1 Laser Optoacoustic Devices

Before the advent of lasers, nondispersive infrared gas analyzers, sometimes called optoacoustic analysers, were universally used. For the sake of better illustration, the operation of laser devices is analyzed in comparison with those gas analyzers.

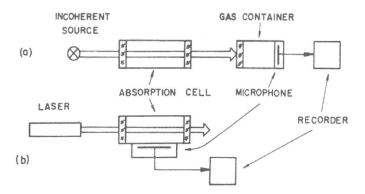

Fig. 2.1. OA instruments (**a**) with an incoherent radiation source; and (**b**) with a laser radiation sources

In a nondispersive OA device (Fig.2.1a) the gas composition is determined by measuring the absorbed power by via the detection of small relative changes in power of radiation that has passed through the cell with a selective OA detector placed behind the cell. In other words, the measurement is reduced to the conventional absorption method. The chamber of a selective OA detector is filled with molecules identical to those whose concentration is to be determined in the absorption cell. Gas pressure pulsations in the working chamber caused by periodic absorption of modulated radiation flux are detected by a microphone in the detector chamber.

In such a scheme the OA detector simultaneously acts as a wide-range spectral filter and a radiation detector "matched" to the radiation flux only in spectral ranges coinciding with several absorption bands of the molecules to be detected. Thus, in contrast to conventional spectral instruments for gas analysis, selectivity of these gas analyzers can be obtained without spectral resolution. This also allows an increase in signal level at the OA detector output because the noncoherent radiation source usually used in such instruments (filament lamps, gas-discharge tubes, etc.) have rather wide-band spectra and are comparatively low in power. The overall threshold sensitivity of gas analyzers is determined by the ability of the recording system to detect small power variations of the radiation passing through the absorption cell. Differential schemes of nondispersive gas analyzers have been developed with two cells (for measurement and comparison) and a differential OA detector. They reliably detect changes of radiation flux in the measuring channel up to 0.1%. This corresponds to a threshold concentration sensitivity up to 1 - 10 ppm (for CO_2, CO and CH_4 gases) with an absorption-cell length of about 10 to 30 cm.

In laser OA instruments (Fig.2.1b), as in the previously discussed scheme, the laser radiation passes through an absorption cell filled with gas, but there is no detector behind the cell. Since the laser lines coincide with the absorption band of molecules under detection, the gas absorbs some part of the radiation, gets heated and increases the pressure in the closed volume. This increase represents an OA signal carrying information on molecular absorption (concentration). Then, the OA signal is transformed to an electric signal using a microphone placed in the cell. The cell with the microphone was called a spectrophone and later, when the laser was used to detect micro-impurities in gases, a laser OA detector of molecules. So, laser OA instruments, unlike nondispersive ones, can directly detect absorbed power right in the spectrophone chamber. Such a scheme was used for the first time by BELL to study the OA effect [2.1]. But the low intensity of thermal radiation sources made it impossible to detect small concentrations of molecules. At the same time, the high power of laser sources together with highly sensitive techniques to detect weak acoustic vibrations enable certain molecules in concentrations up to 0.01 ppb to be detected.

2.2 Formation of Optoacoustic Signal

To estimate amplitude and shape of the OA signal it is necessary to calculate: a) the number of molecules excited by radiation; b) the gas temperature variation, caused by nonradiative relaxation of molecules and c) the variation of gas pressure.

The primary process which initiates the OA effect in a gaseous medium is the variation in population of molecular energy levels resonating with the radiation: electron levels in the ultraviolet and visible spectral regions, vibrational-rotational levels in the IR region and purely rotational ones from the far-IR region to the microwave region. The sequence in which the OA signal is calculated is similar for all spectral regions. But there are specific features in it. For instance, with electron-state excitation account must be taken of possible photochemical processes. As a result, the portion of absorbed energy causing gas heating decreases. It should also be noted that relaxation of excited particles through a number of intermediate states is rather complex, which makes it more difficult to estimate the temporal parameters of OA signals.

Very generally, the radiation power absorbed in a unit of volume is determined by

$$I_{abs}(t) = I(t)\alpha(\nu,t,I) \quad , \tag{2.1}$$

where $I(t)$ is the radiation intensity, $\alpha(\nu,t,I)$ is the absorption coefficient of the medium per unit length that depends on intensity and time on account of possible nonlinear and transient interaction effects between radiation and medium. In prelaser OA spectroscopy medium absorption did not vary under radiation and it was always assumed that $\alpha(\nu,t,I) = \alpha_0(\nu)$ was the linear absorption coefficient at the radiation frequency ν.

With laser sources it is easy to obtain nonlinear effects due to absorption saturation and multiphoton absorption. Then the dependence of α on intensity and time should be taken into account. Depending upon the purpose of the experiment, it is necessary to measure either the linear absorption spectrum or the nonlinear effects in absorption. This dictates the choice of radiation parameters. To illustrate situations typical in this case we restrict considerations to calculating OA signals induced by IR lasers which is of great practical interest due to the presence of characteristic vibrational-rotational bands of most molecules in the IR region.

2.2.1 Calculation of IR-Radiation Absorption at Vibrational-Rotational Transitions

The process of molecular excitation by IR radiation, as applied to OA with weak noncoherent sources, has been studied theoretically in [2.2,3] and with the use of laser sources in [2.4-6]. Here we consider the simplest model of absorption of IR laser radiation of rotational-vibrational molecular transitions in gases [2.7,8]. Figure 2.2a illustrates this model by two vibrational levels with a rotational stucture. Collisions lead to a Boltzmann equilibrium distribution of molecules over rotational levels with the characteristic rate τ_{rot}^{-1}. At the same time the laser radiation with the rate $W = \sigma_j I / h\nu$ (σ_j being the cross section of the radiative transition $V=0,j$ and $V=1,j'$ between the levels concerned) excites the molecules from the rotational sublevel j of the ground vibrational state $V=0$ to the rotational sublevel j' of the excited vibrational state $V=1$. This changes both the rotational and the vibrational distributions of molecules. The vibrationally excited molecules relax to the ground state from all rotational sublevels in the characteristic time τ_{vib}. For simplicity, we neglect here possible excitation of molecules to higher vibrational levels and vibrational-vibrational exchange between molecules [2.8].

The rate equations describing the rotational distribution of molecules under laser excitation have the form

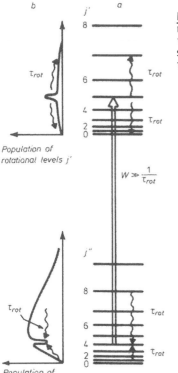

Fig. 2.2. (a) Laser excitation of vibrational-rotational molecular levels in the IR spectral region; and (b) bottleneck effect in absorption saturation of vibrational-rotational transitions

$$\frac{dn_{1,j'}}{dt} = (n_{0,j} - n_{1,j'})W + \frac{q_{1,j'}N_1 - n_{1,j'}}{\tau_{rot}} \qquad (2.2a)$$

$$\frac{dn_{0,j}}{dt} = (n_{1,j'} - n_{0,j})W + \frac{q_{0,j}N_0 - n_{0,j}}{\tau_{rot}} \quad , \qquad (2.2b)$$

where $n_{v,j}$ is the molecular density in the vibrational-rotational state (V,j), N_v is the total density of molecules in the vibrational state V, i.e., $N_v = \sum_j n_{v,j}$. The value $q_{v,j}$ describes the fraction of molecules in the vibrational state V at the rotational sublevel j under consideration. In other words, the factor $q_{v,j}$ describes the rotational distributional equilibrium of molecules in the vibrational state V

$$q_{v,j} = \frac{1}{z^{rot}} g_j \exp(-E_j / K_B T) \quad , \qquad (2.3)$$

where z^{rot} is the statistical sum of the rotational states, g_j is the degeneration of the rotational sublevel j with rotational energy E_j and K_B is the

Boltzmann constant. The density variation of vibrationally excited molecules
is determined by the excitation rate W and the time of vibrational relaxation
τ_{vib}:

$$\frac{dN_1}{dt} = (n_{0,j} - n_{1,j'})W + \frac{N_1^0 - N_1}{\tau_{vib}} \quad . \tag{2.4a}$$

The population of the lowest vibrational state N_0 is determined from the total population of the vibrational levels $V = 0,1$:

$$N_0 + N_1 = N_0^0 + N_1^0 = \text{const} \quad . \tag{2.4b}$$

The values of N_V^0 determine the equilibrium population of the vibrational
state V without excitation.

The time τ_{vib} of vibrational relaxation of excited molecules to the ground
state is described by

$$\tau_{vib} = (\tau_{V-T}^{-1} + \tau_{rad}^{-1} + \tau_W^{-1})^{-1} \quad , \tag{2.5}$$

where τ_{V-T} is the time of vibrational rotational relaxation responsible for
the transfer of the molecular excitation energy to heat, τ_{rad} is the time of
radiative relaxation, τ_W is the relaxation time of molecules onto the cell
walls. For vibrational-rotational molecular states at gas pressures typical
for OA cells (hundreds of Torr or 10^4 Pa) the following relation is valid

$$\tau_{rot} \ll \tau_{V-T} \ll \tau_{rad} \leqslant \tau_W \quad . \tag{2.6}$$

Table 1.2 gives typical values of relaxation times.

Equations (2.2-4) describe the population kinetics of the excited vibrational state and the depletion of the ground vibrational state. Knowing the
level population, it is possible to determine the absorption coefficient per
unit length

$$\alpha(\nu,t,I) = \sigma_j(\nu) (n_{0,j} - n_{1,j'}) \quad , \tag{2.7}$$

and, according to (2.1), the absorbed radiation power per unit volume. Here
we use the cross section of the transition between rotational-vibrational
states. In the literature the transition cross section determined by total
populations of vibrational levels may also be given:

$$\alpha = \sigma(\nu) (N_0 - N_1) \quad . \tag{2.8}$$

From (2.7,8) it follows that both cross sections σ and σ_j are related by the fraction q or relative population of the rotational state j

$$q\sigma_j = \sigma \quad , \tag{2.9}$$

where, for simplicity, we assume that $q_{0,j} \approx q_{1,j'} = q$.

Instead of solving (2.2-4) in the general case and thus calculating $\alpha(v,t,I)$, we consider only three different particular cases which give a clear picture of molecular absorption saturation by IR radiation.

a) Short Laser Pulse $(t_p \leqslant \tau_{rot})$

For $t_p \leqslant \tau_{rot}$ the rotational relaxation during a laser pulse is negligible. Therefore, in the molecular distribution over rotational states a "peak" is formed at the level $V = 1$, and conversely a "hole" at $V = 0$ (Fig.2.2b). In other words, the vibrational-rotational transition $V = 0, j \rightarrow V = 1, j'$ resonating with the laser radiation becomes saturated, and rate of further excitation of the molecules is limited by the so-called "rotational bottleneck" [2.8]. After the pulse is over, the absorbed energy is distributed over all the rotational sublevels by rotational relaxation. And, finally, during τ_{vib} the excited molecules return to their initial state.

In this regime the population of the upper level depends weakly on the shape of the laser pulse. The population fully depends on the pulse energy flux Φ

$$\Phi = \int_0^{t_p} I \, dt \quad , \qquad \text{and} \tag{2.10}$$

$$N_1 \approx N_0 \frac{q}{2} [1 - \exp(-\Phi/\Phi_s^j)] \quad , \tag{2.11}$$

where, for simplicity, we assume that $N_1^0 \ll N_0$ and introduce the absorption saturation energy [2.8]:

$$\Phi_s^j = q \frac{h\nu}{2\sigma} = \frac{h\nu}{2\sigma_j} \quad . \tag{2.12}$$

At the radiation fluence $\Phi = \Phi_s^j$ the absorption coefficient decreases by e times. When $\Phi \ll \Phi_s^j$, absorption is linear. In the other limiting case $(\Phi \gg \Phi_s^j)$, absorption is highly saturated and the maximum population of the upper state can be expressed as

$$N_1^{max} \approx q \frac{N_0}{2} \quad . \tag{2.13}$$

With polyatomic molecules one must be careful with (2.11,13) at $\Phi \gg \Phi_s$ because multistep and multiphoton processes may participate even at a low energy fluence of the laser pulse (Sect.7.3.2).

The average energy E_{abs} absorbed in a unit volume is determined by

$$E_{abs} = h\nu(N_1 - N_1^0) \approx h\nu N_1 \quad , \tag{2.14}$$

where N_1 is the population of the upper vibrational level after the laser pulse desribed by (2.11).

b) Laser Pulse of Intermediate Duration ($\tau_{rot} \ll t_p \ll \tau_{vib}$)

For a laser of intermediate duration, rotational relaxation occurs during the pulse which reduces the "rotational bottleneck" effect and increases the energy absorbed in the gas. On the other hand, the pulse duration is not long, so the vibrational-rotational relaxation of excited molecules in a laser pulse is still negligible. Such a case can be realized in typical experiments on molecules diluted in an atomic inert buffer gas. Since rotational relaxation redistributes molecules during a laser pulse, the population of the vibration-al state $V = 1$ by the end of the laser pulse is approximately described by an expression similar to (2.11)

$$N_1 = N_0 \frac{1}{2} [1 - \exp(-\Phi/\Phi_s)] \quad , \tag{2.15}$$

where Φ_s is the saturation energy of the vibrational transition, i.e.,

$$\Phi_s = \frac{h\nu}{2\sigma} \approx \Phi_s^j / q \quad , \tag{2.16}$$

and $t_p \gtrsim \tau_{rot}/q$. Under high saturation ($\Phi \gg \Phi_s$) the population of the excited vibrational level reaches its maximum

$$N_1^{max} \approx \frac{1}{2} N_0 \quad . \tag{2.17}$$

The energy absorbed with fast rotational relaxation increases by $1/q$ times at most (when $t_p \gtrsim \tau_{rot}/q$).

c) Long Laser Pulse ($t_p \gg \tau_{vib}$)

When $t_p \gg \tau_{vib}$ the population of the upper vibrational level and the absorption factor depend on radiation intensity since during a laser pulse relaxation of vibrational excitation takes place and

$$\alpha = \frac{\alpha_0}{1 + I / I_s} \quad , \tag{2.18}$$

where α_0 is the linear absorption coefficient, I_s is the intensity of absorption saturation under which the absorption decreases twice. The value of I_s is determined by

$$I_s = \frac{h\nu}{2\sigma\tau_{vib}} \quad . \tag{2.19}$$

During a long laser pulse absorbed energy turns into heat during the pulse.

2.2.2 Calculation of Optoacoustic Signals in a Nonresonant Spectrophone

A secondary process occurring in the spectrophone chamber is the transfer of energy of the molecules excited by laser radiation as they return to their initial state of thermal energy, with the simultaneous formation of acoustic vibrations (OA signals). Optoacoustic signals are generated in either nonresonant or resonant spectrophones. In nonresonant spectrophones the modulation frequencies are usually much lower than the natural resonant acoustic frequencies of the spectrophone chamber. As a rule, the characteristic times of measurement with these spectrophones are longer than or comparable to the transient period of thermodynamic equilibrium in a gas. So it is rather easy to estimate OA signals on the basis of classical formulas of thermodynamics for ideal gases. In resonant spectrophones OA signals are recorded at frequencies commensurable with those of acoustic resonances in OA chamber. Therefore, the calculation must be based on the theory of three-dimensional acoustic resonators, whose characteristic dimensions are much larger than the depth thermal waves penetrate from the absorption zone into the walls of the chamber. This section describes the design of nonresonant spectrophones generally employed in laser spectroscopy. Section 5.3.1 considers resonant spectrophones.

OA signals of nonresonant spectrophones were calculated for thermal radiation sources in [2.2,3,9,10] and coherent sources in [2.4-6,11,12]. The best solution for a particular case of harmonic modulation of coherent radiation was given in [2.13].

For an approximate calculation of OA signals we can use the following system of differential equations:

$$\frac{d\Psi}{dt} = h\nu \frac{N_1(t)}{\tau_{V-T}} - \frac{\Psi - \Psi_0}{\tau_s} \quad , \tag{2.20a}$$

$$P = \frac{P}{TV\rho C_V} \int_V \Psi \, dV \quad , \tag{2.20b}$$

where Ψ is the average energy of translational motion of the molecules in a unit volume; $N_1(t)$ is the density of excited molecules in the upper vibrational level; τ_s is the time constant of the spectrophone characterizing the resultant relaxation time in the gas chamber of the spectrophone

$$\tau_s = (\tau_T^{-1} + \tau_L^{-1})^{-1} \quad , \tag{2.21a}$$

$$\tau_T \approx V / (5.76\pi \, Lk) \quad . \tag{2.21b}$$

Here τ_T is the time of thermal relaxation of the heated gas onto the chamber walls, τ_L is the characteristic time of gas leakage from the chamber to additional (or out-of-membrane) volumes. Gas leakage usually occurs in spectrophones during continuous pumping of the mixture through the chamber or due to inclusion of special capillary tubes to compensate for a change of ambient conditions, say, the variation of its temperature. Further, k is the coefficient of thermal gas diffusion $k = K / \rho C_V$, where ρ C_V, K, P, T denote, respectively, density, specific heat, heat conduction, pressure and temperature of the gas, V and L are the volume and length of the chamber.

In the equations above the main approximation eliminates the intermediate step of determining the temperature distribution in the cell volume as the OA signal is calculated. This distribution is usually described by cumbersome expressions derived from solving the heat equations [2.11]. In essence, to calculate an OA signal we use the standard relaxation equation (2.20a) with the principal time constant τ_T (assuming that $\tau_T \ll \tau_L$) which is obtained from the basic solution of the heat equation for a long cylinder. Such a replacement is justified by almost full averaging of the OA signal with respect to the chamber volume of the nonresonant spectrophone (or, in other words, by the equality of pressure in different zones of the chamber), unlike the temperature distribution that is allowed for by integration (2.20b). The validity of such an approximation is substantiated by good agreement between theory and experiment [2.6].

Since this system of equations is rather simple and convenient for practical calculations, it is possible to reveal the basic physical features in spectrophone operation and to outline ways of designing optimum structures at any ratios of excitation duration and characteristic relaxation times in gases. Here we treat only the solutions for the four most typical operating conditions of IR lasers: the pulse, the pulsed-periodic regime with a pulse repetition rate, and the continuous operation having amplitude and frequency modulated radiation. For simplicity, we call them MP, PP, AM, and FM regimes, respectively.

a) Pulsed Regime

When the laser pulse duration complies with the condition $t_p \ll \tau_{vib}$, under linear absorption the OA signal shape is described by the approximate expression:

$$P = \frac{P\tau_{vib}\alpha\bar{\Phi}}{T\tau_{V-T}\rho C_V} [\exp(-t/\tau_T) - \exp(-t/\tau_{vib})] \quad , \tag{2.22}$$

where $\tau_L \gg \tau_T \gg \tau_{vib}$. When the pulse duration $t_p \gg \tau_{vib}$, (2.22) is somewhat modified

$$P = \frac{P\tau_{vib}\tau_T\alpha\bar{I}}{T\tau_{V-T}\rho C_V} \begin{cases} [1 - \exp(-t/\tau_T)] & t \in 0, t_p \\ [1 - \exp(-t_p/\tau_T)] \exp(-t/\tau_T) & , t \geqslant t_p \end{cases} \quad . \tag{2.23}$$

In (2.22,23) $\bar{\Phi}$ or $\bar{I}$ are the mean values of energy flux and intensity in the spectrophone chamber. The determination of $\bar{\Phi}$ or $\bar{I}$ reduces to integrating (2.20b) at a given distribution of $\Phi(r, \varphi, z)$ and $I(r, \varphi, z)$ (r, φ, z being polar coordinates, the z axis is directed along the spectrophone axis) in the chamber

$$\begin{pmatrix} \bar{\Phi} \\ \bar{I} \end{pmatrix} = \frac{1}{V} \int \begin{pmatrix} \Phi(r, \varphi, z) \\ I(r, \varphi, z) \end{pmatrix} \exp(-\alpha z) dV \quad . \tag{2.24}$$

In linear case,

$$\begin{pmatrix} \bar{\Phi} \\ \bar{I} \end{pmatrix} = \begin{pmatrix} E \\ J \end{pmatrix} \times \frac{1 - \exp(-\alpha L)}{\alpha V} \quad , \tag{2.25}$$

where E, J are, respectively, the energy and power of laser radiation.

Comparison between the regime $t_p \ll \tau_{rot}$ and $\tau_{rot} \ll t_p \ll \tau_{vib}$ shows that the signal shape is the same in both cases and the amplitudes begin to differ only when absorptions saturation is achieved. Then the ultimate value of OA signal amplitude as well as its associated saturation energy are smaller by

26

1/q (about 10 to 100 times) in the first case than in the second. As for the ultimate sensitivity, the second regime is preferable although it requires a higher energy consumption. It should be noted that in the pulsed regime with $t_p \leqslant \tau_{vib}$ the amplitude and shape of the OA signal in linear operation do not depend on pulse shape and are fully governed by the pulse energy.

In the pulsed-periodic regime calculation of the OA signal reduces to a repetition of signals like (2.22 or 23) and the determination of harmonic components of the corresponding pulse sequence. This calculation is essentially similar to that for amplitude modulation of continuous radiation (AM regime) given below.

b) Amplitude-Modulation Regime

Under amplitude modulation (AM) the time dependence of the radiation intensity can be presented as

$$I(t) = I_0 [1 + \delta f(t)] \quad , \tag{2.26}$$

where I_0 is the average radiation intensity within the period of modulation, T; $f(t)$ and δ are, respectively, the temporal function and the depth of modulation ($\delta \leqslant 1$). The expression for the first harmonic of OA signal has the form

$$P = K_F \frac{P \tau_{vib} \bar{I}_0 \alpha \tau_T \delta \, \sin[\omega t + \varphi(\omega)]}{T \tau_{V-T} \rho C_V [1 + (\omega \tau_T)^2]^{1/2} \, [1 + (\omega \tau_{vib})^2]^{1/2}} \quad , \tag{2.27a}$$

$$\varphi(\omega) = -\arctan \omega \tau_T - \arctan \omega \tau_{vib} \quad , \tag{2.27b}$$

where $\varphi(\omega)$ is the phase difference between OA and laser signals; $K_F \approx 0.3 - 1.0$ is a coefficient depending on the type of amplitude modulation.

c) Frequency-Modulation Regime

Frequency modulation (FM) in laser systems can be realized simply through a discrete change of wavelengths either from different lasers or in one laser with an appropriate frequency switching device. To determine the value of an OA signal in this case it is sufficient to substitute $\sigma = |\sigma_1 - \sigma_2|$ in (2.27) for α, where σ_1 and σ_2 are the absorption cross sections at the laser frequencies ν_1 and ν_2.

To determine the OA signal amplitude generally one should know the frequency-modulation law and the absorption-band shape. In practice, in most

27

cases of spectrophone operation the profile of the molecular absorption band can be approximated, with sufficient accuracy in collisional spectral line broadening, by the Lorentzian function

$$\sigma(\nu) = \sigma_0 \frac{\Gamma^2}{\Gamma^2 + \Omega^2} \quad , \tag{2.28}$$

with the spectral line half-width $\Gamma = \gamma_\beta P$, where γ_β is the broadening parameter; $\Omega = \nu_L - \nu_0$ is the detuning between the laser frequency ν_L and the center of absorption line ν_0; and σ_0 is the absorption cross-section at the center of the line. For harmonic frequency modulation with modulation depth $\mu \ll 1$

$$\nu(t) = \nu_L [1 + \mu \sin \omega t] \quad . \tag{2.29}$$

Substituting (2.29) into (2.28) we have [2.14]

$$\sigma(t) / \sigma^* = (1 + \alpha_1 \sin \omega t + \alpha_2 \sin^2 \omega t)^{-1} \quad \text{where}$$

$$\alpha_1 = 2ab / (1 + a^2) \quad , \quad \alpha_2 = b^2 / (1 + a^2) \quad , \quad \sigma^* = \sigma_0 / (1 + a^2) \quad , \tag{2.30}$$

$$b = \mu \nu_L / \Gamma \quad , \quad a = (\nu_L - \nu_0) / \Gamma \quad .$$

The harmonic components of the function (2.30) are calculated by Fourier analysis. If, for example, the laser frequency coincides with the center of the absorption line ($a = 0$ and $b = 1$) the relative amplitude values of subsequent harmonics will be: $1(\omega = 0$, dc term); $0(\omega)$; $0.242(2\omega)$; $0(3\omega)$; $0.057(4\omega)$. Frequency modulation was comprehensively analyzed by DEWEY in [2.14]. The absorption-modulation regimes using the Zeeman and Stark effects as the molecules are acted upon by electric or magnetic fields, are similar to the FM case. A specific feature of these modulation regimes is that it is possible to realize a relative frequency shift between absorption band and radiation line on account of the frequency shift of the absorption band about the single-frequency laser line. The fundamental shapes of OA signals calculated with the help of the above expressions are illustrated in Fig.2.3.

2.3 Nonresonant Spectrophones

Optoacoustic signals are generated and subsequently detected in the closed volume of the spectrophone chamber (Fig.2.1b). The one-chamber nonresonant spectrophone, in which laser radiation passes through the working chamber

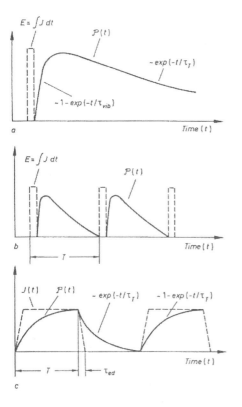

Fig. 2.3. Typical forms of OA signals (——) under (**a**) pulsed, (**b**) pulse-periodic, and (**c**) modulated laser excitation of molecules. The laser pulses are shown by dashed lines

once, is the simplest design among a great number of various types of spectrophone. In the simplest case such a spectrophone is an acoustically closed chamber containing the gas to be analyzed. Its ends have optical windows for through passage of laser radiation.

The OA signals in the spectrophone chamber are detected by acoustic pickups which transform small gas pressure variations to electric signals. The main requirement of these pickups is high sensitivity, a small **level** of intrinsic noise, comparative simplicity and compactness, and linearity over a wide dynamic range. The possibility of optimal fitting of the parameters of the pickup and the working chamber at which the maximum spectroscopic sensitivity is realized is also an essential factor. The type of pickup chosen in many respects determines the specific structural features of the chamber and so the design of the entire spectrophone.

2.3.1 Acoustic Pickups (Microphones)

As OAS was developed, various pickup designs were used to detect acoustic vibrations including the optical Golay microphone, the inductive pickup, the condenser, the electrodynamic, the electret and other types of microphones.

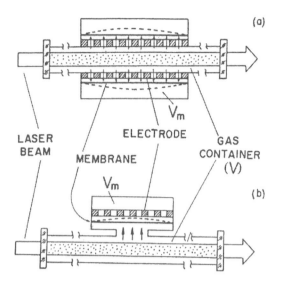

Fig. 2.4. Microphone designs with (a) cylindrical and (b) plane fixed electrodes

V_m

LASER
BEAM

ELECTRODE

MEMBRANE

GAS
CONTAINER
(V)

V_m

(a)

(b)

As practice shows, capacitance-type pickups (condenser and electret microphones) have proved to be the best ones in laser spectrophones as far as the
simplicity of their design, sufficient sensitivity, the possibility of miniaturization and high operating factors are concerned. The frequency characteristics of these microphones embrace almost the entire range of working
frequencies, from a few Hz to tens of kHz. Such microphones with a cylindrical
(a) or plane (b) electrode geometry have gained wide acceptance (Fig.2.4).
An elastic membrane, 1 - 10 μm thick, is a sensitive element in these designs.
In condenser microphones it is usually made of mylar, teflon or metal and
in electret microphones of a special electret film. In plane microphones the
pressure from the irradiated-gas chamber is transferred to the membrane
through an additional channel or merely a hole in the side wall of the chamber (Fig.2.4a). In cylindrical microphones it is transferred through a lot
of holes in the wall of the electrode which is directly wound in a sensitive
membrane. When the membrane sags the pickup varies in capacity. Its plates
form the surface of the fixed electrode and the metallized surface of the
inner (or sometimes external) membrane. As a rule, to protect the membrane
against the action of slow variations in temperature and ambient pressure,
there should be a compensating capillary (not shown in Fig.2.4a) connecting
the working chamber volume V with the back-membrane volume of the microphone
V_m. Then the spectrophone makes recording the constant signal impossible.
  When the microphone is placed in the closed volume of the working chamber
its sensitivity differs from that when it is located in the open space, due

to the influence of gas elasticity. On the other hand, the presence of a microphone membrane, i.e., an elastic wall in the chamber, changes the chamber volume during generation of the OA signal. As a result, their amplitude decreases as compared to the value calculated from (2.20) valid for absolutely rigid walls. To allow for this effect, we shall consider the general equation for the ideal gas state written for small variations of its constituent parameters

$$\frac{\Delta P}{P} + \frac{\Delta V}{V} = \frac{\Delta T}{T} \quad , \tag{2.31}$$

where V denotes the inner volume of the spectrophone chamber with gas. The variation in pressure ΔP is related to the volume variation ΔV as [2.15,16]

$$\Delta P = i\omega Z(\omega)\Delta V \quad , \tag{2.32}$$

where $Z(\omega)$ is the so-called acoustic impedance of the microphone in the spectrophone chamber. Substituting (2.32) into (2.31) and taking into account the form of the expressions obtained in Sect.2.2, it is possible to get the following relation between OA signals for the case of absolutely rigid walls (P) and in the presence of a microphone (P_{mic}) in the chamber, respectively,

$$P_{mic} = \frac{P}{1 + P / i\omega Z(\omega)V} \quad . \tag{2.33}$$

Thus, in the presence of a microphone in the chamber the OA signal amplitude may decrease and the character of its time (or frequency) dependence may change. The microphone's acoustic impedance $Z(\omega)$ may be presented as

$$Z(\omega) = R_{mic} + i(\omega M_{mem} - 1 / \omega Y_{mic}) \quad , \tag{2.34}$$

where R_{mic} is the acoustic resistance due to internal friction of the gas between the membrane and the fixed electrode; M_{mem} is the reduced mass of the membrane and the gas layer vibrating with it; Y_{mic} is the resultant elasticity of the microphone. Usually when operating at lower frequencies smaller than the microphone's resonant frequency ω_{mic} [which is determined from $\omega_{mic} = (1/M_{mem}Y_{mic})^{1/2}$ and lies within several kHz], we may neglect the influence of the first two terms in (2.34), and the acoustic resistance will be determined by the resultant elasticity of microphone, i.e., $Z(\omega) = -i / \omega Y_{mic}$.

In the most general case the resultant membrane elasticity is given by

$$Y_{mic}^{-1} = \left(Y_{mem}^{-1} + Y_p^{-1} + Y_b^{-1} \right) \quad , \tag{2.35}$$

where Y_{mem} is the elasticity of the membrane, Y_p, Y_b are the elasticity of the gas in the premembrane V and the back-membrane V_m volume, respectively. They can be determined by

$$Y_{mem} = \frac{1}{2\pi T^*} \quad ; \quad Y_p = \frac{V}{\gamma P (A_m / 2)^2} \quad ; \quad Y_b = \frac{V_m}{\gamma P (A_m / 2)^2} \quad , \tag{2.36}$$

where A_m is the membrane area, T^* is the membrane tension, $\gamma = C_p / C_V$ is the ratio of the gas thermal capacities at constant pressure and volume.

To increase Y_{mic} it is necessary, first of all, to fulfill the condition $Y_b \gg Y_p$, which can easily be done by increasing the back-membrane volume V_m. A definite recording scheme of the microphone membrane displacement x

$$x(t) = \frac{2P(t)}{A_m Z(\omega) i \omega} \tag{2.37}$$

and, hence, microphone capacity variation $C = (x / d)C_m$, d being the average distance between the membrane and the fixed electrode, reveals other means of increasing the spectrophone sensitivity.

Variations in microphone capacity can be recorded using circuits of two basic types: those where the microphone connects directly to the preamplifier in sequence with high resistance and a polarizing voltage, and connection circuits with frequency conversion. Below we discuss briefly the specific features of their operation.

a) Microphones with Direct Connection

The circuit for directly connecting the microphone (Fig.2.5a) is the simplest parametric circuit, described by

$$\frac{dq_e}{dt} + \frac{q_e}{\tau_e [1 + (\Delta C / C)_{max} f(t)]} = \frac{U_p}{R} \quad , \tag{2.38}$$

where q_e is the electric charge, τ_e is the time constant of the input circuit $(\tau_e = RC_m)$; $(\Delta C / C)_{max}$ is the relative change of microphone capacity with polarizing voltage U_p; f(t) is the function describing the shape of the OA signal. The solution (2.38) can be easily attained by the small-parameter method, Specifically, for $f(t) = \sin \omega t$ the electric signal U with resistance R equals

$$U = \left(\frac{\Delta C}{C}\right)_{max} U_p \frac{(\omega \tau_e)^2}{(\omega \tau_e)^2 + 1} \quad . \tag{2.39}$$

Thus, to generate a maximum electric signal, it is necessary that the condition $\omega\tau_e \gg 1$ should be fulfilled, which in practice can be done by increasing R and using high-resistance preamplifiers. As applied to pulsed operation with $t_p \lesssim \tau_{vib}$, this condition has the form $\tau_e \gg \tau_{vib}$, τ_T.

The microphone sensitivity K_m can be found from the ratio between the output voltage U and the OA signal

$$K_m = \frac{U}{P} = \frac{2U_p}{(d + d_m/\varepsilon_d)i\omega Z(\omega)} = \frac{U_p A_m}{2(d + d_m/\varepsilon_d)(2\pi T* + \gamma P A_m^2/4V)} \quad , \quad (2.40)$$

with $V_m \gg V$ and $\omega \ll \omega_{mic}$, where d is the distance between the membrane and the fixed electrode, d_m, ε_d are the thickness and dielectric permittivity of the membrane (taken into account when the metallized coating is on the external side of the membrane with respect to the electrode).

According to (2.40), microphone sensitivity can be increased by increasing the polarizing voltage, by decreasing the interelectrode distance and by increasing the elasticity of the membrane itself and the gas tension in the spectrophone chamber. The relative matching is approximately determined by

$$n* = \frac{Y_{mem}}{Y_p} = \gamma \frac{P A_m^2}{8\pi V T*} \approx 1 \quad . \quad (2.41)$$

Along with high sensitivity, an important requirement imposed on acoustic pickups is a low level of intrinsic noise including the noise of the preamplifier connected to the pickup. Optimally this noise must be lower than or at least commensurable with the thermal fluctuations in the gas-membrane system which impose a limit on the sensitivity of the laser spectrophone (Chap.6). But in practice, as a rule the electric noise in the preamplifier input circuit is dominant. This noise has been analyzed in [2.5,14]. To define potential ways of decreasing this noise we may use the results of [2.16]. For the simplest preamplifier circuit (Fig.2.5a), this noise has been presented as

$$\bar{U}_N^2 = \left(\frac{4K_B T}{C_m^2 \omega^2 R} + \frac{2eJ_G}{C_m^2 \omega^2} + \frac{0.65}{g_m} 4K_B T + \frac{A_i^2}{f} \right) \Delta f \quad , \quad (2.42)$$

where K_B is the Boltzmann constant, e is the electron charge, J_G is the gate leakage current, g_m is the transistor conduction, and A_i is some constant coefficient. At low modulation frequencies down to 100 Hz, as a rule the thermal noise of the chain of parallel-connected resistance R and microphone capacity C_m as well as the current noise of the input transistor prevail. These noises

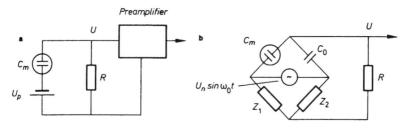

Fig. 2.5 **a,b.** Condenser microphones connected to an electric circuit: (a) with direct connection; and (**b**) bridge diagram with frequency transformation

are described by the two first terms in (2.42). They can be reduced at a given modulation frequency by increasing C_m, R and decreasing J_G. The technical limits for R and J_G approximately range from 10^{10} to $10^{11}\Omega$ and 10^{-11} to 10^{-10} A, respectively.

Thus, taking into account the above results we conclude that it is necessary to minimize T* and to increase C_m in order to raise the nonresonant spectrophone sensitivity. At the same time the average distance from the membrane to the fixed electrode should be stabilized and the attachment of the membrane due to polarizing voltage must be excluded. These conditions are conflicting but can be met, to a greater degree, in cylindrical microphones. In these microphones the free location of the membrane with very low tension makes it possible to minimize the d and T* values, and simultaneously to stabilize the position of the membrane relative to the electrode. Then a considerable increase of the membrane area through increasing its length ℓ_m along the spectrophone's optical axis does not cause the ballast volumes to grow up. The lower limit for d_m is determined from the condition of conservation of stability of the membrane's mechanical properties during its tension and the coating with a metal layer; it is usually of the order of 0.1 - 1 μm. The highest sensitivity of cylindrical microphones can be obtained with a rough electrode surface as well as by placing between the membrane and the electrode small limiting stops made, for example, of a thin fiber-glass wire wound on the electrode with a small pitch [2.6]. This can be explained by the fact that at increased polarizing voltages, up to 300 V, the attachment of the membrane is overcome and the efficiency of its total surface is increased. In such microphones their capacity can easily be up to $C_m \approx 500 - 1000$ pF. The noise level in this case, according to (2.42), will be $\sqrt{\overline{U_N^2}} \approx 20$ nV(Hz)$^{-1/2}$ at the frequency of $\omega / 2\pi = 20$ Hz. It is worthy of note that in this case the intrinsic noise level of the preamplifier is about 10 to 15 nV(Hz)$^{-1/2}$.

In microphones with plane fixed electrodes increasing the membrane radius r_m increases its flexibility at a fixed value of T*, which depends on the properties of the membrane material. But it is difficult here to increase C_m correspondingly because it is complicated to realize a constant distance between the electrode and the membrane over the entire area of the membrane. As far as the microphone design is concerned, it is impossible to increase the ratio r_m / r_C (r_C being the spectrophone chamber radius) without increasing the volume of the ballast channels connecting the working chamber with the microphone. Therefore, from a practical point of view the following limitation is imposed on the choice of a maximum membrane radius: $r_m / r_C \approx 3 - 4$ [2.17]. It should be noted that the behavior of the dependence of the plane-microphone sensitivity on polarizing voltage essentially differs from the same dependence for cylindrical microphones. In a cylindrical microphone, for example, the sensitivity increases linearly with an increase in U_p up to 200 - 300 V, and thereafter it does not change anymore. For a plane microphone, however, in the region of enhanced voltage the microphone "sensitivity", on the contrary, can even rise. This is explained by the influence of the negative elasticity of the electric attractive force. In this case, however, the plane microphone sensitivity is not stable, it is highly dependent on the voltage U_p, and the membrane tends to attach. Yet the cylindrical microphone sensitivity in the region of high U_p is rather stable and depends slightly on the polarization voltage.

Practice shows that with the microphones described a threshold spectrophone sensitivity can be realized which is 10 to 30 times worse than the theoretical limit.

Table 2.1 lists the basic parameters of the microphones under investigation. For comparison, it also gives the parameters of the best commercial microphones. It can easily be seen that the commercial microphones by virtue of a "strong" membrane tension, a low capacity and a small membrane area are not optimum for obtaining high spectrophone sensitivities.

Similar considerations can be applied to electret microphones whose advantage is the absence of polarizing tension. Under standard measurement conditions the sensitivity of the best electret microphones is commensurable with the sensitivity of condenser ones [2.18]. In some cases, with the use of compact commercial microphones with a "highly tensioned" membrane, for example, the spectrophone sensitivity can be increased by several microphones connected in series and placed along the optical axis of the spectrophone [2.19]. Such a gain in sensitivity is due to the fact that the signal increases by N_m times (N_m being the number of microphones), while the noises, being statistically independent, increase just by $\sqrt{N_m}$ times, i.e., the signal-to-noise ratio increases by $\sqrt{N_m}$ times. However, it should be taken into ac-

Table 2.1. Comparison of the parameters of various microphones

Basic parameters	Special microphones		Commercial microphones plane geometry	
	cylindrical	plane	4144 B and K	4 134 B and K
Diameter [mm]	$r_m \approx r_c$, $l_M^* \approx 30-60$ mm	30 – 60	23.77	12.7
Membrane thickness d_M[μm]	0.5 – 3	0.5 – 3	5	5
Distance between the membrane and the electrode d[μm]	0 – 10	10 – 30	22	20
Membrane area A_M [mm^2]	1000 – 2000	1000 – 3500	500	150
Capacity C_M[pF]	500 – 1000	100	55	18
Polarization voltage U_p[V]	200 – 350	50 – 100	200	200
Membrane tension T*[N/M]	1 – 10	10	$1.5 \cdot 10^3$	10^3
Sensitivity K_M [mV/Pa]	100 – 200	50 – 150	50	12,5

* l_M is length of microphone along axis

count that at optimum matching of the parameters of each microphone with the chamber the gain in sensitivity than can be compensated for by a corresponding increase of the total area of microphone membranes and a decrease of their equivalent capacity.

b) Microphones with Frequency Conversion

For highest sensitivity, schemes with frequency conversion (the capacity variation modulates the amplitude, frequency or phase of carrier frequency) are most promising [2.20]. This is due to an essential suppression of low-frequency noise of the preamplifier on account of a rather high carrier frequency (up to $10^5 - 10^6$ Hz) at low values of input resistance R (up to $10^6 \Omega$). This resistance plays an important role in decreasing the influence of stray electric current on the preamplifier input. Figure 2.5b illustrates a bridge circuit with amplitude modulation in which the microphone capacity C_m is balanced by the same capacity C_0: the two resistors Z_1 and Z_2 are chosen to be equal. The bridge is fed by a quartz oscillator with small impedance. In the absence of membrane vibrations, the bridge is balanced and so there is no signal at its output. A balanced bridge also decreases the influence of amplitude fluctuations (U_n) on the carrier frequency ω_0. When the microphone capacity changes, the bridge balances out and at its output a signal is generated, which has the form:

$$U \approx \left(\frac{\Delta C}{C}\right)_{max} \left(\frac{U_n}{4}\right) \sin \omega t \sin \omega_0 t \quad , \qquad (2.43)$$

where ω is the modulation frequency of laser radiation. After amplification, the signal is demodulated and recorded with a phase-sensitive detector.

With frequency modulation, the microphone capacity becomes part of the oscillator, whose frequency varies due to the OA signal. With phase modulation, the capacity is included into the phase-shifting circuit fed from a fixed-frequency oscillator. The signal at the output has the form

$$U = U_n \left(\frac{\Delta C}{C}\right)_{max} \sin\left[\omega_0 t + \varphi_0 + \varphi\left(\frac{\Delta C}{C}\right)_{max}\right] \quad , \qquad (2.44)$$

where φ_0, φ are, respectively, the time-constant and time-varying phase shifts. The phase detector balances the signals before and after the circuit, and produces a voltage proportional to the phase shift between them and hence proportional to the ratio $(\Delta C / C)_{max}$.

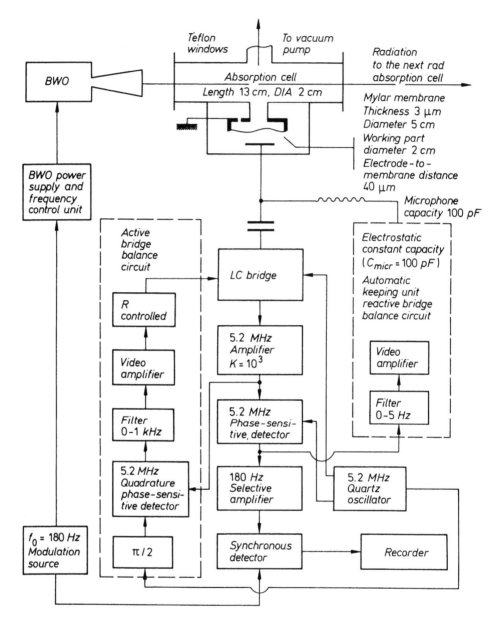

Fig. 2.6. High-frequency circuit diagram of a microphone with phase mudula-
tion (used in [2.21] in operations with a microwave radiation source)

Figure 2.6 depicts a system with phase modulation based on a high-frequency resonant bridge to exemplify the practical realization of the schemes described above [2.21]. The diagram contains two oscillatory circuits connected to the arms of the LC-bridge. One of the circuits is formed by the pickup capacity and inductivity, the other by the etalon capacity, equal to the pickup capacity, and inductivity. The bridge is controlled by a quartz oscillator. To increase the stability of operation, circuits are introduced which balance the bridge by servo tuning the pickup capacity with respect to both reactive and active components [2.22]. As practice shows, such diagrams make it possible to realize a spectrophone's threshold sensitivity only two or three times above the theoretical limit. Yet, compared to the direct-connection case, such schemes are characterized by disadvantages such as relative complexity, unstable operation, rigid requirements imposed on the reference oscillator, and the stability of the capacitance with respect to active and reactive components. It is difficult to satisfy the latter condition in a laser spectrophone due to the frequent change of mixtures to be analyzed in the working chamber.

Experience with different schemes shows that condenser microphones with direct connection to the preamplifier input are very useful, when compact and simple equipment with good operational characteristics at moderate sensitivity is necessary. Furthermore, in the IR range, it is often not necessary to achieve a low noise threshold since the threshold sensitivity is limited by background signals of different origin. In developing highly sensitive laboratory setups, especially with low-power radiation sources, we should use condenser microphones with bridge connection. Plane microphones are convenient for operation with rectangularly geometric laser beams, e.g., gas-discharge TEA or diode lasers as well as in spectrophones of special design, such as those with an extended range of working temperature. Cylindrical microphones, however, are optimum for axially symmetrical beams.

2.3.2 Optimization of Spectrophone Parameters

One of the main requirements of laser spectrophones is to realize a high threshold sensitivity. Therefore, the basic spectrophone parameters can be optimized by maximizing the signal-to-noise ratio for these parameters at a fixed laser power. The expression for the signal-to-noise ratio and hence the results of optimization depend essentially on the regime of laser operation and the nature of the dominant noise of the detecting system. This problem is illustrated below for the most interesting cases: pulsed and continuous operation with threshold sensitivity limited by electronic circuit noise.

The expressions derived in the foregoing sections enable us to express the signal-to-noise ratio at the preamplifier input as

$$\frac{S}{N} = \frac{P_0 N_0 \tau_{vib} U_p A_m f^*}{T \rho C_V \tau_{V-T} (d + d_m / \varepsilon_d)(4\pi T^* + \gamma A_m^2 P / 2V)} \quad , \tag{2.45}$$

where the f* factor is given for pulsed operation by

$$f_p^* = \bar{\Phi}(e^{-t/\tau_1} - e^{-t/\tau_2})_{max} \left(\bar{U}_N^2\right)^{-1/2} \tag{2.46}$$

and for amplitude modulation in CW operation by

$$f_{CW}^* = K_F \bar{J} \tau_T \delta \{\bar{U}_N^2(\omega)[1 + (\omega\tau_T)^2]\}^{-1/2} \quad , \tag{2.47}$$

where τ_1, τ_2 are the time constants determined by the relaxation times τ_{vib}, τ_T and the cutoff frequency of an amplifier.

In the optimization one should account for requirements of particular parameters with respect to measurement conditions and some design limitations in the spectrophone. Specifically, when determining the chamber cross section one should take into consideration the additional condition

$$r_{eff} / r_L = \xi \tag{2.48}$$

which relates the chamber's effective radius $r_{eff} \approx (V /\pi L)^{1/2}$ to the characteristic dimension of laser beam r_L by the coefficient ξ. Its value is determined by calculating the allowable background OA signal arising when stray light strikes the side walls of the chamber. The parameter ξ usually ranges from 2 to 5. Under these conditions the expressions (2.24) for $\bar{\Phi}$ and $\bar{I}$ can be approximated by (2.25) rather accurately.

Below we analyse nonresonant spectrophone parameters with and without matching the acoustic pickup with the spectrophone chamber.

a) Nonmatched Acoustic Pickups

For nonmatched acoustic pickups the spectrophone-chamber walls are supposed to be absolutely rigid, i.e., it is assumed in (2.33) that $Z = \infty$. In the AM regime for weakly-absorbing media ($\alpha L \ll 1$) the maximum value of the OA signal can be obtained if $\omega\tau_T \leqslant 1$, which can be realized by choosing a rather small modulation frequency ω. Then the spectrophone sensitivity does not depend on the chamber geometry and the modulation frequency ω; it is propor-

tional to the gas pressure P. But, since $\tau_T \propto r_{eff}^2$, the condition $\omega \tau_T \leqslant 1$
implies restricting the upper limit of r_{eff} (as well as P). In a first approximation the value of the OA signal slightly depends on the chamber length
L which allows L, and hence the chamber volume, to be minimized. At small ω
the time of measurement increases substantially and, besides that, low-frequency noise and vibrations have a pronounced effect. Therefore, in practice,
the relation $\omega \tau_T \geqslant 1$ is usually realized. In this case the sensitivity does
not depend on the pressure P and is inversely proportional to the square of
the chamber radius r_{eff}^2 and the modulation frequency ω. Since the electronic
circuit noise (2.42), inversely proportional to ω, is prevalent, the signal-
to-noise ratio is independent of ω approximately in the range up to 100 Hz.
In the end, the sensitivity should be increased because of the decrease in
r_{eff} until

$$\omega \tau_T \approx 1 \tag{2.49}$$

is realized.

In pulsed (P) operation with $\tau_1 \gg \tau_2$, (2.46), the amplitude of the OA
signal does not depend on gas pressure and is inversely proportional to the
square of the chamber radius r_{eff}^2. It should be noted, however, that at very
small values of r_{eff}, when $\tau_T \leqslant \tau_{vib}$ is fulfilled, the signal-to-noise ratio
no longer depends on r_{eff}, and its further reduction has no meaning. Then,
to estimate r_{eff},

$$\tau_T \approx 35 \tau_{vib} \quad , \tag{2.50}$$

at which the OA signal amplitude is smaller than its maximum value by just
10%, is recommended.

b) Matched Acoustic Pickup

The difference between matched and nonmatched acoustic pickup consists in
the necessity of taking the matching of the chamber and microphone parameters
into account through (2.41). So the initial problem can be solved by maximizing (2.24) for $\bar{\Phi}$ and $\bar{I}$, with (2.41,48) fulfilled at the same time.
In practice, it is realized by focusing a laser beam with simultaneously
decreasing the chamber cross-section and increasing its length for a given
chamber volume. The requirements (2.41,24) are conflicting and can be simultaneously fulfilled only in the ideal case of highly collimated beams when
an independent decrease in chamber cross-section and increase in its volume
would be possible due to an increase in length. In the limit this produces a

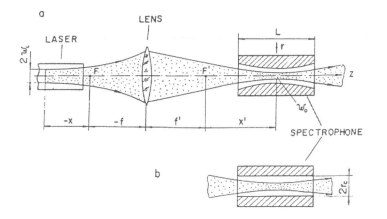

Fig. 2.7. One-component imaging optics for focusing laser Gaussian beams into the chamber of (**a**) a Gaussian-profile; and (**b**) a cylindrical spectrophone

waveguide chamber which cannot generally be used in the IR spectral region
because it is difficult to create appropriate, highly sensitive pickups, and
the background OA signals due to the walls have a pronounced effect.

By virtue of laser-beam divergence, the choice of the parameters r_{eff} and
V is, in practice, not free and is based on minimizing the parameter V for a
fixed chamber length L by providing an optimum focus of the laser beam (Fig.
2.7a). For the most important case of a Gaussian beam the radiation intensity
in the spectrophone chamber is given by

$$I(r) = \frac{2J}{\pi w_0^2 [1 + (\lambda z / \pi w_0^2)^2]} \exp\left(- \frac{2r^2}{w_0^2 [1 + (\lambda z / \pi w_0^2)^2]}\right) \quad , \tag{2.51}$$

where r is the radial coordinate, z is the coordinate along the optical axis,
and w_0 is the beam waist (at $1/e^2$) (Fig.2.7). To obtain maximum sensitivity
the chamber must have a Gaussian profile along the optical axis. Using (2.51),
it means in this case that the optimum waist of the laser beam in the spectrophone chamber is given by

$$w_0^{opt} = \sqrt{\lambda L / 2\pi\sqrt{3}} \quad . \tag{2.52}$$

The length is now fixed, the chamber volume is minimum and equal to

$$V_{min} = \xi^2 L^2 \lambda \sqrt{3} / 3 \quad . \tag{2.53}$$

An optical system must provide laser-beam focusing into the chamber, and
the "confocal" parameter of the beam must be equivalent to $L / \sqrt{3}$. For Gaussian

optics it is easy to get an expression for the focal distance, which matches the waist w_L in the laser cavity and w_0 in the spectrophone working chamber

$$f' = \pi \frac{w_0 w_L}{\lambda} \sqrt{(x\lambda / w_L^2 \pi)^2 + 1} \quad . \tag{2.54}$$

In most practical cases the second factor in (2.54) is close to unity, and thus the position of w_L relative to the lens is slightly dependent on the working section x (Fig.2.7a), i.e., the alignment is not influenced by small variations of the distance from the laser to the lens. In some cases, for example, at small λ and large w_L when the optimal matching of waists, (2.54), is possible, only at rather high value of f' (up to 1 m) a two-component optical system may prove useful to reduce the overall dimensions of the instrument.

These dependences reveal connections between the optimum values for the working-chamber parameters. In particular, the chamber length can be determined by substituting (2.53) into (2.41):

$$L \simeq A_m / 2\xi \; (\sqrt{3}\, P \; / \; 2\pi T*\lambda)^{1/2} \quad . \tag{2.55}$$

Then, using the value found for L one can determine w_0 from (2.52) and the chamber radius r_{eff} from (2.48). It should be taken into consideration that the choice of a Gaussian-profile chamber is justified only when maximum sensitivity is required. To make the chamber design simpler, it is suggested to shape it in the form of a cylinder (Fig.2.7b). The sensitivity is then reduced by no more than 20%.

The influence of the membrane area A_m on the signal-to-noise ratio manifests itself both in the pickup sensitivity K_m and in the noise which depends on A_m. Generally it is necessary to increase A_m, with (2.41) being fulfilled at the same time.

Other types of microphones. In laser Schlieren microphone a low-power He-Ne laser beam is deflected by a reflecting diaphragm mounted on a OA helmholtz resonator. The periodical pressure variations in the resonator distort the surface of the diaphragm so that the reflected laser beam is alternatively focused and defocused. The deflection is converted into an amplitude modulation of the beam by an iris located at a distance from the resonator, and detected with a photodiode. The light beam can be modulated at a high frequency and the signal from the photodiode processed with a lock-in amplifier so that noise with a power spectral density proportional to the inverse of the frequency is significantly reduced in the final OA signal. It appears that the laser Schlieren microphone [2.23] will find the application in OA experiments where small-volume OA cells are employed, or where conventional microphones cannot be employed because of high temperature or chemical corrosion problems.

The acoustic transducer of a unique fiber-optic spectrophone [2.24] is one arm of an allfiber interferometer. The sensing element is designed by winding 20 m of single-mode fiber around a 2.54 cm diameter hollow core formed by wrapping a 75 μm sheet of aluminized mylar into a cylinder. The fiber ends from the cylinder are fused into one arm of an allfiber Mach-Zehnder interferometer. The feature of this sensor is its large sensitive area, made possible by the inherent geometric flexibility provided by optical fibers. These feature may prove useful, for instance, in OA measurements of aerosol absorption.

However, the sensitivity of the fiber-optic spectrophone or of laser Schlieren microphone are not higher than the sensitivity of the condenser microphone. The fiber-optic spectrophone performance can even be improved in liquids where the acoustic impedance matching between fiber sensor and liquid is much better than that for the air.

3. Optoacoustic Spectroscopy of Condensed Media

There are two main OA methods to detect absorbed energy in condensed media: direct and indirect detection of acoustic vibrations. In the first case the acoustic vibrations are detected directly in the sample under study, an acoustic pickup is linked up with it. The main characteristics of this measuring technique are considered in Sects.3.1,2. In the second case the acoustic vibrations are detected in the gas or liquid contacting with the sample. As a rule, the introduction of an additional ('immersion') medium between the sample and the pickup improves their acoustic matching; it also facilitates the study of surface absorption in solids. Analysis of the indirect measuring method and a comparison with the direct one are given in Sects.3.3 - 5.

3.1 Optoacoustic Spectroscopy of Liquids with Direct Detection

Generally the magnitude of the OA effect in condensed media is smaller than in gases. For liquids this is due to their higher density (by about three orders) which results in smaller heating of liquids as compared to gases at the same absorbed energy, as well as in smaller displacements of the particles in the medium as the acoustic wave propagates with amplitude P:

$$\Delta x = P(2\pi \rho f V_s)^{-1} \quad , \qquad (3.1)$$

where f is the acoustic vibration frequency, V_s is the velocity of sound in liquids, and ρ is the liquid density. These specific features also explain why it is inefficient to use microphones for detecting the OA phenomenon in liquids, although they may have proved to be very useful in laser spectrophones for gas analysis. To confirm this we refer to [3.1] where a very low level of OA signals is measured in a number of liquids including water and alcohol. The OA signals were detected by an electrodynamic microphone placed directly in the cell with liquids irradiated by a high-power filament lamp modulated at a frequency of 200 Hz.

The use of much more powerful laser sources increases the OA effect in liquids and enables it to be used in analytical problems. Pulsed high-power laser excitation of the medium and piezoelectric pickups to record OA signals have proved especially efficient. In this case it is possible to measure the absorption coefficient in many liquids at a level of 10^{-6} cm^{-1} with the laser pulse energy of the order of 10^{-3}J [3.2]. Such sensitivity becomes commensurate with that of optoacoustic spectroscopy (OAS) in gases. It is quite sufficient to study the optical properties of most liquids whose absorption coefficients, including the best solvents, usually range from $10^{-4} - 10^{-6}$ cm^{-1}. When the acoustic waves are excited by CW-modulated laser radiation the threshold sensitivity for absorption is about 10^{-5} cm^{-1} with radiation power of around 1 W.

In comparison with the conventional spectrophotometric technique, OAS has the following merits for analysis of liquid substances: 1) a high sensitivity which allows weakly-absorbing media to be studied; 2) a small influence of scattered radiation; 3) the simple correction of spurious reflection on the windows of the measuring cell; and 4) a small effect of slight spatial displacements of the laser beam. As compared to other laser methods for analysing nonfluorescent samples, including the thermolens [3.3-6], the interference [3.7] and frequency modulation [3.8] methods, the main advantage of OAS is its technical simplicity and hence its high accuracy.

The pulsed laser was first used by WHITE [3.9] to generate sound waves in liquids. The generation of acoustic waves during liquid transparent media have been irradiated with a pulsed laser was also oberseved in [3.10,11]. In [3.10] it was assumed that this phenomenon was caused by either electrostriction or thermal expansion due to heating. In [3.11] the initiation of sound waves was connected with a breakdown in the liquid. In [3.12] the experimental and theoretical results agree well for pulsed excitation of sound waves in water containing an absorber of ruby-laser radiation. For other works carried out before 1973 we refer the reader to [3.13]. A number of later works deal with theoretical considerations of sound vibrations generated in liquids under laser radiation [3.14-16]. The utilization of the OA effect in liquids in laser spectroscopy was discussed in [3.2,17-32].

3.1.1 Formation of Optoacoustic Signals

When high-power radiation interacts with a liquid, sound vibrations may be produced. The physical mechanisms differ for different phenomena, such as thermal expansion, induced scattering, electrostriction, steam generation, optical breakdown. It should be noted that not all these mechanisms of sound generation have been studied in detail yet. Nevertheless, the OA effect due to a purely thermal mechanism is of practical value.

As a rule, the thermal mechanism of sound generation is dominant at a comparatively low intensity of the laser radiation. For absorbing media the radiation intensity is restricted by the processes of liquid boiling and steam generation in the focal region of the laser beam, typical for pulsed excitation. From the liquid aggregate state the continuity condition in the area of heating limiting the time-average laser radiation intensity I is:

$$I < (T_B - T_0)\rho C_p / \alpha t_p \quad \text{for} \quad t_p \ll \tau_T \quad , \tag{3.2}$$

where ρ and C_p are the liquid density and specific heat; T_0 is the initial temperature of the liquid; T_B its normal boiling temperature; t_p the laser-pulse duration; τ_T the time of thermal relaxation, and α the absorption coefficient in the liquid.

For weakly absorbing media the radiation power is restricted by optical breakdown and electrostriction. Typical threshold intensities at breakdown occurs for most liquids range from 10 to 10^3 MW / cm^2 [3.13]. The most serious problem is to eliminate the linear electrostriction. It may be seen from preliminary estimates [3.13,14,29] that under the typical measurement conditions of most spectroscopic tasks the thermal mechanism of sound generation in absorbing liquids under pulsed excitation is dominant, with the laser-pulse duration complying with the condition $t_p \leqslant 1$ ms and the radiation continuously modulated in the range from 10^2 to 10^9 Hz. In case of weakly absorbing media, especially with the use of pulsed lasers, electrostriction may limit essentially the OAS sensitivity [3.3].

Let us consider sound excitation when radiation is absorbed inside the medium, the total absorbed energy is small and when the substance does not evaporate. Then the sound is generated due to expansion of the quickly heated volume. This process is isobaric and in most cases it is adiabatic. The adiabaticity criterion of the process here is $\ell_T \ll r_L$, ℓ_T being the thermal diffusion length in the liquid. Under pulsed excitation $\ell_T = (kt_p)^{1/2}$ and under continuous, $\ell_T = (2k/\omega)^{1/2}$, k is being the thermal diffusivity, and ω the modulation frequency. For water $k = 1.5 \times 10^{-7}$ m^2 / s. This condition is usually realizable in practice. For example, with the laser-pulse duration $t_p \leqslant 10^{-3}$s or the modulation frequency $f \geqslant 10^3$ Hz, for most liquids $\ell_T \leqslant 10^{-3}$ cm, being much smaller than the radius of the laser beam usually used. When this condition is not satisfied, account must be taken of temperature waves. First approximations in theoretical analysis often neglect sound attenuation

47

in the liquid, i.e., the influence of heat conduction is neglected and the liquid is assumed to be inviscous.

Under the above assumptions from the linearized hydrodynamics equations and the energy balance in a homogeneous medium we get the following equation for pressure change in a sound wave [3.13]:

$$\frac{1}{V_S^2} \frac{\partial^2 P}{\partial t^2} - \Delta P = \frac{\beta_V}{C_p} \frac{\partial Q}{\partial t} \quad . \tag{3.3}$$

Here $\beta_V = V^{-1}(\partial V / \partial T)_p$ is the coefficient of thermal volume expansion, C_p is the specific heat, Q is the laser-radiation energy flux absorbed per unit time and converted into heat.

Equation (3.3) can be solved using Laplace's transformation and Green's function. As applied to sound excitation in a rather small closed volume of the OA cell, the final expressions are cumbersome and inconvenient for further analysis. So we shall use here some results of the approximated calculations from [3.13,14], which nevertheless adequately explain the main features of the OA effect in liquids.

a) Pulsed Excitation

In the pulsed regime the basic characteristics of an acoustic pulse (its amplitude and temporal shape) materially depend on the geometric parameters of the experiments, particularly on the form of the region of heat generation, on the ratio of its transverse dimension r_L (the laser beam radius) and the effective length $\ell_s = V_s t_p$. The value ℓ_s determines the depth of sound propagation with velocity V_s during energy deposition, i.e., the pulse duration t_p.

With $r_L \ll \ell_s$ the region of energy deposition has time to expand during the laser pulse, so generating a sound. This process is described by solving the boundary problem of expanding region of absorption. With $r_L \gg \ell_s$, the radiation absorption results in tensions which then lead to the sound pulse described by the solution to the initial-value problem.

Let a laser pulse with intensity I and duration t_p travel through transparent and absorbing media along the x axis. There are two types of boundary conditions:

a) free surface (air-liquid boundary)

$$P(x = 0, t) = 0 \quad ,$$

b) rigid surface (cell-wall-liquid boundary)

$$\partial P(x = 0, t) / \partial t = 0 \quad .$$

Provided that the observation point is far from the surface the one-dimensional solution of (3.3) in case (a) is

$$P(x, t) = \frac{V_s \beta_v I}{2C_p}
\begin{cases}
0 & t < x / V_L \\
e^{-\alpha(x-V_s t)} (1 - e^{\alpha V_s t_p}) & t < x / V_s \\
e^{-\alpha(x-V_s t)} - e^{-\alpha(x-V_s t + V_s t_p)} & x / V_s < t < x / V_s + t_p \\
e^{\alpha(x-V_s t)} (1 - e^{-\alpha V_s t_p}) & t > x / V_s + t_p
\end{cases}
\tag{3.4}$$

and in case (b)

$$P(x, t) = \frac{V_s \beta_v I}{2C_p}
\begin{cases}
0 & t < x / V_L \\
e^{-\alpha(x-V_s t)} (1 - e^{\alpha V_s t_p}) & t < x / V_s \\
2 - e^{\alpha(x-V_s t)} - e^{-\alpha(x-V_s t + V_s t_p)} & x / V_s < t < x / V_s + t_p , \\
e^{\alpha(x-V_s t)} (1 - e^{\alpha V_s t_p}) & t > x / V_s + t_p
\end{cases}
\tag{3.5}$$

where V_L is the velocity of light. Analysis of the expressions shows that for the free surface only the compression and dilation pulses following each other at distance t_p are excited while for the rigid surface, only the compression pulse is excited.

The amplitude of the excited sound pulse depends on $\alpha V_s t_p$. If $\alpha V_s t_p \gg 1$ (the case of a strongly absorbing medium) it is possible to attain the maximum amplitude

$$P_{max} \simeq \frac{V_s \beta_v I}{2C_p} \tag{3.6}$$

determined by the intensity of the exciting radiation. When $\alpha V_s t_p \ll 1$ (the case of a relatively weakly absorbing medium) the amplitude

$$P_{max} \simeq \frac{\alpha V_s^2 \beta_v}{2C_p} I t_p \tag{3.7}$$

depends on the total energy flux $\Phi = I \cdot t_p$ of the pulse. Sound can be excited more effectively for a rigid surface since the substance can expand only to one side when heated.

When observed along the light-beam axis and with $\alpha V_s t_p \ll 1$, in a first approximation the OA signal is shaped like the laser-pulse envelope [3.32,33].

49

When observed from one side of the laser beam, the signals from elementary sources across and along the laser beam reach the observation point with time delays that determine the OA signal duration and shape.

At optical excitation of weakly absorbing media ($\alpha L \ll 1$) in the closed volume of an OA cell mostly cylindrical sound waves are formed [3.28,29,34, 35]. The maximum pressure at distance r from the laser beam can be approximately evaluated from [3.29]

$$P_{max}(r) = \frac{\beta_V V_S E \alpha}{\pi r_L^{1/2} C_p r^{1/2}} \times \left\{ \begin{array}{ll} (2\tau_p)^{-1} & \text{for } r_L \ll \ell_S \\[2mm] V_S / r_L & \text{for } r_L \gg \ell_S \end{array} \right.$$

$$\begin{array}{l} (3.8a) \\[4mm] (3.8b) \end{array}$$

where E is the laser-pulse energy.

b) Continuous Excitation

It is assumed that continously excited sound due to modulated laser radiation is described in the plane $x = 0$ (the liquid-surface boundary) by

$$I(r, t) = I(r)(1 + \delta\cos \omega t) \quad . \tag{3.9}$$

Like pulsed operation, the nature of sound excitation in liquids depends on the ratio between the dimensions of the region of heat release, r_L and the sound wavelength $\lambda_S = 2\pi V_S / \omega$.

There are three types of heat radiators [3.36]. If r_L, $\alpha^{-1} \ll \lambda_S$ this heat radiator is an acoustic point monopole at the liquid surface with its radiation, coinciding with the dipole radiation (in a homogeneous medium). If $r_L \gg \lambda_S$, α^{-1} the sound radiation is highly directional along the laser-beam axis with an effective width of $2V_S / r_L \omega$. If $\alpha^{-1} \gg \lambda_S$, r_L (the case of weakly absorbing media) the sound is radiated perpendicular to the laser beam.

In the intermediate region between the latter two cases, the radiation pattern in the plane including the beam axis has a maximum in the direction of an angle from the surface proportional to $\alpha V_S / \omega$, and its angular width equals $2\sqrt{3} \, \alpha V_S / \omega$.

Specifically, for a laser beam with Gaussian intensity profile the solution of (3.3) for absorbing media in the near zone $\ell \ll \alpha r_L^2$ (being the distance between the liquid surface and the point of observation) has the form [3.36]

$$P \approx \frac{\beta_V V_S^2 \omega \alpha}{C_p (\alpha^2 V_S^2 + \omega^2)} \, I \, e^{-\left(\frac{r}{w_0}\right)^2} e^{-\alpha_S \ell} \quad , \tag{3.10}$$

where α_s is the sound absorption coefficient in the liquid. In deriving (3.10) the resonant acoustic properties of the cell filled with the liquid were not taken into account. In the case of excitation of a weakly absorbing medium in a cylindrical OA cell with radius r_c, the pressure near the walls is determined by [3.37]

$$P(r_c) \approx \frac{\beta_V \alpha J V_s}{2\pi C_p r_c} \frac{1}{J_1^*(\omega r_c / V_s)} \quad , \tag{3.11}$$

where $J_1^*(\omega r_c / V_s)$ is the Bessel function of first order. For small $\omega r_c / V_s$, $J_1^*(\omega r_c / V_s) \approx \omega r_c / 2V_s$ and the value of the OA signal is inversely proportional to the modulation frequency ω. As ω increases above $\omega > 2V_s / r_c$, a resonant increase in OA signal is possible [3.38]; its maximum amplitude is limited by acoustic losses in the OA cell. Thus, for the most interesting case of weakly absorbing media the relationship between the OA signal amplitude and the liquid's absorption coefficient is linear

$$P = K\alpha J \text{(or } E) \tag{3.12}$$

where K is a constant factor for a particular measuring setup accounting for the mutual geometry of the measuring OA cell, the laser beam and the acoustic pickup as well as for the properties of the liquid under study, and

$$K \propto \frac{V_s^n \beta_V}{C_p} \qquad n = 1, 2 \quad . \tag{3.13}$$

The OA method can be used in analytical spectroscopy of weakly absorbing liquid media since in a liquid its signals depend linearly on the absorption coefficient.

3.1.2 Liquid Measuring Cells

In optoacoustics the liquid to be studied is placed in a measuring cell comprising a closed chamber with optically transparent windows for the passage of laser radiation, with an acoustic pickup in it.

a) Types of Acoustic Pickups

Piezoelectric pickups are the most effective for detecting OA signals in liquids, because of their high sensitivity (approximately 0.01 - 1 mV / Pa), their linearity over a wide dynamic range, a wide frequency band (10 - 100 MHz) and, very important, their good acoustic matching with the liquid medium. The latter means that an acoustic wave propagates with slight attenuation through the boundary between the liquid and the solid ceramics composing the pickup. The transmission coefficient for normal incidence of an acoustic wave on the surfaces of the two media 1 and 2 can be determined from

$$
T = \frac{4Z_1Z_2}{(Z_1 + Z_2)^2} \quad ,
$$

(3.14)

where $Z = \rho V_s$ is the acoustic impedance of the medium. For most gases (g), liquids (ℓ) and solids (s) the value of this parameter equals, respectively, $Z_g \approx 10 - 100 \; g \cdot cm^{-2} \; s^{-1}$; $Z_\ell \approx (1 - 2) \cdot 10^5 \; g \cdot cm^{-2} \cdot s^{-1}$; and $Z_s \approx (6 - 60) \cdot 10^5 \; g \cdot cm^{-2} \; s^{-1}$. According to (3.14), the transmission coefficient for a typical liquid-gas boundary will be $T_{\ell-g} \approx 10^{-4}$ and for a liquid-solid boundary $T_{\ell-s} \approx 0.1 - 0.5$. It is evident that one should use solid pickups in analyzing liquid media since $T_{\ell-s} \gg T_{\ell-g}$.

b) Cell Designs

The simplest cell design with good sensitivity contacts the liquid directly with the pickup surface by placing the acoustic pickup in the liquid. To isolate the pickup electrically, as for hydrophones, its surface may be coated with some nonconducting substance [3.21-24]. In [3.23] the liquid was inside a cylindrical pickup whose form matched the geometry of the narrow laser beam. An OA cell has been developed in the form of an elliptical cylinder in which essentially all the acoustic energy generated by a laser beam propagating along one axis is focused into a cylindrical acoustic transducer located along the other axis [3.39]. Preliminary measurements on a liquid-filled cell of this design show a high sensitivity and a notably clean pulse response. The disadvantage of such a pickup arrangement is that through direct contact with the pickup surface the liquid may become contaminated. This is especially undesirable in the analysis of highly pure or chemically active media.

Cells where the pickup is located outside the liquid chamber, particularly on the external side of the wall, are free of this disadvantage. Such an arrangement is conditioned by sufficient acoustic matching of different media in the liquid-wall pickup system. In practice, there ought to be good acoustic contact at the wall-pickup boundary to reduce spurious attenuation of OA signals. This can be realized by polishing the contacting surfaces and introducing between them an immersion adapter from any plastic material with similar wave resistance, e.g., indium, or even standard vacuum grease.

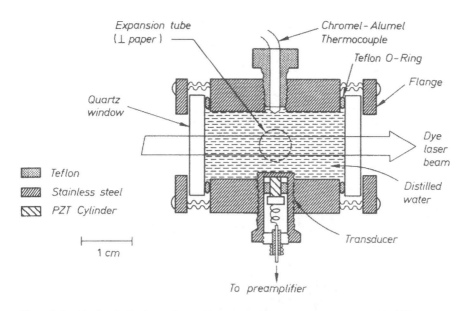

Teflon

Stainless steel

PZT Cylinder

├────────┤
  1 cm

Fig. 3.1. Typical design of an OA cell for analysing liquids [3.28]

Chemical inertness to the liquids under study and low adsorbability are basic requirements of the material for OA cells. Therefore, they are usually made of stainless steel (with teflon seals) and in some cases of quartz or pyrex [3.2,17,40,41]. Figure 3.1 illustrates a typical stainless steel cell (2.5 cm long and 1.2 cm in diameter). When used in the visible and near IR spectral regions their windows are made of quartz. The cylindrical pickup is made of ZTL (zirconate-titanate-lead) ceramics, pressed to a thin polished diaphragm, about 1 mm thick, through which the sound vibrations from the liquid act on the pickup surface. To minimize the sound reflection at the pickup boundary a lead disk is pressed to the external side of the pickup. To improve the acoustic contact between the pickup and the diaphragm a standard vacuum seal is used. Other designs of OA cells were described in [3.28,29].

c) Pulsed Signals

Typically a pulsed OA signal in the above-described measuring cell is a train of damped vibrations with the decay time of several hundreds of microseconds (Fig.3.2). According to (3.5) the first single unipolar acoustic pulse contains information on the value of absorbed energy. Its delay depends on the propagation time of the sound wave from the point of excitation to the pickup. In a closed cell with a liquid, which serves as an acoustic cavity, due to

53

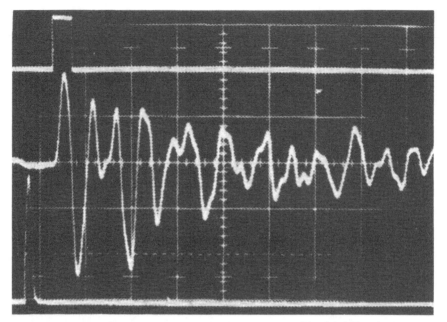

Fig. 3.2. Typical shape of a pulsed OA signal for liquid CCl_4 + benzene (middle oscillogram); shape of a dye laser pulse of $1-\mu$ s (bottom); time interval of integration (top); scanning of 10 μ s/division [3.2]

multiple reflections from the walls the original directed acoustic pulse converts to a train of different-polar nondirectional pulses.

Information on the absorption coefficient can be obtained through selective detection and integration of the first acoustic pulse using the standard gating technique (Fig.3.2). Thus, it is possible to 1) increase the sensitivity of pulsed OAS due to signal accumulation under laser pulse repetition; 2) eliminate background OA signals due to radiation absorption by the cell windows and walls as well as subsequent resonant pulses which strongly depependent on the cell parameters and external conditions of measurement. To avoid background signals from the cell windows, propagating through the cell walls, it is advisable to isolate the windows and the acoustic pickup from the cell walls acoustically. The spectrum of pulsed OA signal, illustrated in Fig.3.2, is concentrated mainly in the band from 100 to 500 kHz.

The volume expansion coefficient β_v and hence, according to (3.13), the OA signal depend greatly on the liquid's temperature (for water at $T \approx 4^{\circ}C$ β_v even changes sign [3.40]). Therefore, for tasks other than acoustic isolation, the OA cell should be also thermo-stabilized.

d) Features of Continuous Operation

When operating with CW lasers one can use almost the same cell designs as in pulsed operation. In continuous operation the OA cell geometry should be carefully considered since as the modulation frequency coincides with one of the resonant acoustic frequencies of the liquid chamber, the increase in signal can increase the OA cell sensitivity. Figure 3.3 shows typical resonant characteristics of a cell with a cylindrical pickup of 7.5 cm long and 1.5 cm in diameter. Low-frequency resonance is usually conditioned in such cells by the influence of the liquid inlet hole which together with the main cell volume forms an acoustic resonator. In the region of low sound frequencies (not presented in Fig.3.3) in the absence of resonances the OA signal decreases with increasing frequency approximately according to a ω^{-1} [3.26-27], as follows from (3.11).

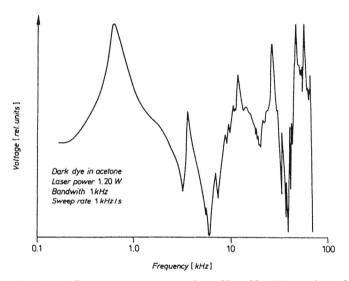

Dark dye in acetone
Laser power 1.20 W
Bandwith 1 kHz
Sweep rate 1 kHz/s

Voltage [rel. units]

Frequency [kHz]

Fig. 3.3. Frequency response of an OA cell with a piezoelectric cylindrical pickup [3.23]

In conclusion, it may be said that the threshold sensitivity of OA cells can be limited by direct incidence of scattered laser radiation on the surface of piezoelectric pickups due to their characteristic pyroelectric properties. The influence of scattered radiation can easily be eliminated by special diaphragms introduced between laser beam and pickup. Such diaphragms must be optically nontranslucent to radiation but able to transmit acoustic vibrations well (acoustically translucent).

In designing OA cells for liquids their sealing ability must be kept in mind. The foregoing results are given for nonsealed cells in which the process of OA signal formation is preferably isobaric. Though it is difficult to realize in practice, hermetically sealed cells can increase the OA sensitivity, since the low compressibility of liquids results in their thermal expansion under certain conditions, considerably increasing pressure on the piezoelectric pickup.

3.2 Optoacoustic Spectroscopy of Solids

Under powerful laser radiation elastic sound waves arise in a solid transparent sample due to the thermoelastic tensions it develops from fast heating of the irradiated part of its volume. The effect of sound-wave generation in solids under laser radiation was originally observed by WHITE [3.42] and others [3.43,44]. In one of the first theoretical works [3.45] the OA effect under pulsed excitation of a solid was analyzed. Experimental techniques were described in [3.46-49]. In [3.50-56] the OA effect in solids was first applied in laser spectroscopy to measure small coefficients of optical materials.

3.2.1 Formation of Optoacoustic Signal

The amplitude of acoustic waves due to the OA effect in solids is proportional to thermoelastic tensions arising from the transformation of absorbed laser energy into thermal energy of the medium. Approximate values for these thermo-elastic tensions were given in [3.53,57].

For simplicity, it is assumed that a solid is isotropic and unlimited in extent. Here we consider only longitudinal elastic waves since the contribution of transverse waves for the class of problems under consideration may, in a first approximation, be neglected. It is assumed that the laser beam is cylindrically symmetric, has a Gaussian intensity profile and travels along the z axis. Under the above assumptions the value of the elastic tension ε_i in a solid sample can be determined by solving the following system of equations [3.57]

$$u = \nabla \Phi^* \tag{3.15a}$$

$$\nabla^2 \Phi^* - \frac{1}{v_s^2} \frac{\partial^2 \Phi^*}{\partial t^2} = \frac{1 + \sigma_p}{1 - \sigma_p} \beta_v T \tag{3.15b}$$

56

$$\nabla^2 T + \frac{2\alpha J(t)\,\exp(-2r^2/w_0^2)}{\pi w_0^2} = \frac{1}{K}\frac{\partial T}{\partial t} \qquad (3.15c)$$

$$\varepsilon_{rr} = \frac{\partial^2 \Phi^*}{\partial r^2} \quad , \quad \varepsilon_{\theta\theta} = \frac{1}{r}\frac{\partial \Phi^*}{\partial r} \quad , \qquad (3.15d)$$

where u is the displacement value; Φ^* is the deformation potential; V_s is the longitudinal sound velocity; σ_p is the Poisson's coefficient; α is the absorption coefficient; k is the thermal diffusivity $k = K/\rho C$, K, ρ, C denote, respectively, the thermal conductivity, the density and the thermal capacity of the sample; and w_0(or r_L) is the radius of the Gaussian beam (at the level $1/e^2$); ε_{rr} and $\varepsilon_{\theta\theta}$ are the radial and azimuthal elastic tensions respectively.

For the OA effect in solids the ratio between the following times is important: the thermal relaxation time τ_T, the deformation relaxation time τ_D, the time of energy relaxation in a molecular system τ_{rel} and the laser pulse duration t_p. For most solids $\tau_{rel} \approx 10^{-13} - 10^{-14}$s. From (3.15) it follows that $\tau_D \approx w_0/V_s$. For example, for typical measurement conditions $w_0 = 0.25$ cm, $V_s = 5 \cdot 10^5$ cm/s and $\tau_D \simeq 5 \cdot 10^{-7}$s. The thermal relaxation time $\tau_T \sim r_L^2/k$ in this case is of the order of 10^{-3}s. Since this value is much longer than τ_D, the influence of thermal conduction may be neglected.

Equation (3.15) can be solved using the integral Hankel and Laplace transforms. Generally expressions to determine ε_i turn out to be very cumbersome. In [3.53] an approximated expression was derived for laser pulses with an exponentially increasing time development $J(t) = J[1 - \exp(-t/\tau)]$, τ being some time constant. It is assumed that $(r/w_0)^2 \gg 1$, i.e., the pickup is far from the laser beam; the condition $r^2 \gg kt$ is realized in practice for short pulses or for a high-frequency modulation; the condition $V_s^2 \gg k^2/r^2$ is fulfilled well with $t \gg \tau$. Then the radial and azimuthal tensions are determined from

$$\varepsilon_{rr} = -\varepsilon_{\theta\theta} = -\frac{\alpha\beta_v J(1 + \sigma_p)t}{2\pi C\rho(1 - \sigma_p)r^2} \quad . \qquad (3.16)$$

Now let us consider the numerical example: $\alpha = 10^{-4}$ cm^{-1}; $J = 0.1$ W; $\beta_v = 10^{-5}$ (K)$^{-1}$; $r = 2$ mm, $C = 1$ J/g$\cdot$K; $\rho = 3$g/cm^3; $\sigma_p = 0.3$; $t = 5 \cdot 10^{-3}$ s [3.53]. Substituting these values into (3.16) gives $\varepsilon_{rr} \simeq -1.25 \cdot 10^{-12}$. For a standard piezoelectric pickup this value of ε_{rr} corresponds to output voltages of several µV [3.53].

A one-dimensional theory for the generation of ultrasonic waves by light absorption in a composite structure was developed in [3.58]. The model consists of a backing material, an absorbing bulk or surface film, and a sample. For instance, the initial calculations were made for two cases of the device: (a) backing: sapphire, film: molybdenum, sample: fused quartz, and (b) backing: air, film: molybdenum, sample: sapphire. The amplitude of the elastic wave in the sample at 20 MHz is 40.8 dB greater in the case (a) than in case (b). Agreement between this theory and the experiment in [3.59] is quite good.

3.2.2 Characteristics of Solid Measuring Cells

It should be noted that many conclusions drawn in Sect.3.1 for OAS of liquids are also valid for OAS of solids. Below we consider the specific features of some schemes of measurement only.

As for an OA analysis of liquids, the use of piezoelectric pickups directly in contact with the surface of the sample is optimal for recording acoustic waves in solid samples. For such pickups joined to the end face of the sample the electric signal output depends on thermal and optical features of the sample [3.54]

$$U \approx \frac{KJ\beta_L}{i\omega\ell\rho_s C_s} \times \begin{cases} -2, & \ell \gg \ell_T, \; \ell \gg \ell_\alpha \quad &\text{(a)} \\ 1, & \ell_T \gg \ell \gg \ell_\alpha \quad &\text{(b)} \\ 1 - e^{(-\alpha\ell)} \pm 6/\ell\alpha[(1 - \alpha\ell/2) - e^{(-\alpha\ell)} \\ \quad \times \; (1 + \alpha\ell/2)], & \ell_\alpha \gg \ell \gg \ell_T \quad &\text{(c)} \end{cases} \qquad (3.17)$$

Here ℓ is the sample length, $\ell_\alpha = 1/\alpha$ is the optical absorption depth, ℓ_T is the thermal diffusion length; ρ_s, C_s are the density and thermal specific heat capacity of the sample, respectively; β_L is the linear expansion coefficient of the sample. The value of K is determined from

$$K = 2e^p_{31}\ell_p \frac{1 + \sigma_p}{\varepsilon^\alpha_{33}A_p\pi} \quad , \qquad (3.18)$$

where e^p_{31}, ε^α_{33} are the piezoelectric and dielectric constants, A_p is the pickup area; ℓ_p is the thickness of pickup. In (3.17c) the plus and minus signs correspond, respectively, to the position of the pickup on the front and back surfaces of the sample. Expressions (3.17) are typical of (a) thermally and optically thick samples; (b) thermally thin and optically thick samples; and (c) thermally thick and optically thin samples.

Accordingly, when the pickup is located on the end face, the output signal is inversely porportional to the modulation frequency ω, which is true for the region of low frequencies, and the sample length ℓ. Also it is linearly dependent on the absorption coefficient α in the region of relatively low values when $\alpha\ell < 1$. In [3.54] these conclusions were supported by erperiments and it was demonstrated that, as the radiation power varied by six orders, the OA signal was linear.

The sample must provide good acoustic contact with the pickup. Therefore, the sample should be specifically prepared for measurements. This preparation consists of polishing a section on its surface, whose dimensions must be commensurable with the overall dimensions of the acoustic pickup in use. With such a scheme, however, it is difficult to perform high-sensitive OA analysis of samples which are different in volume or small in size. Scattered radiation is eliminated by introducing an opaque immersion packing between the pickup and the sample as well as by using a differential scheme of measurement with two pickups [3.51]. Further the pickup in acoustic contact with the sample records the useful and background signals, while the other not in acoustic contact records only the background signal of scattered radiation. As a result, the background signals at the differential circuit output are compensated. With modulated laser radiation the OA signal amplitude depends weakly on the modulation frequency in a wide frequency band, approximately from 150 Hz to 40 kHz [3.53,56].

An increase in OA signal at higher frequencies is caused by the resonant properties of the pickup and the sample, and square-wave modulation being more effective than sinusoidal modulation. It follows from (3.16) that the pickup signal depends quadratically on the distance between the pickup and the laser beam r. With $r \gg w_0$, the signal is independent of the laser-beam diameter.

To obtain separate information on volume and surface absorption poses a serious problem in studying solids. In the scheme described this problem can be solved by phase selecting the OA signal since these signals arrive at the pickup from different zones of the sample at different times. It is possible, in particular, to obtain a $180°$ phase shift of OA signals from surface absorption in relation to the OA signal from volume absorption by selecting a proper mutual geometry of sample, laser beam and pickup [3.50]. It is also possible to determine the output signal as a function of the distance between the pickup and the laser beam. The corresponding dependences for volume and surface absorption behave differently in this case (Fig.3.4), because surface absorption gives rise preferably to a spherical sound wave while volume absorption has a radially symmetric sound wave. The degree of contribution of each type of absorption can be estimated from the resultant curve.

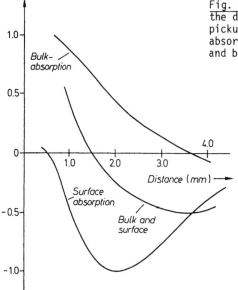

Fig. 3.4. Dependence of the OA signal on the distance between the piezoelectric pickup and the laser beam with surface absorption only, bulk absorption only, and both combined [3.50]

The CW-laser OA technique makes it possible to measure the absorption co-efficient at a level of 10^{-5} cm^{-1} with radiation power of about 1 W.

The generation and detection of short acoustic pulses (10 ns widths) in opaque plates using low-energy N_2-laser pulses ($\lesssim 1$ mJ) for the excitation was described in [3.60]. For time-resolved detection of the longitudinal, shear and surface waves a thin-film piezoelectric transducer (ß-polyvinylidene difluoride PVF$_2$) was used. This OA material testing technique is useful for rapid ultrasonic measurements or for detecting deep subsurface flaws.

An OA signal generated by pulsed laser excitation can be analyzed, for in-stance, by the discrete Fourier transform technique [3.61,62]. Some peculiari-ties of the spectrum allow one to discriminate the OA signal from window absorption and various electrical and mechanical noise sources [3.63].

In comparison with other method for analyzing weakly-absorbing solid samples (calorimetry, interference, etc.), the merits of OAS are high response and the possibility of analyzing samples with a rather small length up to several millimeters. However, it should be said that along with acoustic detection of elastic waves in a sample, the absorbed energy can also be detected by piezoelectric calorimetry, whereby a properly fixed piezoelectric pickup de-tects the pressure from the sample via thermal expansion [3.64].

3.3 Optoacoustic Spectroscopy of Condensed Media with Signal Detection in a Gas Medium (Photoacoustic Spectroscopy)

The energy absorbed in condensed media can also be measured by indirectly re-cording OA signals in the gas contacting the sample. As already noted in Chapt.1, this approach is also known as photoacoustic spectroscopy (PAS) (Fig.3.5). The sample is placed in an acoustically closed cell with optically transparent windows, using a sensitive microphone and a gas not absorbing or slightly absorbing radiation in the spectral range concerned. In most practi-cal tasks air is used as such a gas. The sample is subjected to modulated or pulsed radiation.

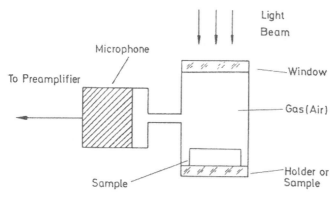

Fig. 3.5. Basic scheme of OAS of condensed media

Optoacoustic signals in a gas form in the following way. During radiation absorption nonradiative transitions are responsible for heating the sample as a whole or only its surface. Part of the thermal energy is transferred through the sample surface to the gas contacting the sample. Periodic heating of the gas causes its pressure to vary, which is detected by the microphone as an OA signal. The OA signal spectrum is recorded by the signal from the microphone as a function of wavelength of the radiation falling up on the sample. The characteristic feature of OAS compared to the standard spectro-photometric technique is that the OA spectrum depends both on the "purely" optical properties of the sample and on its thermodynamic parameters. This feature, to some extent, impedes correct interpretation of measurements. Con-sequently, to obtain strict correspondence between the OA spectrum and the optical absorption spectrum it is necessary to fulfill special requirements placed particularly on the modulation frequency and the sample dimensions. At the same time this peculiarity somewhat widens the analytical scope of the

OA method. Specifically, owing to this it can measure some thermodynamic parameters of condensed media.

Since the OA signal is directly proportional to the energy absorbed in the sample, scattered light causing serious problems in ordinary spectroscopy does not play a significant part in OAS. This makes OAS effective in spectral analysis of highly scattering and highly absorbing objects, inconvenient for ordinary spectroscopy, such a various powders, biological media, paper, fiber, etc. Significant analytical possibilities of OAS and its competive abilities compared to the conventional technique have so far been demonstrated in many applications including physics, chemistry, biology and medicine [3.65,66].

Some problems of practical importance connected mainly with the analysis of highly absorbing substances are solved in OAS by nonlaser radiation sources, for example, with a powerful xenon lamp. More powerful laser sources increase the sensitivity of the method substantially, opening the way for OAS analysis of both weakly absorbing and highly reflecting materials, i.e., objects in which the absorbed energy is relatively low. Besides, laser techniques make it possible to study highly absorbing objects, accessible also for nonlaser OAS, but for example, at high (to tens of kHz) modulation frequencies, but then the sensitivity of the method drops essentially. High frequencies are necessary, in particular, to measure the depth profile of absorption or the absorption coefficients of optically opaque samples (up to 10^5 cm^{-1}).

In this monograph much attention is given to the practical possibilities of OAS with coherent radiation sources. Anyone interested in the potentialities of OAS with noncoherent sources should consult [3.65,66].

3.3.1 Fundamentals

The main task in analyzing the OA effect is to determine the dependence of the OA signal on the optical and thermal properties of the sample, the substratum and the gas, and also on the geometry of cell and sample. Various aspects of OAS are analyzed theoretically in [3.67-94]. Some of these studies offer theoretical models providing satisfactory agreement with experiment for many applications, indicating that the basic mechanisms of OA signal generation have already been clearly understood.

Below we consider briefly some theoretical principles of OAS [3.74]. This reference offers a theoretical model for the OA signal generation in a gas due to the action of an acoustic piston, that is, a thin near-surface gas layer heated by the sample. This is a so-called one-dimensional model for a measuring cell in the form of a cylinder with diameter D and length ℓ (Fig.3.6). The cell is filled with a nonabsorbing gas (g) and has a sample

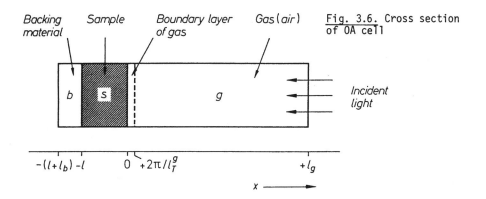

Backing material

Sample

Boundary layer of gas

Gas (air)

Fig. 3.6. Cross section of OA cell

Incident light

b $\quad$ s $\quad$ g

$-(l+l_b)-l \qquad 0 \quad +2\pi/l_T^g \qquad\qquad +l_g$

$x \longrightarrow$

(s) located on a special substratum (b). Below, k is the thermal diffusivity coefficient $(k = K / \rho C)$; $l_T = (2k / \omega)^{1/2}$ is the depth of penetration of thermal waves in the medium or the length of thermal diffusion; $l_\alpha = 1 / \alpha$ is the optical absorption depth. It is assumed that (i) the sample is illuminated by sinusoidally modulated radiation with intensity I_0 and modulation frequency ω so that $\lambda_s = V_s / \omega \gg l_g$; λ_s, V_s being the sound wavelength and velocity in the gas, respectively; (ii) the total radiation energy absorbed by the sample transforms to heat; (iii) the substratum does not absorb radiation; and (iv) the heat diffusion length in the gas l_T^g is much smaller than the gas column length, i.e., $l_T^g \ll l_g$.

The absorption of radiation by the sample changes its temperature $T_s(x,t)$, and is described by the thermal diffusion equation

$$\frac{\partial^2 T_s}{\partial x^2} = \frac{1}{k_s} \frac{\partial T}{\partial t} - \frac{\alpha I_0}{2K_s} \exp(\alpha x)[1 + \exp(i\omega t)] \quad \text{for } -l \leqslant x \leqslant 0 \quad . \tag{3.19}$$

The thermal diffusion equation for the gas and the substratum are identical but with the right-hand side equal to zero. The boundary conditions for solving these equations determine, on one hand, a continuous thermal flow and the temperature at the boundaries $x = 0$ and $x = -l$; and on the other hand, the constant temperature on the cell walls $x = +l_g$ and $x = -l - l_b$.

The solution of the thermal diffusion equation for sample, substratum and gas with allowance made for boundary conditions gives the temperature distribution in the cell as a function of time and the distance from the sample surface. Whence it follows that the temperature perturbations in a gas are concentrated in the near-surface gas-sample layer with a thickness of $2\pi l_T^g$. For example, for the air under normal conditions this thickness is about 0.1 cm at the modulation frequency of $\omega / 2\pi = 100$ Hz. The boundary gas layer expands under heating acting as an acoustic piston and generates an OA signal in the cell, which can be determined from

$$P(t) = \gamma \frac{P2\pi \ell_T^g}{\ell_g} \frac{<\overline{T(t)}>}{T_s(0)} \quad , \tag{3.20}$$

where $<\overline{T(t)}>$ denotes time-average periodic temperature perturbations in the surface layer; $T_s(0)$ is the initial temperature on the sample surface, $\gamma = C_p/C_v$ is the ratio of thermal capacities at constant pressure and volume. The complete expression has the form [3.74]

$$P(t) = K_1 K_2 e^{i(\omega t - \pi/4)} \quad , \qquad \text{where} \tag{3.21a}$$

$$K_1 = \gamma P I_0 / 2\sqrt{2} \, \ell_g T \tag{3.21b}$$

$$K_2 = \frac{\alpha \ell_T^g}{K_s(a^2 - \sigma_s^2)} \frac{(a-1)(b+1)e^{\sigma_s \ell_s} - (a+1)(b-1)e^{-\sigma_s \ell_s} + 2(b-a)e^{-\alpha \ell_s}}{(g'+1)(b+1)e^{\sigma_s \ell_s} - (g'-1)(b-1)e^{-\sigma_s \ell_s}}$$

$$\tag{3.21c}$$

where

$$b = K_b \ell_T^s / K_s \ell_T^b \quad , \quad g' = K_g \ell_T^s / K_s \ell_T^g \quad , \quad a = (1-i)\alpha \ell_T^s / 2 \quad , \quad \text{and}$$

$$\sigma_s = (1+i)/\ell_T^s \quad .$$

Usually in most practically important cases one can assume $b \sim 1$ and $g' < b$ in (3.21).

The OA spectrum depends mainly on the ratio between three parameters having dimensionality of length: the sample thickness ℓ_s, the optical absorption depth $\ell_\alpha = 1/\alpha$ and the thermal diffusion length $\ell_T^s = (2k_s/\omega)^{1/2}$. To clarify the picture physically it is advisable to consider several solutions of (3.21) with regard to transparency, thermal properties and sizes of the sample.

a) Optically Transparent Sample ($\ell_\alpha > \ell_s$)

For $\ell_\alpha > \ell_s$ the radiation is absorbed in an effective distance larger than the sample length. Depending on the thermal diffusion length, the solution of (3.21c) has the form

$$K_2 = \begin{cases} (1 - i)\alpha \ell_s \ell_T^g \ell_T^b / (2 K_b) & \text{for } \ell_T^S \gg \ell_s, \quad \ell_T^S > \ell_\alpha & (3.22a) \\ \text{the same} & \text{for } \ell_T^S > \ell_s, \quad \ell_T^S < \ell_\alpha & (3.22b) \\ -i\alpha(\ell_T^S)^2 \ell_T^g / (2 K_s) & \text{for } \ell_T^S < \ell_s, \quad \ell_T^S \ll \ell_\alpha & (3.22c) \end{cases}$$

Following the terminology adopted in [3.74], (3.22a,b) and (3.22c) correspond to thermally thin and thick samples, respectively. For (3.22a,b) the signal is proportional to $\alpha \ell_s$ and varies as ω^{-1}. The signal magnitude is influenced by the thermal properties of the substratum. For (3.22c) the signal is proportional to $\alpha \ell_T^S$ and varies as $\omega^{-3/2}$. The signal magnitude is here influenced by the thermal properties of the sample rather than the substratum.

b) Optically opaque sample $(\ell_\alpha \ll \ell_s)$

For $\ell_\alpha \ll \ell_s$ the radiation is absorbed in a length shorter than the sample thickness. From (3.21) it follows that

$$K_2 = \begin{cases} (1 - i)\ell_T^g \ell_T^b / (2K_B) & \text{for } \ell_T^S \gg \ell_s, \quad \ell_T^S \gg \ell_\alpha, & (3.23a) \\ (1 - i)\ell_T^g \ell_T^s / (2 K_s) & \text{for } \ell_\alpha < \ell_T^S < \ell_s & (3.23b) \\ -i\alpha(\ell_T^S)^2 \ell_T^g / (2 K_s) & \text{for } \ell_T^S \ll \ell_s, \quad \ell_T^S < \ell_\alpha & (3.23c) \end{cases}$$

Cases (3.23a) and (3.23b,c) correspond to thermally thin and thick samples, respectively. For (3.23a) the acoustic signal is independent of α and varies as ω^{-1}. Its magnitude is influenced only by the thermal properties of the substratum. For (3.22b) the signal is influenced only by the thermal properties of the sample rather than the substratum. The third case (3.23c) is of practical use to determine the absolute absorption coefficient and to study the spectra of highly absorbing substances. Here the signal is proportional to $\alpha \ell_T^S$ and varies as $\omega^{-3/2}$.

If the sample with $\ell_T^S > \ell_s$ is rigidly attached to a substratum with high thermal capacity the OA signal will be proportional to ℓ_T^b / K_b. For thin films fixed at the ends, however, when the gas serves as a substratum the OA signal will be proportional to ℓ_T^g / K_g. The more general behavior of OA signals is analyzed in [3.80].

The above relations are obtained for homogeneous samples. But it is not difficult to generalize them to a practically important case of nonuniform samples which arise, for example, in studying two-layer structures of the coating-substratum type [3.88]. It is of particular interest for many practical applications to study weakly absorbing optical materials by OAS, which can be realized only with high-power lasers. An important task then is to obtain differentiated information on bulk and surface absorption [3.77,78]. It has been shown that $P \propto \omega^{-1}$ for purely surface absorption and the phase shift between the OA signal and the exciting pulses equals $\Delta\varphi^{\circ} = 90^{\circ}$, while for purely bulk absorption they will be $P \propto \omega^{-3/2}$ and $\Delta\varphi = 45^{\circ}$, respectively.

In the measuring cell the pressure may increase not only by the vibrations of the gas layer on the material surface but also by the mechanical vibration of the material surface due to its thermal expansion. As shown in [3.81], the contribution of this effect to the OA signal should be taken into account at small absorption coefficients and high modulation frequencies and in analysing liquid media it becomes prevalent. The complete expression for OA signals for thermally thick samples, with allowance made for mechanical expansion, has the form

$$ P = -\frac{i\gamma PI_0}{\omega \ell_g^2 \rho_s C_s} \left\{ \frac{\alpha}{\sigma_g T(g' + 1)(a + 1)} + \beta_L [1 - \exp(-\alpha \ell_s)] \right\} . \qquad (3.24) $$

It should be taken into account that the OA model described has some limitations. Specifically, it is not valid when the length of thermal diffusion in the gas ℓ_T^g becomes comparable with the length of the gas column in the cell ℓ_g, i.e., at comparatively low modulation frequencies or small values of ℓ_g. In these nontypical cases it is advisable to use the results of [3.76]. The one-dimensional character of the theoretical model makes it impossible to reveal the dependence of the OA signal on the cross-sections of the beam and the cell. This dependence, for example, may be responsible for a reduction of the OA signal as the beam shifts to the cell wall [3.84,92]. An analysis of a three-dimensional theoretical OA model can be found in [3.89-91].

For poor thermal contact of the sample with the environment one should consider the finite value of surface thermal resistance, which somewhat modifies the form of the final expression for the OA signal (3.21) [3.83].

The heat transfer by the interstitial gas between two metallic surfaces in contact has been investigated by OAS in [3.95]. The sample was composed of a thin (12.5 μm) metallic foil brought into contact with a thick, aluminum polished substrate. The thermal conductance was determined by interpreting the phase data obtained by varying the laser-source modulation frequency. These experimental results confirm the hypothesis of most theoretical formulations of the OA effect, that the surface's thermal resistance is generally negligible.

In practical applications it is necessary to choose certain measurement conditions based on the dependence of the OA signal on the modulation frequency. According to the expressions presented, using the value of slope of the straight lines ℓnP vs $\ell n\omega$ it is possible to choose the conditions which enable study of either the optical or thermal properties of substances. The choice of the working modulation frequencies is rather critical. For example, in studying the optical properties of highly absorbing samples at low modulation frequencies, there is danger of thermal saturation ($\ell_\alpha \ll \ell_T^S$) when the OA signal ceases to depend on optical absorption [3.72]. The technique of modulated-radiation intensity is the most widely applied. But in some cases, for example, in measuring relaxation times, the pulsed operation of the radiation source may be useful [3.79,82].

3.3.2 Measuring Cells

The key element in OAS is the measuring cell, whose cross-section is given in Fig.3.5. Here we consider the main requirements used as the basis for estimating and designing such cells [3.96-99]. To attain maximum sensitivity and minimize background signals, easy fabrication and simple and handy operation are the principal and most general requirements.

a) Geometric Dimension

The basic parameters governing the geometric dimensions of a cell are the length of the gas column in the cell, ℓ_g, and the cell diameter D. Other parameters of the cell (modulation frequency, type of gas, etc.) and the sample being fixed, the optimal value ℓ_g can be found from

$$\ell_g \approx \ell_T^g = \sqrt{2k_g/\omega} \quad . \tag{3.25}$$

When this condition is fulfilled, a maximum output signal can be obtained [3.76]. With ℓ_g larger than the thermal diffusion length ℓ_T^g, the OA signal amplitude decreases as ℓ_g^{-1}. Qualitatively this is because at equal energy values in the gas the pressure varies more in the cell with a smaller volume of gas. With $\ell_g < \ell_T^g$, the OA signal amplitude begins to decrease due to fast removal of heat from the gas through the cell windows. Then the signal also depends on the thermal properties of the entrance window. It should be noted that the dependence of the OA signal phase is strong in the region $0 < \ell_g / \ell_T^g \leqslant 1$. Specifically, as ℓ_g / ℓ_T^g increases, the phase grows and reaches its maximum value of $\pi / 2$ in the region $\ell_g / \ell_T^g \geqslant 1$.

To make the measurement scheme of OAS more universal, i.e., so it can be applied to a great number of analytical tasks, the cell must be designed to be applied over a wide frequency range. Then the choice of ℓ_g is based on a compromise between the desire to provide a wide range of working frequencies and a high sensitivity of the cell in this range. According to (3.25), the value of ℓ_g for air at frequency $\omega / 2\pi = 100$ Hz is about 0.1 cm. The value of ℓ_g is usually chosen from 0.1 to 0.5 cm. Besides, at low values of ℓ_g the OA signal may decrease due to thermoviscous attenuation [3.96].

b) Gas Type

Following the expressions presented in Sect.3.3.1, in most practical cases the OA signal is proportional to $\gamma (K_g)^{1/2}(P)^{1/2}/T$, K_g being the coefficient of thermal conductivity, and γ the ratio of gas thermal capacities. Therefore, it may be possible to increase the OA signal by filling the cell with gases of high thermal conductivity at increased pressures and decreased temperatures. For thermally thick and optically opaque samples, in particular, the OA signal is proportional to $\gamma \sqrt{K_g}$ with $\ell_g > \ell\frac{g}{T}$ and $\propto \gamma$ at low ℓ_g. So the use of He instead of N_2 or air increases the OA signal in practice by 3 times (γ_{He} $\cdot \sqrt{K_{He}} / \gamma_{N_2} \sqrt{K_{N_2}} \approx 3.4$ at high ℓ_g).

The OA signal is found to be enhanced several times by the introduction of a volatile liquid (for instance, diethyl ether) into the cell [3.94]. The enhancement is higher at a lower optical absorption coefficient and a larger chopping frequency, and is proportional to the vapour pressure of the liquid. The enhancement of the OA signal over that normally observed in air (adequately described by the theory [3.74]) could be attributed to the physically adsorbed layer on the surface of the sample. Periodic heat flow to the surface of the solid could cause a periodic evaporation and condensation of part of the physically adsorbed layer.

The low absorption in the gas, responsible for the background signals, is an important requirement of the gas especially when laser sources are used to study weakly absorbing media. Practice shows that most problems not requiring maximum possible sensitivity of the OA cell in the visible or the IR, particularly in the region of the CO_2 laser ($\lambda = 9.2 - 10.8$ μm), can be solved with a cell having ordinary air, whose absorption coefficient (in the ranges) is no higher than 10^{-5} to 10^{-6} cm^{-1}.

c) Cell Material

The basic requirement in choosing a material for the cell elements (walls and windows) is to produce a maximum signal, to suppress background signals and to enable working with substances of various chemical properties. Background signals may be caused by direct absorption of radiation in the cell windows as well as by scattered radiation by the cell walls and the acoustic pickup elements. Windows should be chosen with minimum absorption and large enough so that the scattered radiation can be reflected back from the cell. These requirements can be met in the visible and near-IR ranges with quartz and in the middle-IR region with BaF_2, ZnSe, etc.

Scattered radiation is important in studies of substances with low absorption or high reflection, i.e., when the useful signal is small.

To reduce the background signals from the walls it is necessary to choose highly reflecting or highly transparent materials. The first category includes polished metals and alloys (such as stainless steel, duralumin), the second, quartz, organic glass, etc. Stainless steel is preferable over duralumin since it is chemically inert and does not grow dim. Besides, all the inner surfaces must be clean to elimate the possibility of background signals from surface impurity.

d) Types of Acoustic Pickups

As in analyzing gas media (Chap.2), the basic requirement of the pickup in OAS is its high sensitivity, a simple design and a possibility of acoustic matching with the measuring-cell parameters. The uniformity of the frequency spectrum over a wide range is a specific requirement. Condenser and electret microphones satisfy these requirements to a high degree. Usually they are located next to the measuring cell and connected with it by an acoustic channel, eliminating the influence of scattered radiation. To overcome undesirable loss of sensitivity, the volume near the microphone and the volume connecting the acoustic channel should be minimized.

The requirements of uniform frequency spectrum and high sensitivity are contradictory. In commercial microphones the first requirement is fulfilled to the detriment of the second by using highly strained membranes. Therefore, when such microphones are used in OAS, optimum matching (2.41) in the gas-cell microphone system is not fulfilled, and the elasticity depends completely on microphone parameters. Here it is possible to minimize the cell volume without decreasing the microphone's sensitivity. Typical values of the cell volume thereby range from 0.1 to 1 cm^3. However, when we wish to realize the ultimate cell sensitivity over a rather narrow frequency range, special microphones with low-strain membranes should be used and acoustic matching taken into consideration. Typical values of sensitivity for such microphones are given in Table 2.1.

When designing a cell one should take every measure to eliminate the effect of acoustic noise and vibrations. This can be done with good acoustic sealing, rather thick walls and acoustic isolation of the cell from the vibrations in the laboratory, for example, through vibration-isolating cell holders.

The sensitivity of cells for OAS of absorbing media is usually characterized by the signal-to-noise ratio measured under certain conditions. For most cells, in which carbon black is used as a standard absorption sample, the value of this ratio is of the order of 10^3 (radiation power of $J = 1$ mW, modulation frequency: $\omega / 2\pi = 100$ Hz, radiation wavelength: $\lambda = 0.55$ μm and detection bandwidth: $\Delta f = 1$ Hz).

3.4 Optoacoustic Spectroscopy of Solids in Liquids

In OAS of solids in liquids the sample is inserted in a liquid with a small absorption coefficient [3.100,101]. When heated by laser radiation the sample generates acoustic waves, which after traversing the solid-liquid boundary interface are detected in the liquid with an appropriate pickup (Fig.3.7a). As in direct detection of acoustic waves in condensed media, with indirect detection of acoustic waves in the liquid contacting the sample piezoelectric

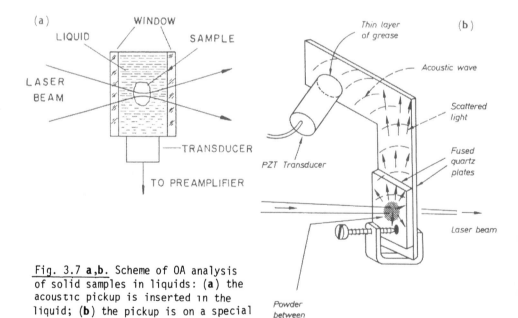

Fig. 3.7 **a,b.** Scheme of OA analysis of solid samples in liquids: (**a**) the acoustic pickup is inserted in the liquid; (**b**) the pickup is on a special soundguide [3.28]

ceramics are usually used as pickups. An intermediate liquid medium introduced in this method improves acoustic matching of the pickup with the sample. The advantage of such a scheme is that it is possible to analyze, with high sensitivity, substances in which it is difficult or almost impossible to have direct acoustic contact with the pickup. Such a situation, for example, arises in studying polycrystalline materials, materials with a characteristic grain structure like powders, as well as when small samples (below 1 mm) are used.

The basic requirement for the liquid is a small absorption coefficient in the spectral range under study. This is essential for eliminating the formation of acoustic waves in it due to the inherent absorption which materially impedes detecting the acoustic waves from the sample. In the visible spectral range ordinary or heavy water, whose absorption coefficient in this range is 10^{-4} to 10^{-5} cm^{-1}, can be used as such an immersion liquid [3.40]. In some cases we should take into account that the liquid should be chemically inert towards the sample.

The mechanism of acoustic wave production in liquids may differ according to the optical and thermal properties of the sample as well as its geometrical dimensions. This may be due, for instance, either to thermal volume expansion of the sample as a whole, or local thermal tensions across the laser beam, or heat transfer from the sample to the liquid, or a combination.

This method is characterized by a considerable amount of scattered radiation, especially at different refractive indexes in the liquid and the sample. This imposes specific restrictions on the method's threshold sensitivity as scattered radiation falls on the surface of the piezoelectric pickup, due to its inherent pyroelectric properties. The contribution of the scattered radiation can be minimized by several methods: 1) matching refractive indexes of liquid and the sample, 2) screening the pickup with an opaque sealing, 3) using the differential measuring technique with two pickups, and 4) removing the pickup from the sample with subsequent acoustic connection through a sound conductor of complex configuration. In such a sound conductor the scattered radiation is essentially reduced and the sound vibrations are not subjected to much loss (Fig.3.7b). With the latter scheme [3.101], for example, it was possible to measure the absorption spectra of some oxides of rare-earth elements in the form of powders in ethylene glycol at a concentration of about 0.01 mg / cm^3. The liquid and powder were placed between quartz plates and irradiated by a pulsed dye laser. Under pulsed laser operation there is an additional possibility of attenuating background signals from scattered radiation, that is, temporal selection of acoustic waves from radiation absorption in the sample.

3.5 Comparison of Various Methods

To summarize the following methods of OA analysis of condensed media are the
principal ones: through direct detection of acoustic waves with a pickup attached to the sample, and indirect detection of acoustic vibrations in the
gas or liquid in contact with the sample.

According to the basic scheme of formation and detection of OA signals,
these methods may simply be called as SP (sample pickup), SGM (sample gas
microphone) and SLP (sample-liquid-pickup). It is pertinent to compare the
potentials of these methods and to define the optimum fields for their application.

The most sensitive method is SP which allows weakly absorbing substances
to be studied both in liquid and solid phases with the absorption coefficients
of down to 10^{-6} cm^{-1}. The sensitivity thresholds of SGM and SLP compared to SP
are much smaller $10^{-2} - 10^{-3}$ cm^{-1} and $10^{-4} - 10^{-5}$ cm^{-1}, respectively, due to indirect energy transfer from the absorption zone to the pickup, and inevitable
losses at the boundaries between various media. The SP and SGM methods are
suited for analyzing both liquid and solid media, whereas SLP is useful in
studying solid samples.

The specific feature of SGM is thermal transfer of absorbed energy from
the sample through its surface to the adjacent gas, whereas in SP and SLP
the energy is transferred to the pickup via acoustic-wave energy. Therefore,
SGM proves to be more inertial. Since the thermal expansion of the sample
across the laser beam is almost instantaneous, the time resolution in SP and
SLP is limited mainly by the time the sound wave takes to be propagated from
the absorption zone to the pickup. But it should be taken into account that
when the ultimate sensitivity is to be realized, the frequency band of detection, which determines the total time of measurement, must often be narrowed.

Due to its high sensitivity to absorption in the surface layer of the substances, SGM is well suited to study surfaces (optical coating, subsurface
structure, etc.) as well as highly absorbing samples with absorption coefficients of up to 10^6 cm^{-1}. Thus, when combined, these methods overlap a huge
dynamic range from $10^{-6} - 10^6$ cm^{-1} with respect to the absorption coefficient.

As for its sensitivity, SLP is suited to studying samples whose absorption
coefficients approximately correspond to the middle of this range, i.e.,
$1 - 10^{-3}$ cm^{-1}. Analysis of samples with a comparatively high absorption coefficient is possible if their optical density is reduced via a decrease in
length [3.100]. In this method there are no strict limitations on dimension
and form of the samples, so it can be applied to spectral analysis of micro-

Table 3.1. Comparison of OA methods to analyze condensed media

Method of detection	Aggregate state of samples	Range of detectable absorption coefficients [cm^{-1}]	Time of response [s]	Main field of analysis	Specific requirements
Pickup joined to sample (SP)	Solid and liquid phases	$10^{-1} - 10^{-6}$	up to 10^{-6}	Weakly-absorbing media	Overall dimensions of sample no smaller than few mm. Sample must be prepared for analysis
Microphone is in the gas contacting the sample (SGM)	Solid and liquid phases	$10^{5} - 10^{-2}$	$1 - 10^{-2}$	Highly absorbing and scattering samples	Sample must be sealed in gas chamber
Pickup is in the liquid contacting the sample (SLP)	Solid phase	$1 - 10^{-3}$	up to 10^{-6}	Amorphous media and micro-samples	Scattered radiation must be eliminated

samples. In the SP method, on the contrary, only comparatively large samples, with characteristic dimensions no smaller than several millimeters, can be analyzed. The form and internal structure of such samples should provide good acoustic matching with the pickup.

Generally at the same sensitivity piezoelectric detection of OA signals is preferable to microphone detection, since the former has better operating characteristics, a fast response and a simpler measuring technique over a wide temperature range (Chap.5). Besides, with SP and SLP the sample need not be sealed in the gas chamber as it must with SGM. The effect of scattered radiation, however, is stronger in these methods. Table 3.1 compares these three methods.

4. Optothermal Spectroscopy

The energy absorbed in a sample can be thermally detected at the first step of OA signal generation from sample heating. By analogy with the OA method, this technique may be called optothermal (OT), and it is similar to OA in application and the general measuring techniques. Sometimes it is difficult to separate these methods from each other. In OAS of condensed media with indirect recording of signals, for example, in a gas, nonapparent physical features of OT appear such as heating of the gas when heat is transferred to it from the irradiated sample. When sound waves in a solid sample are directly detected with a piezoelectric pickup joined to it, output signals can also arise from thermal expansion of the sample due to the OT effect. In some cases OA and OT can be used together and favorably complement each other. Therefore, it is pertinent to consider here briefly the characteristics and technique of OT spectroscopy (OTS) and compare its analytical potential with OAS. Originally the possibility of detecting absorbed energy thermally in gases was first noted in [4.1] and later OTS was experimentally realized in [4.2,3] with molecules excited by continuous radiation and in [4.4] with pulsed excitation. As applied to solids, OTS, also known as calorimetry, is widely used to measure absorption in crystals with low absorption [4.5] as well as to analyze highly scattering and absorbing samples using noncoherent radiation sources [4.6] analogously to OAS.

Since OTS of substances in different aggregate states differs, OTS of gases and condensed media are considered separately.

4.1 Optothermal Spectroscopy of Gases

4.1.1 The Optothermal Effect

Optothermal detection of absorbed energy in gases is based on measuring the thermal effects in the gas cell. According to gas pressure, there are two main channels of thermal energy release in the cell. At high gas pressures,

74

when nonradiative relaxation of excited molecules through collisions with
each other or with the buffer gas molecules is prevalent, the absorbed energy
is released in the gas volume, so increasing the translational energy of the
molecules and, finally, heating the gas totally as in OAS. At low gas pressures,
however, when direct relaxation of excited molecules on the cell walls is prevalent, the excitation energy of the molecules during their collisions with
the wall is transferred to the wall, so heating it. The transition pressure
P^* can be determined approximately from the condition of equality of the thermal relaxation time on the cell walls, $\sim\tau_T / \varepsilon_A$ according to (2.21b), and the
time of vibrational-translational relaxation, τ_{V-T}:

$$P^* \approx \frac{1}{r_c} \sqrt{5.76 \ k^* \varepsilon_A (P\tau_{V-T})} \quad , \tag{4.1}$$

where r_c is the cell radius, k^* is the coefficient of gas thermal diffusion
at a given pressure, say, 1 Torr. The coefficient ε_A characterizes the relaxation probability of excited molecules during their collisions with the cell
walls. The value ε_A depends on the molecules, the wall material and particularly their purity, varying greatly from 10^{-3} to 1, and for complex polyatomic
molecules it is usually about 1 [4.7]. As an illustration for a SF_6 molecule
($\varepsilon_A \approx 1$, $P\tau_{V-T} \approx 158$ $\mu s \cdot$ Torr [4.8]) in a cell with radius $r_c = 1$ cm the boundary
pressure will be $P^* = 0.3$ Torr according to (4.1).

In the low-pressure region, where $P < P^*$, there are two main mechanisms to
transfer molecular excitation energy to the cell walls. If $\Lambda \leqslant r_c$ (r_c is approximately the distance from the excitation zone to the cell wall, Λ is the
free path of molecules), the diffusive mechanism of excited-molecule transfer to the wall is prevalent, the molecules collide with each other, but
do not relax. In this case, in particular, vibrational-vibrational (V-V) energy exchange between molecules takes place. With $\Lambda > r_c$, the energy transfer is
collisionless, i.e., the molecules move to the wall without colliding with
each other. The boundary pressure $\tilde{P}$ for these two regimes can be from the
condition $\Lambda \approx r_c$. For most gases with $r_c = 1$ cm, for example, the value of $\tilde{P}$
ranges approximately from 5 to 20 mTorr.

4.1.2 Types of Thermal Pickups

The type of optimum thermal pickup used to measure the variations in gas temperature or the temperature of the pickup surface itself during relaxation of

75

excited molecules on it[1] is determined primarily by the dominant relaxation channel, laser operation (pulsed continuous), and in some cases by the aim of research. Most thermal detectors for IR radiation can be used with advantage as thermal elements in OTS [4.10-12], as bolometers, pyroelectric pickups, thermistors, thermocouples, etc. Pyroelectric pickups have proved the best, as concerns fast time response and adequate sensitivity [4.2-4].

Such pickups operate by changes in spontaneous polarization of the pyroelectric material as its temperature varies, so generating current [4.13]:

$$i = A_e \gamma_p \frac{dT}{dt} \quad , \tag{4.2}$$

where A_e is the area of the electrodes, γ_p is the pyroelectric coefficient ($\gamma_p = dP_s / dT$, P_s being the mean value of spontaneous polarization). At present, three main groups of pyroelectric materials can be used: monocrystalline pyroelectrics ($LiNbO_3$, KH_2PO_4, KD_2PO_4, $Sr_{0.63}Ba_{0.27}Nb_2O_6$, etc.); polycrystalline pyroelectrics ($BaTiO_3$, ZTL, etc.) and organic polymeric pyroelectrics (compounds of polyvinylfluoride — PVF_2, polyvinylidenefluoride, etc.). All these materials are suited both for recording variations in gas temperature and direct detection of excited molecules. The choice of one or another type of material is governed by specific research tasks. For example, when maximum sensitivity is to be required monocrystalline materials should be used. To obtain fast response up to 10^{-7}s at a sensitivity sufficient for many applications, one should use a very thin (up to 1 µm) polycrystalline film with a columnar structure where some microcrystals are normally oriented to the film surface [4.4]. It should also be kept in mind that the pyroactivity of various materials varies greatly with the temperature range of operation. The properties of pyroelectric materials were described in detail in [4.13] and references therein. The value of the pyroelectric coefficient γ_p for many materials varies approximately between $2 \cdot 10^{-3}$ C / m^2K and 10^{-5} C / m^2K. Such pickups indicate temperature variations at the level of $\Delta T_{min} \approx 10^{-5} - 10^{-6}$ °C.

The corresponding electric diagram of the pickup with external circuit is similar to that in Fig.2.5, where the source of polarization voltage is substituted by an appropriate EMF ($\varepsilon = -\gamma_p A_e \Delta T / C_p$, A_e being the area of pickup electrodes, and C_p the electric capacity of the pickup). This is a typical scheme of an electrostatic source with EMF loaded by the intrinsic leakage resistance R_ℓ. For pickups with a planar geometry

---

[1] The applicability of this method was proved earlier in [4.9], in which the energy absorbed by a molecular beam was measured with bolometric pickups.

$$\text{a)} \quad R_\ell = \frac{2d_p}{A_e \sigma_c} \quad , \qquad \text{b)} \quad C_p = \frac{\varepsilon_d A_e}{d_p} \quad , \qquad (4.3)$$

where d_p, σ_c, ε_d denote the thickness, electric conductivity and dielectric constant of the pyroelectric material. The output signal and its OT recording parameters depend essentially on the operating regime.

4.1.3 Pulsed Regime

Under pulsed laser operation the amplitude and the time shape of the electric output of a pyroelectric pickup depend on many factors including the shape and duration (t_p) of the laser pulse, mutual geometry of beam, cell and thermal detector as well as the ratio between the following times: the characteristic time $\tilde{\tau}$ of molecule transfer from the excitation zone to the cell walls (with $\Lambda \ll r_c$ $\tilde{\tau} = \tau_T$, with $\Lambda > r_c$ $\tilde{\tau} = r_c / \bar{V}_T$, $\bar{V}_T$ being the average thermal velocity of molecules), the pickup time constant τ_p, the external electric-circuit time constant $\tau_e(\tau_e = RC_p$, C_p being the pickup capacity, and R the active load resistance) and the pulse duration t_p. According to gas pressure, the thermal pulse $J_T(t)$ on the pickup surface is formed as a result of a thermal wave propagation in the gas (with $P > P^*$) and diffusive $(\tilde{P} < P < P^*)$ or collisionless (with $P < \tilde{P}$) motion of excited molecules from the absorption zone.

The relationship between the output voltage on the load resistance U(t) and $J_T(t)$ varies with the values of the times $\tilde{\tau}$, τ_p and τ_e (with $t_p \ll \tilde{\tau}$):

$$U(t) \propto \begin{cases} \dfrac{dI_T(t)}{dt} & \text{with } \tau_e \ll \tau_p \ll \tilde{\tau} & (4.4a) \\[2mm] I_T(t) & \text{with } \tau_e \ll \tilde{\tau} \ll \tau_p \ \text{ or } \ \tau_p \ll \tilde{\tau} \ll \tau_e & (4.4b) \\[2mm] \int I_T(t)dt & \text{with } \tilde{\tau} \ll \tau_p \ll \tau_e \quad . & (4.4c) \end{cases}$$

With thermal pulse differentiation (4.4a) it is possible to realize the ultimate time resolution of OTS. It is very important for studying the time-of-flight distribution of excited molecules, for increasing the spatial resolution of OTS, as well as for selection and identification of different types of molecules by means of their various times of arriving at the pickup (with $\Lambda > r_c$).

In (4.4b) the output signal is similar to the thermal pulse. This may occur either for a short time constant of the electric circuit r_e or for a short time constant of the pickup itself τ_p. This regime is convenient for studying the time-of-flight distribution of excited molecules from the irradiation zone to the pickup surface.

Thermal pulse integration (4.4c) is used with pulsed laser operation mainly for correct measurement of the absorbed energy of laser radiation. Fast response and a small contribution to acoustic noise are specific thermal pickup requirements under pulsed operation. Acoustic noise is formed by the reaction of the pickup to accompanying acoustic vibrations in the gas arising from both the piezoelectric effect intrinsic in many pyroactive materials and the secondary piezoeffect caused by thermal deformations of pyroactive materials.

Figure 4.1a illustrates typical thermal pulses in SF_6 irradiated by a pulsed CO_2 laser. The measurements were taken in an OT detector with length $L = 1$ m and diameter $D = 27$ mm. In the middle was the above-described thin-film pyroelectric pickup with time constant $\tau_p = 5$ µs. For the conditions of measurement (4.4b) is fulfilled ($\tau_e = 100$ ms, $\tilde{\tau} \approx 1$ ms), i.e., the pickup records the real shape of the thermal pulse. The signal shapes and their evolution with increasing gas pressure (broadening and maximum displacement) are conditioned by the diffusion transfer mechanism of vibrationally-excited molecules from the excitation zone extending along the optical axis to the pickup surface and their subsequent relaxation on this surface. The required narrowing of thermal pulses can be achieved by diaphragming the zone near the pickup. At pressures below several mTorr the thermal pulse shape no longer changes with pressure because the transportation of excited molecules becomes collisionless.

In the region of high gas pressures, starting with about 0.3 Torr, one can observe for SF_6 molecules a train of pulses instead of a smooth thermal pulse on the leading edge. This can be explained by the fact that the pickup records the energy of the sound waves arising from increasing V-T relaxation of excited molecules in the gas (Fig.4.1b).

Figure 4.2 shows the metrologically important dependence of the pyroelectric pickup output signal on pressure, under integrated thermal pulses for multiphoton absorption of a CO_2-laser pulse ($\Phi = 0.11$ J/cm^2) in SF_6. The linearity of the OT detector with such a pickup points to the possibility of correct measurement of absorbed energy over a wide range of gas pressure. The threshold sensitivity of the OT detector allows measurement of absorption of most molecules up to pressures of 0.1-1 mTorr at laser-pulse energies up to 0.1-1 J.

In measuring weak signals a serious problem in OTS arises from the background caused by heating the thermal pickup surface by scattered radiation. In pulsed operation this problem can be easily solved by time selection of the thermal pulses reaching the pickup, with a time delay $\tilde{\tau}$ relative to the light pulses.

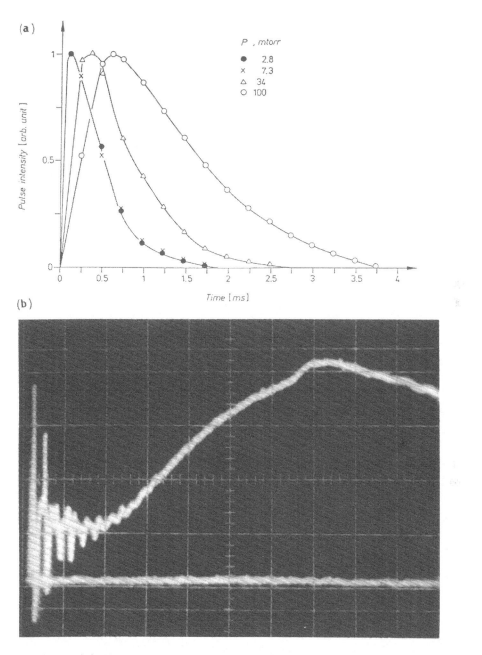

Fig. 4.1. (a) Time profiles of thermal pulses in SF_6 at different pressures (all the curves are normalized to a maximum of 1). The radiation source is a pulsed CO_2 laser, t_p = 50 ns, radiation line P(22), ν = 942,4 cm^{-1}[4.22]. (b) Signal shape from the pyroelectric pickup at SF_6 pressure of 2 Torr [4.4]

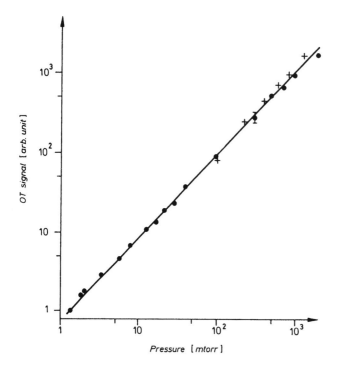

Fig. 4.2. Dependence of the signal from the pyroelectric pickup on SF6 pressure, with thermal pulses integrated for multiphoton absorption. (x): results of direct calorimetry of absorbed energy by comparison of the input and output radiation energy of the cell. [4.4]

4.1.4 Continuous Regime

With continuous laser operation the radiation should be modulated for highly sensitive detection of absorbed power. As in OAS, the modulation frequency ω is chosen from the condition $\omega\tilde{\tau} \approx 1$. In the region of high gas pressures $P \gg \tilde{P}$, $\tilde{\tau} \approx \tau_T$, and the behavior of OT and OA signals is almost the same.

There are no rigid requirements on fast response of the thermal pickup. To exclude OT signals attenuation it is necessary that the condition $\tau_P < \tau_T$ should be fulfilled. Because of this, if we wish to obtain the ultimate sensitivity at rather high values of τ_T it is advisable to use bolometers, whose sensitivity is closest to the theoretical threshold for thermal detectors.

The influence of acoustic noise is not so material as for pulsed operation, the OT detector is effectively at high gas pressures. Sensitivity at different gas pressures was estimated in [4.2-4]. In the low-pressure region $P \ll P^*$ as well as under pulsed operation, the OT detector sensitivity is independent of gas pressure (also the type of gas). In the high-pressure region P >> P*, the sensitivity behaves as it does in OAS. The variation of OT detector sensitivity at P ≃ P* is related to the influence of V-T relaxation in the gas. Using pyroelectric pickups at low pressures it is possible to measure the absorption coefficient at a level of 10^{-5} cm^{-1} with the radiation power of about 1 mW. In the high-pressure region the sensitivity of OTS is somewhat lower but can be increased by using more sensitive pyroelectric pickups with a higher pyroelectric coefficient, a higher dielectric constant and a smaller thickness.

Compared to OAS, the sensitivity of OTS at high gas pressures is lower by one or two orders. However, the small influence of absorption in the windows should be considered as the advantage of this method. We should also note that pyroelectric pickups have low sensitivity to acoustic noise and vibrations, and they operate reliably as opposed to microphones.

The effect of scattered radiation can be overcome by thorough alignment of the OT cell; blackening its internal walls; introducing special diaphragms both across the cell and in front of the pyroelectric pickup; increasing the coefficient of reflection from the pickup surface, etc.

The OT method is most efficient at low gas pressures in nonlinear and multiphoton laser spectroscopy, including the technique of molecular beams [4.23, 24].

A new method of detection absorption in gas flow utilizing the OT method was considered in [4.25]. It is based on probing the flow by transverse laser radiation with subsequent measurement of local temperature variations with a thermal transducer placed across the flow. The advantage of this method is a high spatial resolution (up to 0.1 mm) at high sensitivity ($\simeq 10^{-7}$ cm^{-1}).

4.2 Optothermal Spectroscopy of Condensed Media

In OTS of condensed media absorbed energy in the sample is detected thermally by recording the temperature variations in the sample itself. By analogy with OAS, due to its low sensitivity this method with low-power non-coherent sources of radiation holds much promise for analyzing highly absorbing substances, and with lasers for studying optical materials with a small absorption coefficients. Thermal waves penetrate smaller into solids and liquids than gases. For NaCl, for example, they penetrate only 0.3 mm with a modulation frequency of f = 10 Hz. Because of this, for effective recording of thermal effects in the sample it is necessary to place the pickup in the immediate vicinity of the laser beam or to operate at low modulation frequencies or, in general, under continuous irradiation of the sample. It should be noted that in the first case it is difficult to eliminate the direct irradiation of the pickup surface to avoid undesired background signals. In the investigation of condensed media, thermocouples and thermistors are usually used as thermal pickups.

4.2.1 Laser Measurement of Weakly Absorbing Media

Lasers providing up to 10 - 100 W radiation power, combined with OTS, allow rather small absorption coefficients of up to 10^{-5} - 10^{-6} cm^{-1} to measure in solids [4.14,15]. This is essential for many tasks including the fabrication of fiber optics with low losses, and high-quality optics for high-power lasers.

For the most part, there are two methods of measuring the absorption iso-thermic and adiabatic. In the first case the sample is heated by the radia-tion passing through it until its temperature T_e reaches a constant value determined by the condition of thermal equilibrium. Then the radiation is cut off, and the cooling time constant τ_c is measured. These two parameters allow the absorption coefficient α to be found from [4.14]

$$\alpha = C\rho\pi r_s^2 T_e J^{-1} \tau_c^{-1} \quad , \tag{4.5}$$

where ρ and C are, respectively, the density and specific heat of the sample; r_s is the radius of the cylindrical sample; J is the radiation power inside the sample.

In the second technique the heating rate dT/dt of the sample is measured after the laser is switched on. The absorption coefficient is then calculated from [4.15]

$$\alpha = \frac{C\rho\pi r_s^2}{J} \left(\frac{dT}{dt}\right)_{t=0} \quad . \tag{4.6}$$

Samples may be disk- or rod-like [4.16,17] with polished surfaces. The in-fluence of scattered light can be eliminated by introducing an opaque thin diaphragm between the thermocouple and the sample surface.

It is of great importance, especially in reducing background signals from the windows in OA cells, to obtain differentiated information on the values of bulk α_V and surface $\varkappa_s$ absorption (the coefficient $\varkappa_s$ is determined from the ratio of the absorbed power in the surface layer to the total radiation power at the sample input). In OTS these parameters are measured separately by using samples in the form of comparatively long rods (up to 40 - 60 mm) and different versions of time selection for thermal waves from various zones of the samples. Such selection, for instance, can be realized using samples of different lengths [4.18], measuring the temperature at different points on the sample [4.19], using the slope of the heating curve [4.20] and other techniques [4.21]. According to the type of selection, the characteristic time of measurement in OTS ranges from tens of seconds to several minutes. In this context measures should be taken to stabilize the laser power and to use the thermostated sample. To reduce environmental effects a differential circuit is usually used with two similar samples and thermocouples joining to them, only one of which is irradiated by the laser light.

4.2.2 Analysis of Highly-Absorbing Media

Ophothermal analysis of highly-absorbing condensed media can be carried out both with laser sources and intense noncoherent sources. The sample temperature is usually measured with a thermistor contacting or inside the medium. The temperature threshold of such a thermal detector is approximately 10^{-3} - 10^{-4} K. To minimize the background signal from scattered radiation the thermistor surface is coated with a reflecting layer. The time constant of a thermistor is rather long (0.1 - 1 s) and so the sample is usually continuously irradiated. In certain cases, when studying thin samples and when special transistors with time constants down to several microseconds are used as thermal pickups, modulation of the radiation is possible.

As far as the applications and the measuring techniques are concerned, OTS and OAS (Sect.3.3) are very similar. Advantages of OTS are small acoustic noise, less strict requirements on sample geometry and the possibility of operation in vacuum. More detailed information on the features of OTS in analyzing highly-absorbing media can be found in [4.6].

Recently, the OT method has been employed to demonstrate the feasibility of OT microscopy with pyroelectric detectors [4.26]. The application of pyroelectric thin-film transducers for OTS studies was described in [4.27]. The sensitivity of pyroelectric detection are demonstrated by spectroscopic studies of neodymium-oxide-doped polycrystalline (methyl methacrylate) films. At energy fluences of less than 10mJ / cm^2, sensitivities of 10^{14} molecules / cm^2 are achieved with a conventional amplitude-modulated light source as well as with a pulsed laser. A detector response time of less than 100 ns can be reached with pulsed excitation.

5. Principles of Laser Optoacoustic Instruments

The analysis of weakly-absorbing media provides one of the most important application of laser OA instruments because of high sensitivity. In practice, it may be difficult to realize the ultimate sensitivity due to the influence of different background OA signals. So measures taken to reduce background signals may affect the choice of the scheme of measurements and of OA cell designs.

5.1 Basic Sources of Background Signals

Optoacoustic signals whose generation is not connected with the radiation absorption in the medium under investigation are background signals. They may be caused by radiation absorption in the walls and the windows of the OA cell, scattered radiation entering the acoustic detector, the presence of extraneous impurities with their absorption bands in the spectral range under study, etc. In most cases the background OA signals as well as the signals from radiation absorption in the sample are proportional to the power of the radiation entering the OA cell. Therefore, irrespective of the place and origin of these signals, it is convenient to estimate their level in the form of an equivalent background absorption factor α_b. The value of α_b characterizes such absorption in the medium at which a useful OA signal equal to the background signal arises. Thus, the parameter α_b determines the ultimate sensitivity of OA instruments in the presence of background signals. Experimentally α_b is estimated away from the sample in the OA cell or with the sample replaced by a standard sample that does not absorb radiation.

In most cases the nature of basic background signals is one and the same for the substances in different aggregate states. This point has been briefly discussed in the foregoing chapters. It allows general classification of background signals of OA analysis both of gas and condensed media. More attention is given to the problem of gas analysis with the highest threshold sensitivity achieved.

5.1.1 Background Signals from Windows

The background OA signal from a window is caused mainly by heating from passing radiation, which is accompanied by thermal expansion as well as by partial transfer of heat to the medium contacting with the windows. The value
of the OA signals thus formed in the medium can be estimated. Surface absorption is dominant in the IR spectral range in which such hygroscopic materials
as NaCl, KBr, etc. are used. It can be calculated with the results given in
[5.1-3]. Specifically, according to [5.1,4], the OA signal from radiation
absorption in two cell windows equals

$$P_B = \frac{2^{5/2}(K_g C_g)^{1/2} \, I k_F \alpha_s \ell_s}{3\pi\omega\ell_g (K_w C_w)^{1/2}} \quad , \tag{5.1}$$

with $\omega < 2V_s/L$ and $C_w K_w / C_s^2$, $(\alpha_s \ell_s)^2 \ll 1$, where α_s is the absorption coefficient of the surface layer with thickness ℓ_s; C, K are, respectively, the
thermal capacity and thermal conductivity of gas (g), windows (W) and surface
layer in the windows (s); L is the cell length; ω is the modulation frequency;
I is the radiation intensity; k_F is the form factor of the modulation function,
V_s is the sound velocity in gas.

If the role of the volume absorption coefficient α_v in the generation of a
background OA signal is dominant, its value depends little on the window thickness ℓ_w. If the role of surface absorption with $\ell_T^w \ll \ell_w (\ell_T^w$ being the length
of thermal diffusion in windows, and ℓ_w the window thickness) is dominant,
the results are the same. Conversely, if $\ell_w \leqslant \ell_T^w$, with decreasing ℓ_w the background signal can even increase due to the effect of absorption in the outer
window. From this it follows that in choosing a material for the windows in
the spectral range under investigation, along with the smallness of the absorption coefficients α_v and α_s, we should see that the OA background signals from
volume and surface absorption are almost the same. For illustration, it is
noted that the thermal diffussion length in most optical materials, with modulation frequency $\omega / 2\pi = 10$ Hz, is about $\ell_T^w = 0.01 - 0.05$ cm.

The background OA signal in gases with allowance made for the thermal expansion of the window heated by radiation can be estimated from (3.24). Here
the value is proportional to the window thickness. For liquids the value can
be estimated from the results in Sect.3.4.

5.1.2 Background Signal from Scattered Radiation

Inevitable scattering of radiation on the walls of the cell and the surface of the acoustic detector gives rise to a background OA signal, caused by heating of the walls and their subsequent expansion and transfer of some heat to the adjacent medium. This means that such signals are formed in the same way as for windows. This type of background may arise in many designs of OA cells particularly in cells with metal walls, where the optical absorption depth is rather small, i.e., $\ell_\alpha \ll \ell_T^W$ ($\ell_\alpha = 1/\alpha$ and ℓ_T^W being the length of thermal diffusion in the wall material). The value of a background OA signal in a cell filled with gas, for example, can be calculated from (3.23b). The effect of reflection from walls is taken into account by the substitution $J \to J(1 - \bar{R})$ where J is the total radiation power falling on the cell walls, and $\bar{R}$ is the average reflection coefficient. For transparent walls it is necessary to use (3.22c) to calculate the OA signal. Since heat waves hardly penetrate into metals (for brass $\ell_T^W = 1.2$ mm with $\omega/2\pi = 10$ Hz), with wall thickness $\ell_w \gg \ell_T^W$, the dependence of the background OA signal on wall thickness may be neglected.

Background signal due to scattered radiation change the tension of the membrane and cause its thermal expansion in condenser microphones, as well as the secondary piezoeffect resulting in thermal expansion, and the pyroeffect of the piezoelectric detector. Specifically, according to the results of [5.5], the sensitivity of such standard piezoelectric ceramics as PZT-4 type with silver coating facing the incident laser radiation at a modulation frequency of 100 Hz is about 10 mV / W in the visible and near IR ranges. Using this data it is possible to estimate the permissable level of scattered radiation which results in a signal not exceeding the level of electric noise or that of the useful signal. It must be taken into account that the scattered radiation signal, being inversely proportional to the modulation frequency, is linear as the radiation power increases approximately up to 0.1 W. It depends little on the laser beam cross section and increases with a decrease in dimensions of the piezoelectric detectos [5.5].

Radiation entering the walls of the OA cell and the detector may be caused by (Fig.5.1):

a) misadjustment which may be most significant if the cell cross section is small;

b) diffraction and scattering at input elements, for example, at diaphragms, mountings, the edge of mechanical obturator blades;

c) Frenel reflection in the windows or from the sample surface;

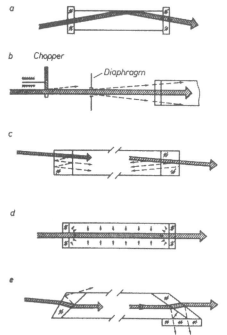

Fig. 5.1 **a-e**. Causes of laser radiation entering OA cell walls: (**a**) mis-
alignment; (**b**) diffraction and scattering at input elements; (**c,e**) Fres-
nel reflection; and (**d**) scattering in
the windows and the medium

d) various types of scattering in the windows and medium. The presence of
microscopic nonuniformity and various inclusions, rough surface treatment of
the sample, etc., may be responsible for radiation scattering in condensed
media. Rayleigh and Mie scattering whereby the presence of aerosols as well
as fluorescent radiation can also cause radiation scattering in gas media.

5.1.3 Background Signal in the Sample

Background absorption in a medium under investigation may be caused by:

a) the presence of foreign impurities with absorption bands at the same
frequencies as the laser;

b) hot absorption bands of different gases;

c) multiphoton effects;

d) weak absorption at forbidden transitions and continuous absorption.

As applied to liquids, the threshold sensitivity of OAS can be limited by
background signals due to the electrostriction effect. According to [5.6,7,
77], under pulsed laser operation the corresponding equivalent background
absorption is about 10^{-5} - 10^{-6} cm^{-1}. For gases this value is smaller by three
or four orders of magnitude. The background signal due to electrostriction
is characterized by a weak spectral dependence, a specific time shape of the

pulsed signal and, because of absorption in the medium, a dependence of the
amplitude on the laser-beam diameter more pronounced than for OA signals.

5.1.4 Comparison of Background Signals

It is difficult to carry out true quantitative estimates of background signals
because they depend on many parameters, regarding which there is not always
reliable information. Table 5.1 compiled from results of [5.8] presents approximated values of such background OA signals in laser spectrophones. These
values were obtained in a spectrophone chamber 20 cm long and 0.6 cm in diameter; the gas type is air under normal condition; the walls are made of brass;
the windows borosilicate glass ($\lambda = 0.63$ µm), quartz ($\lambda = 3.39$ µm) and NaCl
($\lambda = 10$ µm); the window thickness is 4 mm; the laser operation is continuous
with power modulation at $f = 20$ Hz.

According to Table 5.1, the dominant factor in limiting the threshold sensitivity of laser spectrophones is radiation absorption in the spectrophone
windows ($\alpha_b = 10^{-6} - 10^{-7}$ cm^{-1}) and absorption of the radiation reflected from
the windows in the walls ($\alpha_b \approx 10^{-7}$ cm^{-1}). For a spectrophone of small cross section account should be taken of the radiation falling on its side walls. For
illustration, Table 5.2 compares background signals from windows made of different IR materials in the spectral region near $\lambda = 10$ µm under the conditions

Table 5.1. Background signals (α_b) in laser spectrophones

| Sources of background signals | α_b[cm^{-1}] | | |
	$\lambda = 0.63$ µm	$\lambda = 3.39$ µm	$\lambda = 10$ µm
Absorption in windows			
Bulk absorption	$10^{-8} - 10^{-9}$	$10^{-8} - 10^{-9}$	$10^{-8} - 10^{-9}$
Surface absorption	$10^{-7} - 10^{-8}$	10^{-7}	$10^{-6} - 10^{-7}$
Thermal expansion	$10^{-8} - 10^{-9}$	$10^{-8} - 10^{-9}$	$10^{-8} - 10^{-9}$
Experimental data (total effect)	10^{-7}		$5 \cdot 10^{-7}$ [5,9] $(6-30) \cdot 10^{-7}$ [5.8]
Absorption in walls			
At single contact of the beam with the walls	10^{-4}	$5 \cdot 10^{-5}$	10^{-5}
Single reflection from the Brewster output window	$4 \cdot 10^{-7}$	10^{-7}	$2 \cdot 10^{-7}$
Absorption of the Gaussian beam wing			$10^{-8} - 10^{-9}$
Rayleigh scattering in gas	$2 \cdot 10^{-9}$	$3 \cdot 10^{-12}$	$3 \cdot 10^{-14}$
Mie scattering	10^{-7}	$5 \cdot 10^{-8}$	$5 \cdot 10^{-8}$
Scattering in windows	10^{-8}	10^{-9}	10^{-10}
Fluorescence of molecules under analysis	$6 \cdot 10^{-2} \alpha$ $(\tau_{rad} = \tau_{v-T})$	$5 \cdot 10^{-5} \alpha$ $(\tau_{rad} = 10^{3} \tau_{v-T})$	$6 \cdot 10^{-5} \alpha$ $(\tau_{rad} = 10^{3} \tau_{v-T})$

Table 5.2. Background signals from windows

Material	Ge	BaF$_2$	KBr	KCl	NaCl	NaCl*	ZnSe	KRS-5
$\alpha_b \cdot 10^{+6}$ cm^{-1}	12.6	8.8	0.6	0.3	0.2	3.8	0.4	0.6

* with protective coating

specified above. The measuring technique consisted in successive introduction of these materials into the laser beam in the spectrophone working chamber filled with spectrally pure nitrogen, i.e., this technique is typical for PAS.

For most materials the OA signal drops as ω^{-1} with increasing frequency which theoretically (Chap.3) is characteristic when surface absorption is dominant. The values of α_b vary greatly from sample to sample, and this is explained by small changes in the state of the sample surfaces and by the presence of different inclusions on them. From the results in Table 5.2 it follows, in particular, that any coating to protect hygroscopic materials from moisture does not give a positive effect due to its inherent strong absorption.

Reflection or radiation scattering from the sample are dominant factors in limiting the sensitivity of cells in OAS. In a direct analysis of condensed media the entry of radiation onto the surface of the piezoelectric detector has the main effect, about $\alpha_b \approx 10^{-4} - 10^{-5}$ cm^{-1} [5.10].

5.2 Elimination of Background Signal

Fundamental methods decrease background OA signals in measuring cells:
   a) modifications of the cell design;
   b) special conditions of laser excitation and measurement and
   c) using compensating circuits.
We now consider these methods briefly.

5.2.1 Modifications of Cell Design

In almost all OAS methods (except the OAS with indirect detection) the OA signal from absorption in the medium varies slightly with cell length whereas the background OA signal from windows varies inversely with length. Therefore, one way of reducing these signals is to increase the length of the cell, whose upper limit is determined by limitations of the overall dimension and the maximum possible volume of the medium. Our measurements have shown, for example, that in using a CO_2 laser it is possible to reduce the absorption in NaCl window to $5 \cdot 10^{-8} - 10^{-7}$ cm^{-1} on account of increasing the laser spectrophone length up to 125 cm. In some cases, e.g., in air pollution measurements, we may ignore windows altogether if we use special designs of resonant OA cells (see below).

Under pulsed operation the background signals from windows can be suppressed substantially (up to two orders and more) by means of diaphragms introduced between the microphone and the window [5.11,12]. Such diaphragms introduce substantial acoustic resistance to longitudinal acoustic waves from windows. A specific feature of the spectrophone given in [5.11] is that the sensitive membrane of the cylindrical microphone is placed inside the chamber and the OA signal is preferentially resonant with the longitudinal waves of the chamber. Under CW laser operation with modulation such diaphragms also allow effective suppression of background OA signals from windows in resonant spectrophones (Sect.5.3) [5.13].

The simplest way of reducing background from walls is to decrease the energy absorbed in them by polishing the metal walls and coating them with a mirror layer or, simply, by fabricating the walls of optically transparent materials. In the latter case most scattered radiation is removed from the cell, and just a small zone of absorption near the inner surface of the wall comparable with the thermal diffusion length in the wall participates in the generation of background signal. The use of plexiglass instead of Al for the cell walls in OAS with simultaneous screening of the microphone decreases the background in the visible region by more than an order of magnitude [5.14]. If there is no appropriate optically transparent material of large dimensions or its technological processing is not easy, a similar effect can be obtained by coating the metal walls inside with layers of such a material. The layer thickness must be several times larger than the thermal diffusion length. To get rid of secondary reflection it is advisable to blacken the surface of the metal walls.

In addition, it is possible to attenuate the background OA signal from a wall by reducing the relative fraction of radiation falling on it. For example, to decrease the diffraction effect it is reasonable to increase somewhat the cell diameter as well as to eliminate different mountings and diaphragms in the path of the laser beam. In certain cases, for example, in gas analysis, this may decrease spectrophone sensitivity. So the diameter should be increased until equality between the signal-to-background and the signal-to-noise ratios, where noise means noises of microphone and preamplifier, is realized.

Generally to reduce Frenel's reflection in the windows it is advisable to use windows with antireflection coating. This method, however, is effective only in a narrow spectral range. Reflection on the windows at a large cell diameter can be eliminated easily when they are located transverse to the optical axis. Because of multiple reflection this leads to almost complete removal of the radiation from the cell. At small cell diameters it is useful to employ Brewster windows and polarized radiation. To decrease the effect of reflections at the faces of the input window it should be located transverse to the optical axis and its thickness minimized so that the directional distribution of scattered radiation is concentrated along the optical axis. Another way is to increase the thickness and wedge angle of the window and its angle of inclination to the optical axis so that the reflected radiation cannot enter the chamber (Fig.5.1d). To reduce the reflection from the output window it is advisable to make an additional window near it to remove the reflected radiation from the spectrophone chamber.

The background OA signals from the windows and the walls of the cell are proportional to the gas pressure in the cell. Therefore, with weak pressure dependence of the absorption in gas the signal-to-background ratio can be increased by reducing the gas pressure.

5.2.2 Special Measurement Conditions

Optoacoustic signals produced in the medium under investigation and background signals, according to the geometry of the region where they are generated, can be conventionally divided into two main types: local and distributed signals. The ways of eliminating background signals vary with the type of signal.

For illustration, we consider the case of gas analysis with laser spectrophones. Here, the absorption in the medium is distributed. The signals from absorption in the windows and on the walls in a limited zone can be attributed to local background signals. Distributed background signals are caused by different effects of scattering and background absorption in gas media.

Methods to eliminate local background signals can be based on the time delay τ_{del} of their arrival at the acoustic detector relative to the useful OA signal from absorption in the medium. Under pulsed operation this method is reduced to a time selection of signals. For effective elimination of background signals from windows, for example, the following conditions should be met: $\tau_{del} \geqslant 3\tau_{vib}$ and $t_p \leqslant \tau_{vib}$ with $\tau_{del} \simeq L/2V_s$, L being the cell length, and V_s the sound velocity in gas. Under pulsed-periodic operation a similar effect can be achieved through gating with the interval t_g complying with the condition $t_g \leqslant \tau_{del}$. At CW modulation the time selection consists in phase selection. With volume absorption prevalent in the windows, the background signal from the windows varies with frequency as $\omega^{-3/2}$ and with the useful OA signal as

ω^{-1}, so one way to increase the signal-to-background ratio here is to use high modulation frequencies. It is also reasonable to use the resonant properties of the spectrophone chamber.

Similar ways to eliminate background OA signals can be applied to analyse condensed media with direct detection of acoustic vibrations. But we should take into account that for liquids at the corresponding cell geometry the background signals from windows can reach the detector faster than the useful signals, due to their propagation across the cell body. The analysis of condensed media is somewhat specific in eliminating the background signals due to radiation scattered onto the piezoelectric detector surface. Here on the contrary, useful OA signals can be selected due to their arrival at the detector with a time delay relative to light pulses. To increase the reliability of selection the detector should be somewhat removed from the excitation zone. As indicated above, it is possible to exclude scattered radiation effects also by introducing screening diaphragms in front of the detector and special soundguides of complex configuration.

In OAS with indirect detection the absorption in the medium may be local and the background signal, on the contrary, distributed. Such a situation arises, for example, in studying the coatings on optically transparent samples or the absorption in microsamples inserted in a liquid. Here the background signals can also be eliminated by the methods described above. For example, the volume absorption in optically transparent materials can be isolated from the surface absorption on the basis of phase and frequency techniques of selection (Sect.3.3). The absorption in a solid inserted in a liquid can be isolated from the absorption in the liquid through time selection, with the sample placed near the detector [5.15].

It is more difficult to overcome the effect of background signals when useful and background OA signals have a distributed character. In such a situation the signal-to-background ratio can be increased by changing to special conditions of measurement:

a) detecting OA signals at the second harmonic regime when absorption saturation is only in the medium under investigation;

b) frequency modulation of laser radiation;

c) two-frequency excitation of molecules at coupled transitions with OA signals recorded at sum and difference frequencies;

d) selective modulation of absorption through the Stark and Zeeman effects, etc. The peculiarities of these operating conditions are considered in the corresponding chapters of Part II, in more detail.

5.2.3 Compensating Measuring Schemes

Most background OA signals are systematic, i.e., their value is constant or varies slightly with time. This allows background signals to be eliminated by the measuring schemes to be considered below. According to the OA cells and the acoustic detectors in use, there are four basic types of measuring schemes intended mainly to eliminate or compensate for background OA signals.

Scheme A. Only one cell (or one sample) and one acoustic detector are used in the simplest scheme (Fig.5.2a). The background signals can be eliminated electrically at the amplifier output. For this purpose preliminary information on background signal is required. It can be obtained through special measurements, for example, during input of a nonabsorbing gas into the laser spectrophone or by recording only scattered radiation during analysis of condensed media, etc. In practice, such a method makes it possible to attenuate background signals by a factor of about 30 to 50. Its efficiency is limited by the degree of stability of the output signal from the OA cell, for example, due to fluctuations of the laser radiation as well as by uncontrolled changes of the background signals in the measuring process due to absorption of aqueous vapor or other impurities on the window surface.

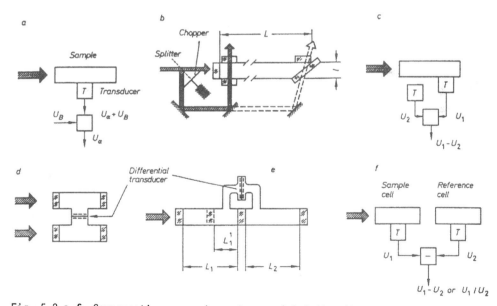

Fig. 5.2 a-f. Compensating measuring schemes: (a) 1 OA cell and 1 acoustic detector; (b) as in (a), with successive radiation transmission in 2 perpendicular directions through the OA cell; (c) 1 OA cell and 2 detectors; (d,e) 2 cells and 1 differential detector. The cells can be parallel (d) or successive (e); (f) 2 cells and 2 independent detectors

It is possible to apply the time-difference principle to eliminate background signals from windows with successive radiation transmission in two mutually perpendicular directions through the OA cell (Fig.5.2b) [5.4]. Within the first half of the modulation period the radiation is passed along the OA cell of length L, and during the second half perpendicularly to the optical axis when the optical path l in the cell becomes comparable with its diameter. The OA background signal from the windows is normally identical during different half-periods and will be compensated. The scheme can be calibrated to zero by filling the cell with a nonabsorbing gas and controlling the radiation power in one of the channels, for example, by rotating the polarization element. A mirror modulator or a semitransparent plate optically commutates laser beams with subsequent alternating modulation of the beams, as shown in Fig.5.2b. In practice, such a scheme attenuates background signals approximately 50 to 100 times [5.8]. To reduce undesired secondary reflections it is advisable to direct the reference beam through the Brewster window, as shown in Fig.5.2b by the dashed line. Here efficiency can be limited by nonuniform reflection over the mirror-modulator surface, by the dependence of the splitting factor of a semitransparent plate on wavelength, as well as by incomplete compensation for laser intensity fluctuations by virtue of the time-difference principle in the OA cell. The main advantage of this scheme is its simplicity and that just one highly sensitive acoustic detector is used.

Since the absorption in windows is weakly dependent on wavelength, a very effective way of eliminating background signals is frequency modulation of laser radiation. For example, the two-frequency discrete modulation through periodic beams propagation, having different wavelengths at the cell input with a mirror modulator, makes it possible to decrease the nonselective background from windows approximately 100 to 200 times [5.8].

Scheme B. One cell (or one sample) and two detectors are used (Fig.5.2c), so effectively eliminating the effect of background OA signals due to scattered radiation falling onto the surface of a light-sensitive acoustic detector, for example, a piezoelement (Sect.5.1.2). One of the detectors is usually in acoustic contact with the sample while the other is acoustically isolated from it and detects only scattered light. Subsequent elimination of signals from the two detectors at the preamplifier input makes it possible to attenuate background signals [5.16]. This scheme also compensates for the ambient conditions (variations of air temperature and pressure, acoustic noise and vibrations caused, for example, by the modulator, etc.), since the response of both detectors to variations of these parameters is the same.

Scheme C. Two cells (samples) and one detector operating in a differential scheme are used (Fig.5.2d,e). It is possible to use schemes with parallel
94

(d) and successive (e) cell arrangements. The first scheme is a close analogue of the differential selective OA detectors widely used in nondispersive gas analyzers [5.17], in which both spectrophone chambers are filled with identical mixtures, and only one of them contains the molecules to be detected. Differential detectors detect signals by time-difference comparison during alternating radiation transmission through spectrophones and by simultaneous comparison of signals in both spectrophones. The latter is simpler, since more favorable operation conditions exist for the detector so just relative changes of pressure in the spectrophones are detected. Such a scheme makes it possible to attenuate the background OA signals both from windows and distributed absorption in gas. Two disadvantages, however, are that it is difficult to inlet separately nonidentical mixtures into separate chambers, and there is a strong influence of laser intensity fluctuations in each channel.

To eliminate radiation fluctuations a longitudinal arrangement of the spectrophones is preferable (Fig.5.2e). In [5.18] such a scheme and BaF_2 windows were used to suppress the background OA signals at the lines of a DF laser by approximately two orders, up to $\alpha_b = 3 \cdot 10^{-9}$ cm^{-1}. In such a scheme the spectrophones should have identical characteristics for effective suppression of background OA signals in the gas. It should be noted that with parallel arrangement of the OA cells, the regime can be easily realized by passing the radiation through one spectrophone and then into the other.

If it is necessary to exclude only the absorption in the windows it is advisable to use spectrophones of different lengths [5.19,20], as shown by the dashed line in Fig.5.2e. With $L_1' \ll L_2$, the useful signal in the first spectrophone is much smaller than the signal in the other, and the background OA signal can be easily equalized provided that the volumes of the chambers are equal. This is realized, for example, by using a big after-membrane volume of the cylindrical microphone [5.19] or by increasing the diameter of the chamber with smaller length [5.20]. So, when a differential acoustic detector is used, the useful OA signal attenuates only slightly and the background signal is eliminated. In this case both chambers can be filled with the same mixture of gases of molecules to be detected.

Scheme D. This differs from the previous scheme in that it has two independent differential detectors instead of one in each OA cell with subsequent electric elimination of amplifier signals (Fig.5.2f). Such a system is more autonomic, and more sensitive nondifferential acoustic detectors can be used. Besides, the presence of two independent detectors simplifies electrical balancing of the cell, as well as measuring the signal ratio from two OA cells that is required in some tasks. But such balancing should be carried out as

often as possible due to unavoidable variations in the sensitivity of separate
channels.

5.3 Resonant Cells

The operation of resonant OA cells is based on the resonant acoustic properties
of the cell volume or the sample itself. Since the resonant OA cells for analyzing gas and condensed media have different designs, they are considered
separately.

5.3.1 Resonant Spectrophone

In a nonresonant spectrophones high sensitivity can be attained at a rather
small volume ($V \approx 1 - 10$ cm^3), a small chamber diameter ($D = 0.3 - 1.5$ cm) and
low modulation frequencies (up to several tens of Hz). Under these conditions
the magnitude of OA signals is the same across the whole volume of the spectrophone chamber. An essential disadvantage is the relatively large background
OA signal from radiation absorption in the windows and walls of the spectrophone chamber. A resonant spectrophone reduces the amplitude of these signals.
A positive effect is thus attained, as a rule, from the corresponding change
in geometry of the spectrophone chamber, mainly due to an increase of its
volume and operation at higher modulation frequencies. The resonant spectrophones also enable, e.g.:

1) windows to be excluded in some cases;

2) gas to flow continuously through the spectrophone chamber;

3) the wall's adsorption of gas to be minimized since the ratio of the wall
surface of the chamber to its volume is comparatively small;

4) some molecular parameters to be determined including the relaxation
times and thermodynamic characteristics, etc.

Since the chamber of a resonant spectrophone is a volume acoustic resonators, a unlike nonresonant spectrophones, the pressure is substantially redistributed according to the acoustic mode. Generally the spatial distribution of pressure in the spectrophone chamber results from the linear combination of orthogonal modes P_j with amplitudes A_j and frequencies ω_j [5.21],

$$P(\omega, \bar{r}) = \sum_j A_j(\omega) P_j(\bar{r}) \quad , \qquad \text{where} \qquad (5.2a)$$

$$A_j(\omega) = \frac{i\omega\alpha(\gamma - 1)\int P_j^* I \; dV}{\omega_j^2(1 - \omega^2/\omega_j^2 - i\omega/\omega_j Q_j)V} \quad , \qquad (5.2b)$$

where P_j^* is the complex conjugate value of P_i normalized so that $\int P_i^* P_j dV$ $= V_c \delta_{ij}$; Q_j is the quality of the system determined from the fraction of the energy lost during one period of acoustic vibration. The solution of (5.2) depends greatly on the configuration of the spectrophone chamber, the excitation geometry and types of excited modes. Generally, the direction of the beam through the spectrophone should be chosen so that the integral value in (5.2b) is maximum. Figure 5.3 illustrates resonant spectrophones which are widely used in laser spectroscopy. The basic parameters of the best ones are summarized in Table 5.3.

The simplest design is a spectrophone of cylindrical geometry (Fig.5.3a,b), in which it is possible to excite longitudinal, azimuthal and radial modes. Then (5.2) has the form [5.21,26]

$$P(r, t) = \sum \left(-\frac{i\omega}{\omega_j^2}\right) \frac{(\gamma - 1)I\alpha \exp(-\mu_j) \cdot L}{(1 - \omega^2/\omega_j^2 - i\omega/\omega_j Q_j)VJ_0^{*2}(\pi a_{0,j})} \times J_0^*(\pi a_{0,j} r/r_c)$$
$$\times \exp(-i\omega_j t) \tag{5.3}$$

where $\mu_j = (w_0/r)^2 (2\pi a_{0j})^{-2}$, w_0 is the effective radius of Gaussian beam. The basic characteristics of resonant spectrophones are the spectrum of resonant frequencies ω_j and quality Q_j. For cylindrical spectrophones

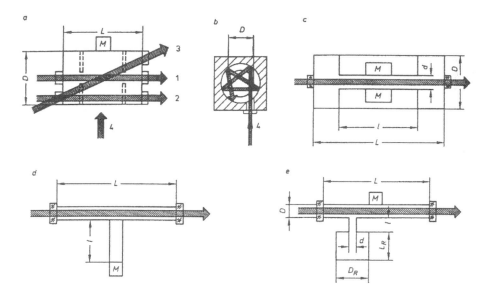

Fig. 5.3 a-e. Resonant spectrophones: (a,b) cylindrical geometry [5.13,22-33]; (c) H geometry [5.34,35]; (d) T geometry; and (e) with a Helmholtz resonator [5.36]

Table 5.3. Comparison of resonant spectrophones

Type of resonant spectrophone	Overall dimensions [mm] (Fig.5.3)	Resonant properties — Mode	Resonant frequency [Hz]	Quality	Background signal from windows — Windows	Spectral range [μm]	α_b [cm^{-1}]	Sensitivity α[cm^{-1}] for J=1W, Δf=1Hz	Ref.
Cylindrical	D=106, L=100	001	4000	890	KCl	10	$9 \cdot 10^{-8}$	$3 \cdot 10^{-8}$	[5.24]
	D=37, L=120	010	5480	100	Quartz	0.3	$7.5 \cdot 10^{-9}$	$4.6 \cdot 10^{-9}$	[5.28]
(Fig.5.3a)	D=155.6 L=65.4	001	2700	560	NaCl	10	$2 \cdot 10^{-7}$	$3.8 \cdot 10^{-8}$	[5.29]
(Fig.5.3b)	V≈2.5 cm^3	200	2200	70	Pyrex	0.48	$2 \cdot 10^{-7}$	10^{-8}	[5.33]
H geometry	D=229, L=234 l=165	001	1822	1800		10	$5 \cdot 10^{-8}$	$1.46 \cdot 10^{-9}$	[5.35]
(Fig.5.3c)	D=15, L=400 d=6, l=60	100	870	28	NaCl	10		$6 \cdot 10^{-10}$	[5.8]
	d=10, l=150	100	1075	20	BaF$_2$	10	$7 \cdot 10^{-7}$	10^{-8}	[5.63]
T geometry (Fig.5.3d)	d=15, L=800 l≈100			30	KRS-5	10	10^{-7}	10^{-8}	[5.37]
Helmholtz Resonator (Fig.5.3e)	L=153, D=9 l=44, d=2.45 D_R=38, L_R=90		305	9	Ge	10		$K_{J\alpha}$=117 V/W·cm^{-1}	[5.36]

$$\omega_{k,m,n} = \pi V_s \left[(k/L)^2 + (a_{m,n}/r_c)^2 \right]^{1/2} \quad , \tag{5.4}$$

where V_s is the sound velocity in gas; k,m,n are, respectively, the numbers of longitudinal azimuthal and radial modes; the coefficient $a_{m,n}$ is determined from $dJ_m^*(\pi a)/da = 0$, with $r = r_c$ (J_m here is the Bessel function of m-th order) [5.22].

To find the quality Q all acoustic energy loss in the spectrophone chamber should be taken into account. Assuming that only the surface losses are dominant, according to [5.26], we get

$$Q_{max} \approx L \left[\ell_v + \ell_T^g (\gamma - 1)(1 + L/r_c) \right]^{-1} \quad , \tag{5.5}$$

where $\ell_v = (2n_v/\rho\omega)^{1/2}$ is the thickness of the viscous boundary layer, $\ell_T^g = (2k/\omega)^{1/2}$ is the thickness of the thermal boundary layer, and n_v is the viscosity factor. Practice shows that (5.4) gives resonant frequencies in good agreement with experiment. At the same time the values of Q from (5.5) are somewhat higher ($Q = 10^3 - 10^4$) when compared to those obtained in practice (Table 5.3), because in deducing (5.5) we neglected other mechanisms of energy dissipation, for example, the effect of the microphone, different types of scattering, etc. Therefore, for precise estimation of Q one must prefer experimental methods on the basis of measuring the resonance half width $\Delta\omega$ at the level of 0.7 ($Q \approx \omega_j/\Delta\omega$) or the damping decrement δ_α under pulsed excitation of resonant vibrations ($Q \approx 1/2\,\delta_\alpha$).

Let us now consider briefly attenuation of background OA signals from windows in resonant spectrophones.

The changeover from nonresonant spectrophones to resonant results in

1) an increase of the signal-to-background ratio,

2) some increase of its sensitivity, and

3) an enhancement of the signal-to-noise ratio where by noise we mean the electric noise of the preamplifier. The increase of the signal-to-background ratio makes sense until the background absorption coefficient α_b is decreased to the level of the minimum detectable absorption α_{min}, whose threshold depends on electric noise. In analytical form, as applied to cylindrical spectrophones, these two conditions can be written as

$$\frac{(S/N)_r}{(S/N)_{nr}} = \frac{\tilde{Q}(\omega_r)r_{nr}^2 \sqrt{1 + \omega_{nr}^2 \tau_{Tnr}^2}}{r_r^2 \omega_r \tau_{Tnr}} \frac{\sqrt{\bar{U}_N^2(\omega_{nr})}}{\sqrt{\bar{U}_N^2(\omega_r)}} \geqslant 1 \quad \text{and} \tag{5.6a}$$

$$\alpha_b \leqslant \alpha_{min} \quad , \tag{5.6b}$$

where the index nr denotes the parameters of a standard nonresonant spectro-phone, and r the parameters of a resonant one. The factor $\tilde{Q}(\omega_r)$ here charac-terizes the degree of increase in OA signal at the resonant frequency ω_r com-pared to an equivalent resonant spectrophone of the same dimensions but free of resonant properties, and $[\bar{U}_N^2(\omega)]^{1/2}$ is the spectral distribution of micro-phone noise. The essence of (5.6a) is that the sensitivity loss due to an in-creasing volume of the resonant spectrophone is compensated for by an increase in OA signal at the resonant frequency with account taken of the spectral distribution of microphone noise. Let us consider a numerical example: for standard conditions ($f = 10$ Hz, $r_{nr} = 5$ mm, $r_r = 50$ mm, $f_r = 1000$ Hz, $\tau_{Tnr} = 0.1$s $[\bar{U}_N^2(\omega_{nr}) / \bar{U}_N^2(\omega_r)]^{1/2} = 10$, condition (5.6a) can be realized for $\tilde{Q} \geqslant 10^3$.

The signal-to-background ratio can be increased by recording the OA signal only at the acoustic modes. The uniformity of gas absorptions plays a dominant role and local absorption by the windows is negligiable. It is advantageous to place the windows and holes for gas inlet and outlet at the nodal points of the acoustic modes. Then the contribution of the absorption in windows to the corresponding mode will be minimum.

In cylindrical spectrophones, as the beam travels along the optical axis radial acoustic modes are excited most effectively. For the first radial mode, ($k = 0$, $m = 0$, $n = 1$), according to (5.3), pressure maxima are formed on the cylindrical wall and near the optical axis. The nodal point where the pressure equals zero is at about 2/3 of the chamber radius from the optical axis. Usually the microphone is located in the middle of the chamber flush with the side wall (Fig.5.3a). In the low-frequency region, as the frequency increases the frequency characteristic of resonant spectrophones behaves similarly to that of nonresonant spectrophones, i.e., it follows the law $P \propto \omega^{-1}$. Yet in the rather high-frequency region it does not obey this law because of the interaction of the different acoustic modes. If the microphone is placed in the middle of the chamber near the wall, the increase of the signal-to-back-ground ratio under axisymmetric excitation (Beam 1 in Fig.5.3a) may be 5 to 8, provided that the ratio of the chamber diameter to its length is about unity. Diaphragms introduced between the windows and the microphone (the dashed line in Fig.5.3a), as proposed in [5.13], increase the signal-to-back-ground ratio up to two orders of magnitude. Their effect is explained by se-lective quenching of OA signals from windows.

As experiments show, in resonant cylindrical spectrophones, together with background suppression, a decrease of total sensitivity by about 5 to 10 times takes place. This means that (5.6a) is not fulfilled. The loss can be partially compensated for by increasing the number of beam passages through the chamber (Sect.5.7) [5.22,26,28].

According to (5.6a) the efficiency of a resonant spectrophone is proportional to the ratio $Q(\omega_r) / r_r^2 \omega_r$, i.e., to produce a large signal lower modes
with the lowest resonant frequency ω_r and chambers with high Q and small r_r
must be used for detection. These requirements are conflicting but, to a great
degree, they are complied with at the first azimuthal mode ($k = 0$, $m = 1$, $n = 0$).
For its effective excitation the laser beam must run near the spectrophone wall
beside the microphone or from the opposite side (Beam 2 in Fig.5.3a).

To reduce the effect of absorption and reflection in the windows at the
same time it has been proposed in [5.29,32] to direct the laser beam at an
angle to the optical axis so that its angle of incidence to the window is
equal to the Brewster angle Θ_B, and the window is located at a nodal points
(Beam 3 in Fig.5.3a). To provide such geometry the ratio of the chamber radius
r_c to its length L should be determined from [5.29]

$$\frac{r_c}{L} = a_{01} \cdot \frac{\tan \Theta_B}{2\beta_{01}} = 0.7967 \, \tilde{n} \quad , \tag{5.7}$$

where $a_{01} = 1.2197$ is the first zero of $dJ_0^*(\pi a)/da$, $\beta_{01} = 0.7655$ is the first
zero of $J_0(\pi\beta)$, and $\tilde{n}$ is the index of refraction. The microphone is placed
along the optical axis flush with the end plane wall. This made it possible
to minimize the geometric distortion of the wall surface and to produce a
signal somewhat larger than that when the microphone is located on the cylindrical wall. Only the first radial resonance is dominant. In such a spectrophone the magnitude of the background signal from the window depends greatly
on the location, the angle of incidence and the polarization of the laser beam.

The excitation of longitudinal modes in a cylindrical design is most effective when the beam is transverse to the optic axis. To increase the sensitivity
then, it is advisable to pass the beam many times through the chamber since
it is reflected from additional mirror walls of the OA cell (Beam 4 in Fig.5.3b)
[5.33]. The microphone is arranged flush with the end wall of the cylindrical
chamber.

Requirement (5.6a) is best fulfilled with chambers of small diameter and
with excitation of longitudinal modes. The background signal from windows can
be eliminated by using a H-shaped spectrophone and an additional volume, damping the signal from the windows, between the windows and the main chamber
(Fig.5.3c). The signal in the gas is recorded at a longitudinal mode of the
chamber now with open ends. To increase the efficiency of such a design some
diaphragms may be arranged at the ends of the chamber [5.11]. Experience shows
that one may use with advantage both plane microphones and cylindrical ones
with outer [5.19] and inner location of the membranes [5.11].

A considerable OA signal can be obtained in a spectrophone of H geometry, with the radial modes excited in the additional cylindrical volume. In accordance with [5.35], for higher efficiency it is reasonable to match the frequency of these modes to the frequency of the longitudinal mode in the chamber with a microphone.

It is also possible to eliminate the background signal in the chambers with small diameter by using spectrophones of T geometry (Fig.5.3d) in which the longitudinal modes are excited in the lateral appendix with a microphone at the end.

As concerns only the increase of the absolute sensitivity of the spectrophone, the chamber with a Helmholtz resonator is of interest (Fig.5.3e) [5.36]. In this design the acoustic vibrations in the chamber with a microphone, through which radiation is passed directly, are amplified via vibration excitation in the additional volume connected with the main chamber through a narrow channel. Such a spectrophone with the dimensions given in Table 5.3 increase the signal at a comparatively low resonant frequency (305 Hz) by about 15 times as compared to the spectrophone without a resonant volume.

In improving resonant spectrophones one should keep in mind that since the resonant frequency depends strongly on the gas composition and the temperature $T(\delta f / f \propto \delta T / 2T, \delta M / 2M$, M being the gas molecular weight), rigid requirements should be placed upon stabilizing the modulation frequency and the gas parameters to increase the accuracy at high values of Q (up to $0.05 - 0.1\%$).

5.3.2 Resonant Cells for Optoacoustic Spectroscopy with Indirect Detection

Resonant cells for OAS with indirect detection can be used when there is no need for a smooth frequency response over a wide range of radiation modulation frequencies. Specifically, these cells

1) increase the sensitivity of OAS in operating at one modulation frequency, especially in the high-frequency region;

2) complement pulsed radiation sources well and

3) eliminate background signals. Some typical resonant OA cells are presented in Fig.5.4.

In the first cell (a) the sample chamber in combination with an additional volume and a microphone form a Helmholtz resonator, whose resonant frequency is calculated from [5.39]

$$\omega_r = \left[v_s^2 \frac{A_c}{V\ell_c} - \left(\frac{4\pi\eta_v}{\rho A_c} \right)^2 \right]^{1/2} \quad , \tag{5.8}$$

where the first term determines the natural frequency of the resonator and

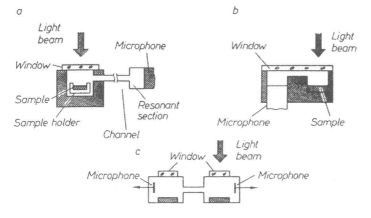

Fig. 5.4 a-c. Resonant cells for OAS with indirect detection:(**a,b**) with Helmholtz resonators; (**c**) a differential cell

the second one allows for the effect of sound attenuation due to gas viscosity in the connecting channel between the two chambers. Here A_c is the cross section of the connecting channel with length ℓ_c, $V = V_m V_s / (V_m + V_s)$, V_m, V_s being the volumes of the chamber with a microphone and the sample, respectively, and n_v is the coefficient of viscosity. The quality of such a resonator can be found from

$$Q = \frac{\omega_r \, \rho A_c}{8\pi n_v} \quad . \tag{5.9}$$

The cell described can be used with both pulsed and CW sources. According to the results from [5.40], the advantages of such a cell under pulsed operation (a dye laser with pulse duration $t_p = 1$ μs) over the nonresonant version are an increase in sensitivity by up to one order, no strong dependence of the OA signal on the properties and geometry of the sample, as well as a lower level of background signal. The parameters of the cell are: $V_m = 0.58$ cm^3, $V_s = 0.28$ cm^3, $\ell_c = 1.43$ cm, $A_c = 2 \quad 10^{-3}$ cm^{-2}, the total volume is 0.86 cm^3, the material is aluminum, $\omega_r / 2\pi = 360$ Hz (the theoretical value from (5.8) equals 430 Hz). The linearity of the output signal is demonstrated by changing the radiation intensity by almost three orders of magnitude. A connecting channel with a small cross section reduces the fraction of scattered radiation falling onto the microphone.

Another version of Helmholtz resonator in an OA cell is illustrated in Fig.5.4b [5.39]. The microphone and the sample are in the same plane and their chambers are connected with a channel near the window surface. Studies of such a design have shown that the experimental values for the resonant frequencies are consistent with those calculated from (5.8). At the same time, at high modulation frequencies the quality determined experimentally is much lower than the calculated one (for example, at 1.5 kHz $Q_{exp} \approx 2.6$ while $Q_{theor} = 24$). The discrepancy may be due to losses in the connecting channel, the effect of the microphone, etc. Under pulsed operation it is easy to identify the type of background signals (wall absorption, encounter of scattered radiation by the microphone, etc.) from their arrival times at the microphone. Results of OAS studies on resonance characteristics of cells were treated in [5.41].

As shown in the previous section, various differential schemes eliminate background signals. Such a scheme with resonant OA cells is illustrated in Fig.5.4c [5.42]. A differential cell is designed as a combination of two resonant cells connected by a narrow channel, i.e., two connected Helmholtz resonators, with the sample irradiated in one of them. The gas vibrations in these resonators caused by absorption of radiation in the sample add up at a resonant frequency while all the acoustic noise from external sources is compensated. As a result, such a design makes it possible to suppress the external acoustic noise by a factor of about 20. It is also possible to eliminate the background signal from windows. For this purpose the radiation must pass, in succession, through two cells.

5.4 Spectrophones with Spatial Resolution

Spatial resolution means the minimum volume of a local zone in the OA cell from which an OA signal can be extracted and detected. The value of spatial resolution is very important at nonuniform radiation absorption in the cell volume. Such a situation arises, for example, in detecting microimpurities in the presence of background absorption from windows, in studying nonlinear and multiphoton effects, in research of spatial nonuniformities in various media, etc. Furthermore, the fact that it is possible to detect OA signals from a rather small region of the cell makes it much easier to create sufficiently uniform distribution of electric and magnetic fields or temperature in this region.

In most of the OA cells described above the OA signals are averaged over the cell volume during the detection time. The means of increasing spatial resolution depend essentially on the laser operation.

Under pulsed operation it is possible to record OA signals from separate local zones along the laser beam by time selection, due to the different

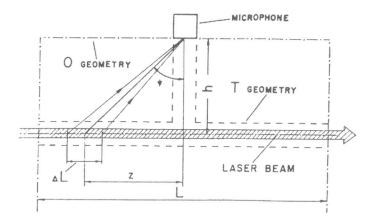

Fig. 5.5. Estimation of spatial resolution under pulsed operation. (O geometry means a cylindrical OA cell)

arrival times of these signals at the microphone (Fig.5.5). The value of the ultimate spatial resolution (ℓ_{sp}) can roughly be estimated from

$$\ell_{sp} \geqslant V_s \tau_R \sqrt{1 + (h/z)^2} \qquad (5.10)$$

where h is the distance between the microphone and the optical axis, z is the coordinate along the optical axis in the cell reckoned from the center of the cell; τ_R is the time resolution of the recording system. The value of τ_R depends on (i) the laser pulse duration t_p, (ii) the time constant of the acoustic detector τ_d, (iii) the time of nonradiative relaxation of excited levels τ_{rel}, and (iv) the mutual geometry of the cell, the microphone and the laser beam. Analyzing gas media under normal conditions $\tau_d \approx 10^{-5} - 10^{-6}$s, $\tau_{rel} = \tau_{V-T} \approx 10^{-5} - 10^{-6}$s, for P = 1 atm, it is easy to realize the condition $t_p \ll \tau_d$, τ_{V-T}. Thus, time resolution is limited mainly by the processes of relaxing absorbed energy or the time constant of the acoustic detector. For condensed media $\tau_d \leqslant 10^{-7}$s, $\tau_{rel} \leqslant 10^{-12}$s and with $t_p \leqslant \tau_d$ the value of τ_R is limited by the time constant of the acoustic detector (piezoelement). A numerical example follows: for $h/z \approx 0$, $V_s = 3 \cdot 10^4$ cm/s (medium - air), $\tau_R \approx 10^{-6}$s, and (5.10) yields that $\ell_{sp} = 0.3$ mm, i.e., the theoretical value of spatial resolution of the fluorescence method [5.43]. The cell with T geometry (Fig.5.5) offers better possibilities for effective time selection.

Increase in spatial resolution due only to time selection may be impeded, for example, by continuous superposition of OA signals that have approximately equal amplitude but are from different zones, or when the duration of the leading edge of an OA signal is increased due to long t_p or τ_{V-T}. In these

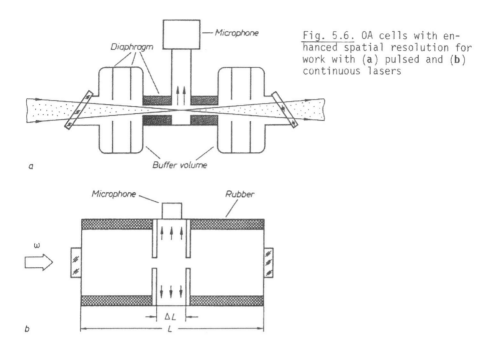

Fig. 5.6. OA cells with enhanced spatial resolution for work with (a) pulsed and (b) continuous lasers

Microphone

Diaphragm

Buffer volume

a

Microphone

Rubber

ω

ΔL

L

b

case one should use the cell in which spatial resolution is obtained through spatial localization of the zone under investigation by diaphragms and additional buffer volumes (Fig.5.6a) [5.12,44]. These elements effectively attenuate the OA signals propagating to the microphone from the "end" zones, and the excitation of natural resonant vibrations in the buffer volumes increases the time delay of these OA signals. Similar cell designs (Sect.5.3) make it possible to eliminate almost completely the effect of a background signal from windows even for short chamber length. At low gas pressure the efficiency of transforming absorbed to acoustic energy decreases. In this cell with diaphragms it is possible to increase spatial resolution also by using OT detection of excited molecules with a pyroelectric detector (Chap.4).

Under continuous operation spatial resolution can be increased by localizing the zone concerned with diaphragms and recording OA signals at the radial acoustic resonance of a cylindrical chamber (Fig.5.6b). To reduce the effect of the lateral chamber it is useful to cover the chamber walls with dampers, rubber for instance, or to change their geometry. However, as the distance between the diaphragms is reduced spectrophone sensitivity decreases. For example, in a spectrophone with the parameters $D = 60$ mm, $L = 200$ mm, gas CO_2, a CO_2 laser as source, a modulation frequency of $\omega / 2\pi = f_{001} \approx 5530$ Hz, $Q = 90$, as the distance between the diaphragms decreases from 180 to 20 mm the OA signal and Q are reduced by about 5 times. High spatial resolution is also typical of resonant spectrophone of T geometry (Fig.5.3d) with the longitudinal mode excited in the lateral appendix. Spatial resolution was about 5 - 10 mm under pulsed operation and 15 - 20 mm under continuous operation.

5.5 Cells with an Enlarged Temperature Range

Cells with an enlarged temperature range essentially widen the possibilities of laser OAS and allow, for example, the spectra of weakly absorbing samples to be measured over a wide temperature range, excited states to be studied and increase the sensitivity and selectivity of the gas analysis, etc.

According to the temperature range of the acoustic detector itself there are two main designs of OA cells: (i) the detector is in thermal contact with the medium, and (ii) the detector is thermally isolated from the medium and its temperature remains constant at the fixed level. Below we consider briefly the specific features of such cells.

The microphones widely used in OA gas and condensed-media analyses have a very limited temperature range since the tension of a thin membrane depends substantially on temperature. Microphones with highly tensed metal membranes, for example, make it possible to cover the temperature range approximately from -50 - +200°C. Higher potentialities are inherent to piezoelectric detectors which cover the range of 4 - 700 K [5.45]. Thus, the use of OA cells where the detector is in direct contact with the medium is most advisable in analysis of condensed media with piezoelectric recording of the OA signal.

More universal, however, is the second design of OA cells where the acoustic detector can operate over a wide temperature range by introducing a matching acoustic channel between the detector and the medium to be heated. The OA signals propagate through this channel to the detector with rather small attenuation while the temperature gradients near the detector attenuate substantially. This enables the detector temperature to be kept at the same level, for example, close to room temperature. The OA cell operates due to OA signals passing through the medium in the matching channel, which has a static temperature gradient. In addition, a forced change of the channel temperature is possible.

Figure 5.7 illustrates a spectrophone designed to study absorption in gases at low temperatures down to 90 - 100 K [5.46,78]. The gas in the central chamber is cooled upon pumping liquid nitrogen vapor through the cavities. The operation of the windows and the microphone is maintained by heating them with a nichrome

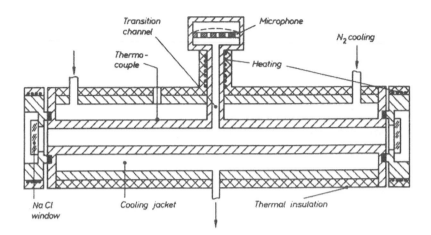

Fig. 5.7. Laser spectrophone for studying gases at low temperatures [5.46]

wire. The OA signals act on the microphone membrane through the holes in the stationary electrode which also serves as a heat filter. The gas temperature can be changed by a copper heat conductor joined to the gas chamber and connected with the corresponding system for temperature variation. A similar microphone design was used in [5.47] to study excited molecular states at temperatures up to 600 K. The gas in the chamber was heated with a nichrome metal coiled around it. The channel with a microphone and the windows were cooled with running water. In such a design, the matching channel 4 cm long and 0.8 cm in diameter, with just one heat filter and natural air cooling of the transition channel, the microphone temperature increased by only $10^{\circ}C$ as the gas temperature in the chamber was increased by $300^{\circ}C$. The potentialities of spectrophones with a widened temperature range were also described in [5.46, 48,49].

A disadvantage of the OA cell design with the acoustic detector removed from the zone of thermal action is that there are volumes near the detector as well as near the windows with a fixed temperature, usually close to room temperature. This may result in an increase of measuring error, for example, due to the effect of molecular thermodiffusion, and in condensation of the constituents on the surface of the windows and the detector. Therefore, these volumes must be minimized.

By analogy, it is possible to construct measuring cells with a wide temperature range for OAS condensed media with indirect detection. The chamber with a sample should be filled with helium, since it has the lowest liquefaction temperature of all gases. For example, when an OA cell with a sample is put into liquid nitrogen [5.50] or a cryostat is pumped with liquid helium

108

vapor [5.51], and the microphone removed from the cooling zone, then we can study absorption in samples at temperatures down 77 K and even 5 K, respectively.

The superconducting microphone has been developed for OAS at a low temperature (4.2 K) [5.79]. The microphone consits of a thin mylar membrane coated with a film of lead whose motion is detected by a SQUID magnetometer. The limiting pressure sensitivity of this set up is 7.6×10^{-9} Pa / $Hz^{1/2}$.

The high-temperature OA cell was designed for the temperature range between room temperature and 1050 K [5.80]. It is made from stainless steel with a quartz window and shaped in the form a Helmholtz resonator. This is the most appropriate design for a high (or a low) temperature OA cell because the inner gas volume is kept small and at the same time the microphone is separated from the hot region. Helium at atmospheric pressure was used as the sound transmitting gas. The technique to get OA spectra of solids at temperatures down to 4.2 K with pulsed dye laser excitation and piezoelectric ceramic detection was described in [5.81]. The sample holder is made of copper. A thin wall (0.5 mm thick) separates the piezoelectric ceramic and the sample. Both sides of the wall are machined to a smooth finish and it is silvered on the sample side. The silvered surface of the copper wall reflects most of the light transmitted by the sample reducing unwanted signals from absorption on the copper wall. The sample holder fits into the sample rod of a CF204 Oxford Instruments cryostat, which allows the temperature of the sample holder to be controlled between 4.2 and 300 K. With this technique a spectrum of a 0.5 mm thick sample of caesium hexachlorouranate (Cs_2UCl_6) was measured at 4.2 K from 445 to 465 nm.

The technique of recording OA signals using an acoustic detector removed from the locally restricted heating zone widens the scope of absorption studies of flames and vapors of various metals [5.52]. If the microphone is only 4 cm from the flame of an oxyacetylene torch this allows reliable recording of pulsed OA signals from resonant absorption of radiation from a pulsed dye laser in Na with a concentration of $5 \cdot 10^{10}$ atoms / cm^3 at the transition $3^2P - 3^2S$ and in Li at the transition $2^2P - 2^2S$. The Na and Li atoms are introduced into the flame as aqueous solutions of the NaI and LiCl compounds [5.52]. The generation of these OA signals at a flame temperature of about 2500 K is caused by quenching the electron-excited states.

The temperature of these OA cells can be calibrated by two main methods:

1) using a sample where the temperature dependence of the absorption coefficient is known e.g., the spectrophone with a CO_2 laser and CO_2 molecules, whose basic parameters are well known as a function of temperature; and

2) through independent measurement of the energy absorbed in the sample.

5.6 Other Cell Designs

5.6.1 Cells with Electric and Magnetic Fields

Optoacoustic cells with electric and magnetic fields are required to study the Stark and Zeeman effects in molecules [5.53] to realize the regime of absorption modulation for increasing the selectivity of gas analysis and decreasing various background signals, from windows for instance; to tune the absorption line maximum to the laser line, etc. To solve such problems ordinary OA cells are modified by placing appropriate elements in them to set up electric or magnetic fields. To produce an OA Stark cell a pair of rectangular parallel metal plates are placed in the cell. A high electric voltage up to several kV is then applied. To increase the electric strength between the plates the gap between them is usually reduced to one or several mm. Uniformity of the electric field is achieved by polishing the inner surfaces of the plates and precise maintenance of the gap size, for example, through dielectric calibrated gaskets between the plates. The advantage of using an OA cell in Stark spectroscopy is its high sensitivity at a relatively small length (up to several cm). This factor essentially simplifies the design of Stark electrodes and makes it possible to set up rather strong local fields with compact devices. The area of Stark electrodes, for example, may be no larger than 10×20 mm^2 [5.54].

The most serious problem in Stark OAS is to eliminate the effect of strong fields on the acoustic detector. When highly sensitive condenser or electret microphones are used the following may be sources of background signals [5.54]:

1) electric stray current on the input circuit of the microphone preamplifier,

2) the motion of the microphone diaphragm under the action of the electric-field force, and

3) gas pressure variation caused by the motion of Stark electrodes under variable electrostatic attraction. The background signals from the first two sources can be eliminated through electric insulation of the microphone. It is somewhat removed from the Stark electrodes and screened against electric stray current, for example, using metal gauze filters. To exclude motion of Stark electrodes they should be made massive with rigid mutual fixing. In [5.55], for example, to increase the electrode rigidity in the submillimeter spectral range, they were in placed in an axially symmetric. The walls of the cylindrical cell and a rigid rod located along its optical axis represent the Stark electrodes in this case. Because of nonuniformity of the electric field

110

between such electrodes it is difficult to carry out quantitative measurements and it is useful to use it only for selective modulation for molecules and transitions with a strong Stark effect. In some cases remote recording of the membrane deflection using the Golay autocollimation scheme [5.56] or the interference scheme may be useful. The OA cells in Stark spectroscopy and the results of preliminary tests have been considered in more detail in [5.54,55, 57,58].

Similar considerations can be applied to the OA cells in Zeeman spectroscopy [5.59-61]. The miniaturization of OA cells makes it possible to set up strong enough local magnetic fields. A superconducting magnet, for example, in combination with a cell 1 cm long having a chamber volume of 2 cm^3 enable a relatively uniform magnetic field with an intensity up to 105 kG in the cell volume 57,58].

5.6.2 Gas-Flow Spectrophones

The gas mixture is replaced in the spectrophone chamber in one of two regimes: cyclic and by gas flowing. In the cyclic regime the mixture to be analyzed is first let into the spectrophone chamber which was preevacuated or cleaned with a pure gas, and then, after the inlet and the outlet are closed and thermodynamic equilibrium is established, measurements are taken. The environmental effect in this case is minimized (acoustic noise, variations in gas temperature and pressure, etc.) and the maximum sensitivity of OAS can be obtained. However, the total time of measurement increases, so hindering effective control of environmental pollution, timely detection of explosive and toxic agent leak, studying the kinetics of chemical reactions, etc. Therefore, to increase the efficiency of gas analysis it is necessary to take measurements with continuous gas flow through the spectrophone chamber.

The basic problems in developing gas-flow spectrophones are:

1) to counteract the reduction in sensitivity caused by the ballast volumes in the gas inlet and outlet systems added to the main chamber;

2) to reduce the effect of gas pressure fluctuations because of flow turbulence;

3) to decrease the acoustic noise penetrating into the chamber from the environment through the gas inlet and outlet;

4) to reduce the measuring errors caused by insufficient flow of the gas near the windows and the microphone as well as by variations of gas pressure and temperature in the chamber. To decrease these factors in nonresonant spectrophones it is necessary to minimize the diameters of the gas inlet and outlet, and to suppress outer acoustic noise by introducing special acoustic

filters. The geometry of the chamber should provide uniform gas flow near the microphone, and the gas volumes near it must also be minimized. Generally, the modulation frequency is chosen by maximizing the signal-to-noise ratio, where noise means gas pressure fluctuations during its flow, near the microphone. Practice shows that these measures enable the effect of background signals in nonresonant spectrophones to be reduced to $\alpha_b \approx 10^{-6} - 10^{-7}$ cm^{-1}, with the flow rate of up to 30 - 50 ml / min [5.37,38,62].

In the gas-flow regime, gas pressure fluctuations are mostly concentrated at low frequency, approximately to 1-2 kHz where the fluctuations decrease with increasing frequencies by a 1 / f law. Therefore, resonant spectrophones, in which the influence of low-frequency acoustic noise is minimized due to relatively high modulation frequencies, have great potentialities in realizing continuous gas flow. To reduce the effect of the gas inlet and outlet dimensions and acoustic gas perturbation near these holes, it is advisable to place them at the nodal points of the pressure distribution over the wall surface of the spectrophone chamber. For example, when a cylindrical spectrophone operates at the first radial mode these points are located on the face walls at about 0.63 r_c from the axis, r_c being the chamber radius [5.24]. Besides, in solving atmospheric tasks it is possible to remove the windows in resonant spectrophones, resulting in an appreciable background signal without considerable loss of sensitivity [5.34,63].

To illustrate the potentialities of these spectrophones, Fig.5.8 shows the dependence of the background absorption coefficient due to turbulent fluctua-

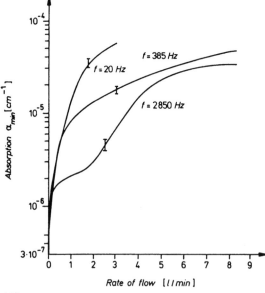

Fig. 5.8. Minimum detectable absorption coefficient in a resonant spectrophone with T geometry as a function of gas-flow rate at different modulation frequencies. The radiation power is assumed to be 1 W [5.37]

tions of the gas pressure in a spectrophone with T geometry (Fig.5.3d) on the rate of gas flow at three different modulation frequencies. The 385 and 2850 Hz frequencies coincide with the longitudinal acoustic resonance in the main spectrophone chamber and in the lateral appendix with a microphone respectively. The length and the diameter of the main chamber are, respectively, 80 cm and 1.5 cm, and those of the lateral appendix are 11 cm and 1 cm. The experiment was performed with a CO_2 laser and mixtures based on CO_2. Operating a resonant spectrophone at the resonant frequencies of the lateral appendix with a microphone decreases acoustic noise from one to two orders with flow rates up to 2 ℓ/min.

Such spectrophones must have consistent parameters — the transmission band of the recording system Δf, the flowing rate v_f and the chamber volume V. These parameters, in turn, determine the relationship between the threshold sensitivity of the spectrophone and its time resolution. In this case it characterizes the ability of the device to analyze without error at rapidly varying concentrations of the component to be detected. In a first approximation the optimum value of Δf at fixed V_f and V can be found from

$$\Delta f_{opt} \approx K_{sp} \frac{V_f}{V} \quad , \tag{5.11}$$

where K_{sp} is a numerical factor depending on the spatial resolution of the spectrophone and the character of gas replacement in the chamber during flow. In the simplest case, for example, when each subsequent portion of gas replaces the preceding one without mixing, and at full averaging of the OA signals over the chamber volume $K_{sp} = 1$. When $\Delta f > f_{opt}$ the spectrophone threshold sensitivity decreases without improvement in its temporal resolution, and when $\Delta f < \Delta f_{opt}$ the sensitivity may increase to the detriment of temporal resolution. According to (5.11), at typical measurement conditions when $V_f = 3$ ℓ/min and V = 0.1 ℓ, $K_{sp} \approx 1$ the optimum band is $\Delta f_{opt} = 0.5$ Hz, which is quite reasonable for many applications.

5.6.3 Optoacoustic Cells for Aggressive Gas and in Vacuum

In ordinary OAS cells the sample must be placed in a closed chamber with the gas and a microphone. This is sometimes difficult to do, e.g., when we have to work with a high vacuum or in chemically aggressive media. Then a cell with an "open" membrane can be used, Fig.5.9 [5.64], although other cell designs described in [5.65-72] (see also Sect.5.3) are also feasible for OAS. In an open membrane cell the sample is placed on a rather thin metal membrane.

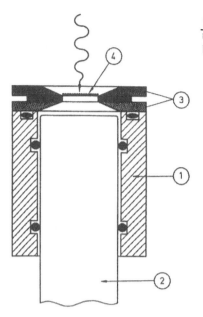

Fig. 5.9. OA measuring cell with an open membrane (1) microphone chamber; (2) microphone; (3) membrane holders; and (4) sample

The laser-induced heat from the sample is transferred through the membrane to the gas in the inner volume of the cell. Then, like in an ordinary cell, the variation in gas pressure is recorded with a microphone. Thus, the sample turns out to be isolated from the chamber with a microphone, and there is no longer any need for windows, so measurements can be taken under the above extreme conditions. When such a cell is used, it is necessary to choose an appropriate thickness of the membrane. For example, with thermally thick membranes ($d_m > \ell_T^m$, ℓ_T^m being the length of thermal diffusion in the membrane) the cell sensitivity decreases because of poor heat transfer. When the membranes are too thin, however, acoustic interference, and environmental noise, as well as thermal expansion of the membrane affect the microphone.

Such a cell is preferable for analyzing highly absorbing thin samples since for low-absorption samples the radiation may fall on the membrane thickness of about several tens of μm which is quite reasonable for certain problems. Such cells hold promise for studying the surface states in vacuum, aggressive media, in analysis of catalytic and photochemical reactions on the surface, etc.

Operation with aggressive gases requires OA cells and particularly microphones made of chemically inert materials. For example, in [5.73] a laser microphone made completely of quartz has been successfully tested. A thin metallized quartz membrane fixed to the quartz cell wall served as a sensitive element of the condenser microphone. It was 60 μm thick, 25 mm in diameter and located about 20 μm from the fixed electrode. The signal was optimum with argon added near the microphone to equalize the pressure on both sides of the membrane. The threshold sensitivity of such a system made it possible to detect ICl at 0.1 Torr with a signal-to-noise ratio of 4, the time constant of 300 ms and the dye laser power of 50 mW, which is equivalent to a minimum detectable absorption factor of about 10^{-5} cm^{-1}. In some cases, for example, when HF is used, it is adequate to make the OA cell of stainless steel and the membrane with Ni coating [5.74].

A new type of open OA cell was described in [5.82,83]. In this cell the temperature variations in the sample are transferred to a window. The thermal expansions of the window are detected by a piezoelectric crystal. Thus, the length expansion of a solid material (a sapphire window) is measured instead of a gas in the closed cell. This cell is especially suitable for liquid samples and gives the possibility to apply OAS to measurements of flowing aggressive liquids.

5.6.4 Combined OA Cells

In some cases it is possible to use a combination of OA cells, designed differently and supplemented with other methods and devices. To widen the dynamic range of a spectrophone with respect to pressure, for example, it is advisable to combine OAS with the fluorescence and OT methods by introducing a window to detect fluorescent radiation and a pyroelectric detector.

5.7 Measuring Techniques of Optoacoustic Instruments

5.7.1 Generalized Scheme

Every variety of laser OA instrument may be reduced to the one block diagram, presented in Fig.5.10. The basic elements of this diagram are a laser and an OA cell. The radiation from the laser is directed to the cell using a matching optical system. When the laser operates continuously a mechanical or electro-optical modulator is used to modulate its radiation. The simplest modulator is mechanical, but it cannot modulate at frequencies above several kHz and, besides, at these frequencies it serves as a source of strong acoustic and vibrational noise. The electrooptical modulator is preferable for high modulation frequencies. To its disadvantages we should attribute a narrow spectral range, a small modulation depth and a small aperture. A modulator

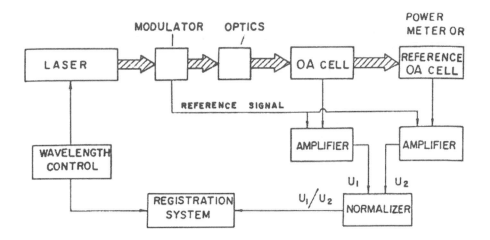

<u>Fig. 5.10.</u> Generalized block diagram of laser OA instruments

is unnecessary for both pulsed operation and for frequency modulation and
absorption modulation based on the Stark and Zeeman effects.

In linear operation the OA signal amplitude is proportional to the power
(energy) of radiation. Therefore, to eliminate the effect of radiation fluctuations on the measuring accuracy, the laser parameters should be stabilized
or, what is much simpler, a second reference channel should be used to record
the radiation power (energy), subsequently normalizing the OA signal to it
via digital devices and analog computers, as shown in Fig.5.10. The optical
and electronic parts of the measuring circuit of OA instruments are considered separately.

5.7.2 Optical Schemes

The optical scheme provides radiation energy transport from the laser to the
medium in the OA cell for optical excitation of acoustic vibrations in this
medium. To increase the sensitivity of OAS in most cases it is necessary to
increase the average radiation intensity (energy fluence) across the OA cell.
With the laser parameters fixed, this can be accomplished using various optical schemes (Fig.5.11). The results compared in Table 5.4 are based on one of
the simplest schemes, in which the radiation is passed right through the OA
cell without an additional matching optical system (Fig.5.11a). The easiest
way of increasing the intensity in a nonresonant OA cell is radiation focusing with a reduction of the cell diameter (Fig.5.11b). The task of an optical
system is then to fit the beam waist in the optical resonator to the optimum
waist in the OA cell (Sect.2.3).

116

Table 5.4. Comparison of the efficiency of various optical schemes for OA instruments

Type of optical circuit	Relative sensitivity
Cell of the laser cavity	
Unfocused beam (Fig.5.11a)	1
Focused beam (Fig.5.11b)	5 - 10
Focused beam with a reflecting mirror at the output (Fig.5.11b)	10 - 20
Resonant spectrophone with a multipass chamber (Fig.5.11c)	10 - 50*
Cell inside the cavity	
Direct beam with optimum transmission of the output mirror in the cavity	5 - 25
Unfocused beam	50 - 100
Focused beam (Fig.5.11d)	100 - 300

* Sensitivity with respect to a one-pass resonant spectrophone

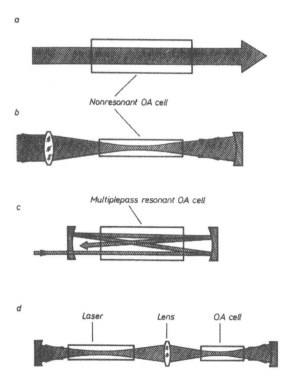

a

Nonresonant OA cell

b

Multiplepass resonant OA cell

c

d

Laser *Lens* *OA cell*

Fig. 5.11 a-d. Optical schemes of laser OA instruments: (a) simplest scheme; (b) two-pass nonresonant OA cell with radiation focusing; (c) multipass resonant cell, and (d) OA cell inside the laser cavity

117

The absorbed power can also be increased by increasing the number of beams passing through the cell. But in nonresonant OA cells a positive effect can already be achieved by twofold passage of the beam through the cell via reflection from an additional mirror (Fig.5.11b). With a greater number of passages it is necessary to increase the chamber diameter to eliminate the background signal caused by radiation of the walls, which may nullify the gain. However, if we use low-power (e.g., semiconductor) lasers, neglecting the effect of a wall background signal, a material increase in sensitivity can be realized due to multiple reflections from the cell walls. For high-power lasers a positive effect is gained by multiple passages of the beam through the large-diameter chamber in resonant cells (Fig.5.11c) [5.22,26,28]. This is explained by substantial attenuation of wall background signal and a slow dependence of the OA signal amplitude on the effective section of laser beam. It is easy to illustrate that the magnitude of ultimate gain in absorbed-power increases in a multipass cell J_{abs}^N as compared to its one-pass version J_{abs}^1 can be estimated from [5.26]

$$\left(\frac{J_{abs}^N}{J_{abs}^1}\right)_{max} \approx \frac{1}{1 - R \exp(-\alpha L)} \quad , \tag{5.12}$$

where R is the mirror reflection coefficient, N is the number of times the beam passes through the cell. For example, in analysis of weakly absorbing media ($\alpha L \ll 1$) and with $R = 0.96$, we have $(J_{abs}^N / J_{abs}^1)_{max} \approx 25$ according to (5.12). To reach a 90% level the number of passes should be no smaller than $N \geqslant -2.3 / \ln R \simeq 56$. The value N achievable in practice ranges from 20 to 60 [5.26,28].

In constructing multipass cells optical systems are used, with mirrors external and internal with respect to the cell. In the first case an antireflecting coating reduces radiation loss in the windows. In the second case to eliminate the background signal caused by heating the mirrors, the mirrors should be thermally isolated from the gas in the cell using a radiation-transparent plate [5.26].

On increasing the power in the cell good results can be obtained when the cell is placed inside the laser cavity. If we use a nonmodified cavity the gain will be $1 / T_m$, T_m being the transmission coefficient of the output mirror in the cavity. A much better result can be obtained by increasing the reflection coefficient of the output mirror and matching the waists in the resonator and in the cell using an additional lens (Fig.5.11d). The main requirement on an OA cell placed inside the cavity is that it introduces a minimum loss. In an analytical form this condition is written as

$$\alpha L + \varkappa_w \leqslant \varkappa_\Sigma \quad , \qquad\qquad\qquad\qquad (5.13)$$

where $\varkappa_w$ is the loss in both windows of the OA cell, $\varkappa_\Sigma$ denotes the total loss in the cavity. Condition (5.13) imposes a restriction from the top on the range of absorption to be measured. For example, with $\varkappa_w \leqslant \varkappa_\Sigma$, L = 20 cm and $\varkappa_\Sigma \approx 10^{-2}$, we get $\alpha \leqslant 5 \cdot 10^{-4}$ cm^{-1} from (5.13). At large α the gain in power increase is reduced, and quenching of laser action may occur. The losses on the windows can be reduced by applying antireflection coats on them, placing them at the Brewster angle as well as by using windowless resonant cells (Sect.5.3).

5.7.3 Electronic Schemes

Electronic schemes process and record the electric signals coming from acoustic detectors. The function of these circuits essentially depends on whether the laser is pulsed or continuous.

The merits of pulsed operation are high values of radiation energy and peak power, lasing in a large number of active media and transitions, and a rather simple elimination of background OA signal from windows. The monopulse regime, however, impedes the processing and automation of measurements. These operating conditions require a comparatively wide frequency range which in-creases the noise and hence reduces OAS sensitivity.

According to the research task, there are two different requirements placed upon the electronic circuit under pulsed operation: (i) recording the pulsed OA signal shape as far as possible without distortion, necessary when measuring relaxation times and in increasing spatial and time resolutions, or (ii) realizing the maximum sensitivity. Without temporal resolution in the first case, the frequency band is chosen rather wide, up to $10^4 - 10^6$ Hz. In the second case, the band is optimized by maximizing the signal-to-noise ratio. To reduce the low-frequency noise it is advisable to transform the pulsed OA signal spectrum to a higher-frequency region on account of quasi-harmonic mo-dulation (Fig.5.12a). This modulation can be realized either in the OA cell itself by using resonant properties [5.11] or with the use of various circuits with frequency conversion (Fig.2.5b).

Lasers operating under pulse repetition offer good possibilities of increas-ing OAS sensitivity. In this case accumulation of pulsed OA signals is possible. At a low repetition rate (below 1 pulse / s) this can be accomplished using the gate integration technique. At a higher repetition rate, due to a comparatively long tail of the pulsed OA signal (Fig.2.3d), even at a high pulse duty factor

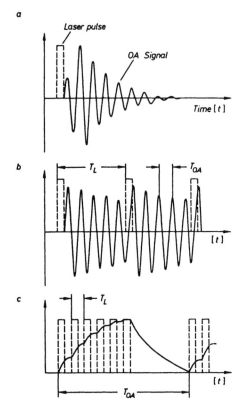

a

Laser pulse

OA Signal

Time [t]

b

T_L T_{OA}

[t]

c

T_L

T_{OA}

[t]

Fig. 5.12 a-c. OA signal recording: (a) quasi-harmonic modulation of pulsed OA signals; (b) modulation at resonant frequency subharmonics; (c) modulation with a pulse sequence. The solid line shows the shape of OA signals; the dashed line denotes laser pulses

(up to $10^4 - 10^6$) we can use the ordinary technique of lock-in integration. The lock-in amplifier and the reference pulses are tuned to that harmonic of the periodic OA signal frequency for which the maximum signal-to-noise ratio can be obtained. Under these conditions one can also use the resonant properties of the OA cell by choosing a repetition rate $f_r = T_{OA}^{-1}$ of the cell (Fig.5.12b) [5.75]. Due to the high Q the amplitude of resonant OA peaks in the period between two successive laser pulses decreases very little [$\propto \exp(-f_r / 2Q\, f_L)$]. This allows resonant cells to be used with lasers having rather a low repetition frequency (f_r / f_L ranging from 4 to 800 according to the magnitude of Q).

For nonresonant OA cells and rather low-power lasers with a high repetition rate, the sensitivity may increase due to additional radiation modulation with low frequency $f = T_{OA}^{-1}$ with phase-sensitive detection of the envelope of the pulse sequence formed at this frequency (Fig.5.12c). The enhancement is due to a comparatively long time constant of the OA cell which enables us to accumulate and integrate single OA signals from each laser pulse right in the cell chamber.

Under CW operation the standard technique of lock-in integration is used for highly sensitive detection of weak OA signals. As shown in [5.76], some possibility to increase OA instrument sensitivity lies in using a digital technique in the initial stages of OA signal processing.

OAS with a correlation technique can be used for a direct delay-time measurement of an OA signal output or the non-destructive measurement of the spectra of separate layers of multilayered samples [5.84,85]. The former situation is realized when correlating a randomly excited signal and the resultant OA signal as a function of the delay time of the reference signal at a constant irradiation wavelength.

6. Analytical Characteristics of Optoacoustic Instruments

This chapter classifies the characteristic parameters of OA instruments (spectrometers or gas analyzers). This is essential for estimating their potentials in different applications as well as for correct comparison of their characteristics with each other and with those instruments operating on other principles.

6.1 Definition of Characteristics

The representation of the OA-instrument parameters depends on the basic equation which relates the output signal to the detectable parameter in the medium under investigation. If weakly absorbing media ($\alpha L \ll 1$) are analyzed and there is no saturation, this equation may be expressed in general form as

$$U = K_i \alpha \times \begin{cases} E, \ i=E\alpha & \text{pulsed operation} \\ J, \ i=J\alpha & \text{continuous operation} \end{cases} , \qquad (6.1)$$

where α is the absorption coefficient; E, J are the energy and power of laser radiation, respectively, and K_i is determined below.

6.1.1 Sensitivity

The first important parameter is the sensitivity K_i characterizing the efficiency of transformation of absorbed radiation power (or energy) to an electric signal of acoustic pickup U. Below, for simplicity, we consider continuous operation with modulation and, when necessary, pulsed operation is considered.

Sensitivity depends on the parameters of the laser, the OA cell, the microphone and the properties of the medium. In analyzing weakly absorbing media the sensitivity mostly depends very little on the length of the radiation-medium interaction. Therefore, it is advisable to present it in units of power absorbed per unit length, i.e.

$$K_{J\alpha} = \frac{U}{J\alpha} \; [V/W \cdot cm^{-1}] \quad . \tag{6.2}$$

In analyzing highly absorbing samples with OAS with indirect detection, for example, it is advisable to estimate the sensitivity similarly as for ordinary detectors, through the radiation power completely absorbed on the surface of an etalon spectrum-nonselective sample (in units of V/W). Sometimes it is useful to divide the sensitivity into two components

$$K_{J\alpha} = K_{J\alpha P} \cdot K_{PU} \quad , \tag{6.3}$$

where $K_{J\alpha P}$ is the coefficient of transformation of the absorbed power to the acoustic signal ($Pa/W \cdot cm^{-1}$), K_{PU} is the coefficient of transformation of the acoustic signal to an electric one (microphone sensitivity in the OA cell) [mV/Pa]. In some applications the absorption coefficient or impurity concentration can be considered as detectable parameters, and the sensitivity can be expressed in terms of the absorption α or the concentration C reduced to a particular power level, usually at $J = 1$ W. To change from the relative concentration C to the absorption coefficient we should use the slightly modified form of (1.2)

$$\alpha = \sigma N_0 C \quad , \tag{6.4}$$

where N_0 is the total number of molecules in a unit volume (under normal conditions, specifically, for gases $N_0 = 2.7 \cdot 10^{19}$ Mol/cm^3 at $P = 10^5$ Pa).

6.1.2 Sensitivity Threshold or Detection Limit

One should distinguish between the sensitivity threshold of the OA cell itself, the minimum detectable absorption coefficient α_{min} or impurity concentration C_{min} in this cell for specific laser parameters (J or E) and the ultimate magnitude of these parameters.

The *sensitivity threshold of an OA cell* is the minimum value of power absorbed per unit length $(J\alpha)_{min}$, detectable in the cell at a given signal-to-noise-ratio (usually assumed to be equal to unity) and frequency band (dimension of $W \cdot cm^{-1} / \sqrt{Hz}$). The form presented is the most universal for this characteristic since it depends neither on the impurity absorption cross section nor the laser power, and it is determined only by the parameters of the OA cell and the equivalent noise of the recording system at the preamplifier input $(\sqrt{\overline{U_N^2}})$. Making the output signal in (6.2) equal to the noise level gives the formula for $(J\alpha)_{min}$

$$(J\alpha)_{min} = \frac{\sqrt{U_N^2}}{K_{J\alpha}} \quad . \tag{6.5}$$

The sensitivity threshold with respect to the absorption coefficient can be determined from the sensitivity threshold of an OA cell for a given radiation power

$$\alpha_{min} = \frac{(J\alpha)_{min}}{J} \quad . \tag{6.6}$$

The concentration sensitivity threshold is determined by (6.4). The ultimate magnitude of these parameters are achieved at radiation power (energy) equal to the saturation intensity (energy fluence) in the medium

$$\begin{pmatrix} E \\ J \end{pmatrix} = A_L \begin{pmatrix} \Phi_s \\ I_s \end{pmatrix} \quad , \tag{6.7}$$

where A_L is the laser beam cross-sectional area, Φ_s and I_s are the saturation parameters determined by (2.16,19).

From a combination of (6.4-7) it is possible to derive a formula estimating the minimum detectable concentration of molecules at absorption saturation in a medium

$$C_{min} = \frac{2\tau_{vib}(J\alpha)_{min}}{A_L h\nu N_0} \quad . \tag{6.8}$$

It is of interest to determine the threshold sensitivity of OAS from the mass of the gas under analysis (absolute threshold sensitivity) since this enables us to compare the sensitivity of OAS with other analytical methods particularly the mass-spectrometric method. The minimum detectable mass of molecules M_{min} can be determined from

$$M_{min} = \frac{V}{V_{gm}} C_{min} \cdot M \quad , \tag{6.9}$$

where C_{min} is the minimum detectable relative concentation of molecules in the mixture (volume); V is the OA cell volume; V_{gm} is the gas gram-molecular volume; M is the molecular weight.

The detection limit is affected by *the background absorption coefficient* α_b. The value of α_b has been defined and comprehensively discussed in Chap.5. If the background signal prevails over the noise of the recording system, the absorption sensitivity threshold is determined by α_b and does not depend on the

124

radiation power. This is valid when $\alpha_b \geqslant (J\alpha)_{min} / J$. At lower powers the absorption coefficient threshold can be found from (6.6).

6.1.3 Accuracy of Measurement

The accuracy of measurement determines the coincidence consistence of the measured absorption (concentration) factor and its true value. It depends on errors in measuring the OA signal value, reproducibility of measurements and calibration error. It is analyzed in more detail in Sect.6.3.

6.1.4 Resolution

Resolution depends on the measuring technique. In linear laser spectroscopy a narrow spectral width allows distortionless measurement of the true values of the spectral lines in the media under investigation. The resolution is limited by the width of the spectral lines, which have the lowest value in gases with Doppler broadening. In nonlinear laser spectroscopy it is possible to measure the sub-doppler structure of spectral lines. However, the resolution is limited by the homogeneous width of spectral lines or the spectral width of the laser output. This point will be discussed in more detail in Chap.7.

6.1.5 Information Capacity

The information capacity is very useful for comparing the analytical potentialities of different instruments. Specifically, if the OA instrument operates as a spectrometer its informative capacity depends on the total information obtained from measuring one spectrogram

$$ H_s = \frac{\nu_1 - \nu_2}{\Delta \nu} \log_2(L_\alpha + 1) \quad , \tag{6.10a} $$

where L_α is the number of resolved values (gradations) of the absorption coefficient α; $\nu_1 - \nu_2$ is the spectral range of spectrometer, ν_1, ν_2 denoting the frequencies at the ends of the spectral range; $\Delta \nu$ is the spectral resolution. In practice, L is determined by the dynamic range of the instrument, the measuring error in absorption $\Delta \alpha$, and the distribution law of this error within the dynamic range.

When the OA instrument operates as a gas analyzer its information capacity is characterized by the amount of detectable components N and the number of resolvable concentration grades of each component

125

$$H_{ga} = \sum_{j=1}^{j=N} \log_2(L_{c_j} + 1) \quad . \tag{6.10b}$$

6.1.6 Dynamic Range

The dynamic range is determined by the ratio of the absorption coefficients measured at both ends of the linear section of calibration characteristic of the instrument: $\alpha_{max} / \alpha_{min}$. According to (2.25), for full pressure averaging over cell volume, the dependence of the output signal on the absorption coefficient α in the neighborhood of its peaks has the form

$$U = K_{J\alpha} \frac{J}{L} [1 - \exp(-\alpha L)] \quad . \tag{6.11}$$

From this it is possible to estimate α_{max}. For a linearity of better than 5%, for example, $\alpha_{max} \leqslant 0.1 / L$.

The limitation on sensitivity is explained by the presence of background signals. Since they are independent of concentration of the component to be detected it is possible to estimate α_{min} from the relation $\alpha_{min} = 20\alpha_b$, allowing an uncertainty of 5%. As a result, under standard conditions of measurement ($\alpha_b = 10^{-7}$ cm^{-1}, $L = 10$ cm) the dynamic range of OAS within the linear regime may be as large as 10^3 to 10^4.

In some applications other characteristics of OA instruments such as spatial and time resolution, detection selectivity, minimum sample volume, temperature and pressure ranges of operation, etc. are of great importance.

6.2 Sensitivity Threshold

Due to its high sensitivity laser OAS is mostly applied to analyze weakly absorbing media. In this connection it is of interest to estimate the ultimate potentialities of OAS in analyzing substances in both gas and condensed phases.

6.2.1 Ultimate Sensitivity

To achieve ultimate sensitivity of OA instruments it is necessary that the following conditions prevail:

a) Maximum Transformation of Absorbed Energy to an Acoustic Signal

This condition can be achieved when nonradiative relaxation of excited particles is dominant and effective transformation thermal energy in acoustic energy.

126

b) Brownian-Noise Threshold

The principal noise threshold is caused by thermodynamic fluctuations in the medium under study. In an ideal case the level of natural noise of the OA signal detector must be lower than or at least comparable to these fluctuations.

c) Absorption Saturation in the Medium

The sensitivity of OAS increases with radiation power until saturation effects manifest themselves in the medium. At absorption saturation the absorbed power no longer increases with increasing radiation power. The upper limit here is saturation intensity (or energy fluence) above which the useful signal increases by just two times (or 1.6). For gas media at normal conditions in the IR spectral range $I_s \simeq 10^3 - 10^5$ W/cm^2. In condensed media the value of I_s is many orders higher and in practice especially under pulsed laser operation, the upper limit is restricted by the effects of optical breakdown, thermal action on the medium or electrostriction. As multiphoton effects become manifest, characteristic, for example, in studying polyatomic molecules in a strong IR field, the absorbed energy may increase if the laser intensity considerably exceeds the saturation intensity in the medium. Then the molecular dissociation threshold at which the analysis becomes destructive by character should be taken as the upper limit of intensity (Sect.7.3).

d) Reducing Background OA Signals

It is possible to increase the sensitivity by increasing the radiation power only in the absence of background OA signals. Otherwise the sensitivity gain due to increased power can be compensated for by the attendant increase in background signals. Let us now estimate the ultimate sensitivity for gas and condensed media.

6.2.2 Analysis of Gaseous Media

When OA signals are recorded in gaseous media with a sensitive microphone the noise level of OAS is substantially determined by the thermal fluctuations in the gas-membrane system when it is in thermodynamic equilibrium with the surroundings. There may be two components of these fluctuations: the membrane fluctuations resulting from the chaotic motion of the molecules colliding with the membrane, i.e., Brownian noise, and the natural thermal fluctuations of the membrane which can take place in a vacuum as well.

According to [6.1], the resultant displacement of the membrane caused is determined by

$$(\Delta \bar{x})^2_f = \frac{4K_B T \left(D_d + \frac{PA_m^2}{G_T T}\right)\Delta f}{\left[\left(4\pi T^* + \frac{PA_m^2}{2V}\right) - M_m \omega^2\right]^2 + \left(D_d + \frac{PA_m^2}{G_T T}\right)^2 \omega^2} \quad , \tag{6.12}$$

where D_d is the damping coefficient of the membrane with mass M_m; G_T is the thermal conductivity between the gas and the OA cell walls; the rest of the notations are as before. The first term in the numerator determines the spontaneous thermal fluctuations of the membrane itself $(\Delta \bar{x}_T)$, the second one determines the Brownian motion of the membrane $(\Delta \bar{x}_{Br})$. By integrating (6.12) we get the root-mean-square value of fluctuations over the whole frequency range

$$(\Delta \bar{x})^2 = \frac{K_B T}{4\pi T^* + (PA_m^2 / 2V)} \quad . \tag{6.13}$$

Expression (6.12) allows studying the spectral distribution of fluctuations, estimating the relative contribution of different terms at various gas pressures and membrane tension, etc. Specifically, under standard operation of the spectrophone the resonant properties of the membrane manifest themselves slightly because of strong damping. The latter is evident in the dominant part of the second term of the denominator in (6.12). In this case the magnitude of $(\Delta \bar{x}^2)^{1/2}$ depends minimally on frequency only in the region of low frequencies, and as the frequency increases it decreases as ω^{-1}. Since the frequency dependence of the OA signal behaves in much the same way, the signal-to-noise ratio depends slightly on frequency.

Let us consider a specific calculation using (6.12) under the following typical conditions: the medium to be analyzed is air under normal conditions $P = 10^5$ N/m^2, T = 293 K, $A_m = 10$ cm^2, $D_d = 0.9$ N $\cdot$ s/m, $\omega = 100$ rdn/s, f = 1 Hz, $T^* = 10$ N/m, L = 10 cm, $r_c = 0.5$ cm (V = 8 cm^3), $G_T = 5.9 \cdot 10^{-2}$ J/s (for a cylindrical chamber $G_T = 8\pi LK$, L being the chamber length, and K the gas thermal conductivity [6.2]), $M_m = 10^{-2}$g. The following numerical values for thermal fluctuations are then $(\Delta \bar{x}_{Br}^2)^{1/2} = 4 \cdot 10^{-4}$ Å and $(\Delta \bar{x}_T^2)^{1/2} \approx 2 \cdot 10^{-5}$ Å. These fluctuations correspond to the noise of several nV at the preamplifier input, and its equivalent gas pressure is of the order of 10^{-7} to 10^{-8} Pa. The results of recalculating these parameters for the sensitivity limit of OAS are presented in Table 6.1. These calculations were carried out in [6.3] using (6.1,

Table 6.1. Threshold sensitivity of OA instruments ($\varkappa_s$ is the surface absorption coefficient, α_v [cm^{-1}] is the bulk absorption coefficient)

Media		Threshold sensitivity $(J\alpha)_{min}$ or $(E\alpha)_{min}$			
		Pulsed operation [J·cm^{-1}]		Continuous operation [W·cm^{-1} / $\sqrt{Hz}$]	
		Theory	Experiment	Theory	Experiment
Gases		$4 \cdot 10^{-12}$[6.3]	$2 \cdot 10^{-10}$[6.3]	$2 \cdot 10^{-11}$[6.3,4]	$6 \cdot 10^{-11}$[6.5] * $6 \cdot 10^{-10}$[6.3] ** $\sim 10^{-10}$ [6.6] ***
Solids	direct detection		$10^{-7} - 10^{-6}$[6.7]		$\alpha_v, 10^{-4} - 10^{-5}$[6.9,10] $\varkappa_s, 10^{-4} - 10^{-6}$[6.8]
	indirect detection			$\alpha_v, 10^{-6} - 10^{-5}$[6.7-8] $\varkappa_s, 10^{-7}$	$\alpha_v, 10^{-3} - 10^{-5}$[6.7] $\varkappa_s, 10^{-3} - 10^{-5}$[6.11,12]
Liquids, direct detection		10^{-10}[6.13]	10^{-9}[6.14,15]	$10^{-7} - 10^{-9}$[6.13,37]	10^{-5}[6.16]

* For schemes with frequency transformation
** For schemes with directly connected microphone
*** For an integration time of 10 min

5,2.22-40) with the following parameters: U_p = 100 V, d = 30 μm (d_M/ε_d<<d). Substituting the threshold sensitivity $(J\alpha)_{min}$ = 2 · 10^{-11} W cm^{-1} / $\sqrt{Hz}$ into (6.6)
shows that with the saturation power of I_s = 10^5 W / cm^2 and a beam cross-section
of A_L = 10^{-2} cm, the minimum detectable absorption coefficient in gaseous media
is about 10^{-14} cm^{-1}. This is consistent with the theoretical limit for most
of the molecules in the IR spectral region, being equal to 10^{-5} - 10^{-6} ppb.
Under typical conditions the sensitivity of the spectrophone itself varies
from 10 to 100 V/W · cm^{-1}.

6.2.3 Analysis of Condensed Media

In OAS of condensed media with direct detection noise limit of the sensitivity is imposed by thermal fluctuations in the system medium-piezoelectric
detector. These fluctuations have two main components: the natural temperature
fluctuations of the detector $(NEP)_{th}$ and the Brownian fluctuations $(NEP)_{Br}$.
They can approximately be estimated with [6.17]:

$$(NEP)_{th} = \frac{(4K_B G_{Td})^{1/2} \, T_{Yp} \ell \, (C\rho)_s}{(\rho C)_p \, \ell_p \, e_{31}^P \, \beta_L (1 + \sigma_p)} \tag{6.14a}$$

$$(NEP)_{Br} = \left(\frac{4K_B T \, A_p \, \ell\omega \, (C\rho)_s}{\ell_p^3 Q_p J_p \omega_0^3 \, \beta_L (1 + \sigma_p)} \right)^{1/2} , \tag{6.14b}$$

where Y_p is the pyroelectric coefficient, G_{Td} is the thermal conductivity
between the detector and the surroundings; $(\rho C)_p$ is the product of the detector
density and thermal capacity, Q_p is the detector quality factor, ω_0 is the detector's resonant frequency. In some cases we should also take into account the
current fluctuations due to dielectric losses $(NEP)_D$ and leakage $(NEP)_L$:

$$(NEP)_D = \frac{\ell(C\rho)_s}{e_{31}^P \beta_L (1 + \sigma_p)} \left(\frac{\ell_p}{4K_B T \omega \varepsilon_{33}^P A_p \tan\delta} \right)^{1/2} \tag{6.15a}$$

$$(NEP)_L = \frac{\ell(C\rho)_s}{\ell_{31}^P \beta_L (1 + \sigma_p)} \left(\frac{\ell_p}{4K_B TR_V A_p} \right)^{1/2} , \tag{6.15b}$$

where tanδ is the tangent of dielectric losses, R_V is the specific volume resistance, and A_p is the detector area.

130

Let us now consider a typical numerical example. For standard conditions with a PZT-type ceramic detector the basic parameters constituting (6.14,15) are $C_{si} = 1.2 \cdot 10^{-8}$ F, $A_p = 10^{-5}$ m^2, $\tan\delta = 0.01$, $\ell_p = 1.78 \cdot 10^{-4}$ m, $R_V = 10^{-11}\Omega^{-1}$ cm^{-1}, $G_{Td} = 7 \cdot 10^2$ J/s$\cdot$°C, $\gamma_p = 0.1 \cdot 10$ C/cm$^2 \cdot$°C, $\rho_s = 5.22$ g$\cdot$cm^{-3}, $C_V = 0.4$ J/g$\cdot$°C, $Q_p \simeq 75$, $\omega_0 = 2\pi \cdot (5 \cdot 10^5)$rdn/s. Substituting these values into the above formulas gives [6.17]: $(\bar{U}_D^2)^{1/2} \approx 11.42$ nV/$\sqrt{Hz}$; $\sqrt{\bar{U}_L^2} = 0.61$ nV/$\sqrt{Hz}$; $(\bar{U}_{th}^2)^{1/2} = 1.67$ nV/$\sqrt{Hz}$; $(\bar{U}_{Br}^2)^{1/2} = 1.37$ nV/$\sqrt{Hz}$. For comparison, it should be noted that the preamplifier noise for a piezoelectric detector similar to a microphone preamplifier is of the order of 13 nV/$\sqrt{Hz}$ in the low-frequency region. Thus, for piezoelectric recording of OA signals the sensitivity threshold to be achievable is less than the theoretical prediction by about one order of magnitude. ($NEP_\Sigma \approx 10^{-6} - 10^{-7}$W).

The results for the threshold sensitivity of OAS in solids are presented in Table 6.1. The analysis of these results reveals that although the threshold sensitivity of OAS obtained in practice is rather high, there is still nonrealized potential for further optimization of parameters.

6.3 Measurement Accuracy

The analytical efficiency depends on metrological characteristics of the OA instrument. So the optimization of basic parameters does not only consist in maximizing the signal-to-noise ratio but also in minimizing the total measuring error. The general questions of gas-analytical measurement metrology, including the specific features of operation of nondispersive OA gas analyzers with incoherent radiation sources, were considered in [6.18-21]. This aspect is briefly discussed below as applied to laser OA detection of molecular trace amounts in gases.

6.3.1 Static Errors

Static errors can be observed in cyclic measurement after the thermodynamic equilibrium and the impurity concentration have been established in the OA cell, at constant or slowly varying environmental conditions. The relative measuring error of the output signal δ_u for statistical independence of error sources can be expressed in the general form

$$\delta_u = \left[U^2 \sum_i K_i^2 + \sum_j (\Delta U_j)^2 \right]^{1/2} \cdot \frac{1}{U} , \tag{6.16}$$

where U is the average value of the output signal. The first term describes the multiplicative noise proportional to the signal from the i-th detectable component ($K_i \ll 1$ is the proportionality factor). The second term gives the additive interference superimposed on the signal. An estimation of δ_U can be made from the general expression for OA signals. In the simplest case the basic equation of OA instruments, (6.1), takes the form

$$U = (K_a + \Delta K_a) \{(K_{J\alpha} + \Delta K_{J\alpha}) (J + \Delta J) [\alpha(\nu) + \alpha_b(\nu) + \alpha_w(\nu)] + U_N$$

$$+ U_{ev} + K_M P_a\} \quad , \tag{6.17}$$

where K_a is the amplification coefficient of the amplifier chain; α_b is the background absorption coefficient in the medium; α_w is the equivalent background absorption in the windows and the walls of the OA cell and due to scattered radiation falling on the detector; U_N is the magnitude of electric and equivalent thermal noise at the amplifier input; U_{ev} is the magnitude of influence of electromagnetic noise and vibrations; P_a is the magnitude acoustic noise in the OA cell chamber. Below, errors are analyzed for binary mixtures, i.e., it is assumed that $\alpha_b = 0$.

The additive noise in laser OA instruments includes the noise of the acoustic detector, various kinds of noise induction and acoustic interference, background absorption in the medium and in the elements of the OA cell, zero drift of a lock-in amplifier, etc. All these types of noise are most essential at low concentrations of the component to be determined. The multiplicative noise includes fluctuations of the laser radiation power, frequency and microphone sensitivity, changes in the gain coefficient of the electron chain, variations in the parameters of the medium (temperature, pressure, and composition), modulation-frequency variations, etc.

Such a division of noise into additive and multiplicative noise has been very useful in the development of methods to decrease its influence. For example, additive noise can be eliminated basically by subtracting it from the signal (electrically or acoustically), the use of compensating systems, differential schemes and selective modulation. To reduce multiplicative noise one should use signal normalization to the reference signal from the etalon sample as well as various ways of stabilizing the main parameters (thermo-, baro-, power-, frequency-, gain-stabilization, etc.). Below we estimate the degree of such noises with respect to the relative measurement accuracy.

a) Laser-Power Variations

In most lasers the radiation-power fluctuations δ_J may be 5% - 10% and higher. Through passive stabilization the power instability during the measurement period can be reduced to 1%. Much more complex and expensive active stabilization systems allow δ_J to be reduced to below 0.1%. Its influence can also be decreased substantially under absorption saturation in the medium when $I \gg I_s$.

b) Laser-Frequency Variations

The laser-frequency fluctuations $\Delta\bar{\nu}$ change the absorption cross section of the molecules. According to (2.28), the relative error $\delta\nu$ is determined assuming a Lorentzian profile of the absorption line, by

$$\delta_\nu = \frac{\Delta\sigma}{\sigma} = \begin{cases} (\Delta\bar{\nu}/\Gamma)^2 & \text{for } |\nu_L - \nu_0| \ll \Gamma & (6.18a) \\[2ex] 2(\nu_L - \nu_0)[\Gamma^2 + (\nu_L - \nu_0)^2]^{-1}\Delta\bar{\nu} & \text{for } |\nu_L - \nu_0| \gg \Gamma & ,(6.18b) \end{cases}$$

where Γ is the line's half-width. The value δ_ν depends essentially on the position of the laser line relative to the center of the absorption line. At standard conditions (e.g., when a low-pressure CW CO_2 laser is used to study gas mixtures at the atmospheric pressure): $P = 10^5$ Pa; $\Delta\nu_L \ll \Gamma$, $\Delta\nu_L$ being the spectral width of the laser output; the mean laser-frequency fluctuation $\Delta\bar{\nu} \approx 2 \cdot 10^{-3}$ cm^{-1}; $\Gamma \approx 0.1$ cm^{-1}, and from (6.18) it follows that at $\nu_L - \nu_0 \ll \Gamma$, δ_ν is minimum (0.04%); its maximum equals 2% at $(\nu_L - \nu_0) \approx \Gamma$. Rather simple systems for automatic frequency tuning provide frequency reproducibility in the order of $10^{-6} - 10^{-7}$ and make it possible to reduce essentially the value of this error. The latter can have a pronounced effect at low gas pressures because of decreasing Γ since $\Gamma \propto P$. When a substance is investigated in the condensed phase, the influence of this error is smaller due to the larger absorption-band width.

c) Temperature Variations

In laser spectrophones the temperature variation strongly effects the molecular absorption cross section σ and the gas' thermal conductivity $(K \propto T^{1/2})$. The temperature dependence of σ manifests itself in the line width since $\Gamma \propto T^{1/2}$. The dependence $N_i \propto \exp(-E_i/K_BT)$, E_i being the energy of the lower excited level of a transition participating in absorption, is a dominating factor especially in case of hot absorption bands. Then

$$\delta_T(N_i) = \frac{E_i}{K_B T} \frac{\Delta T}{T} \quad . \tag{6.19}$$

Measuring the temperature with an accuracy of $\Delta T = \pm 1$ K, $T = 293$ K and $E_i / K_B T \approx 7$ (corresponding to absorption measurement in CO_2 using a CO_2 laser), from (6.19) we get $\delta_T(N_i) = 2.4\%$. At absorption from the ground level this error is smaller, and for most molecules it is no higher than 0.5%.

Changes in gas thermal conductivity K cause the time constant τ_T to vary since $\tau_T \propto K^{-1}$ according to (2.21b). With allowance made for the form of expressing the OA signal (2.27), we have

$$\delta_T(K) = \frac{1}{2(1 + \omega^2 \tau_T^2)} \frac{\Delta T}{T} \quad . \tag{6.20}$$

In the worst case with $\omega \tau_T \ll 1$ and with the allowances made above, from (6.20) it follows that $\delta_T(K) = \Delta T / 2T = 0.2\%$.

d) Pressure Variations

The influence of pressure on spectrophone sensitivity in the high-pressure region manifests itself as variations in the absorption coefficient in the microphone sensitivity K_m, and in the time constant τ_T. According to (1.2) and (2.28), α is independent of P at $|\nu_L - \nu_0| \ll \Gamma$ and $\alpha \propto P^2$ at $|\nu_L - \nu_0| \gg \Gamma$. Thus, in the worst case with $|\nu_L - \nu_0| \gg \Gamma$ and a relative allowance for uncontrolled pressure of variations, $\Delta P / P = 1\%$, the error caused by α variations is $\delta_p(\alpha) = 2\Delta P / P \approx 2\%$.

The form of (2.40) shows that the error caused by microphono-sensitivity variation can be estimated from

$$\delta_p(K_m) \approx \frac{n^*}{n^* + 1} \frac{\Delta P}{P} \quad . \tag{6.21}$$

Thus, with the matching condition for membrane and volume elasticities fulfilled ($n^* \approx 1$) and the above allowance for pressure, $\delta_p(K_m) \approx 0.5\%$. When $n^* \gg 1$, we find the error $\delta_p(K_m) \approx 1\%$. Provided that $\tau_T \propto P$, by analogy with (6.20) it is easy to make the error caused by τ_T variation reach its maximum value $\delta_p(\tau_T) \approx \Delta P / P \approx 1\%$ when $\omega \tau_T \ll 1$.

e) Modulation-Frequency Variations

The error related to the variation of the modulation frequency δ_ω depends on the types of frequency response of spectrophone, microphone and amplifier. For a nonresonant spectrophone the frequency response is $P \propto (1 + \omega^2 \tau_T^2)^{-1/2}$. Therefore, by analogy with $\delta_T(K)$, we then have

$$
\delta_\omega = \frac{2\omega^2 \tau_T^2}{1 + \omega^2 \tau_T^2} \frac{\Delta\omega}{\omega} \quad .
\tag{6.22}
$$

Allowing for a modulation frequency stability $\Delta\omega / \omega$ of 1%, it follows from (6.22) that at $\omega\tau_T \simeq 1$ $\delta_\omega = 1\%$ and at $\omega\tau_T \gg 1$ $\delta_\omega = 2\%$.

For a resonance spectrophone or an amplifier with the quality factor Q and the resonant frequency ω_r, the frequency characteristic is $K_f = K_0[1 + Q^2(\Delta\omega / \omega)^2]^{-1/2}$, $\Delta\omega = \omega - \omega_r$ being the modulation frequency shift of the resonant frequency. From this it follows that at typical value of Q = 300 and $\Delta\omega / \omega \simeq 0.01$ the signal may be reduced essentially up to 70%. So the modulation frequency should be stabilized. For example, for the error δ_ω to be no higher than 5%, the frequency stability should be at least $\Delta\omega/\omega \simeq 10^{-3}$.

In resonance spectrophones detuning between ω and ω_r can also occur due to ω_r variation. Indeed, since according to (5.4) $\omega_r \propto V_s$, and $V_s = (\gamma R_g T/M)^{1/2}$, R_g being the gas constant, and M the molecular weight, the variation of ω_r can be caused by changes in T and M (by the law $\delta\omega_r \simeq \Delta T/2T \simeq \Delta M/2M$). From this it follows that for the error to be no higher than 5%, the gas mixture parameters T and μ should be stabilized at a level no lower than $\Delta T/T \simeq \Delta M/M = 5 \cdot 10^{-4}$.

f) Noise and Drift Errors

These errors result from detector noise, electromagnetic induction noise, acoustic interference and vibrations, output-stage drift, etc., and arise in measuring weak signals. In most cases their influence on the measurement accuracy of the output signal ΔU can be estimated using a two-term formula of the type

$$
\Delta U = \delta U + U_N \quad ,
\tag{6.23}
$$

where U is the output-signal amplitude, U_N is the noise level, δ is the average relative error at a high signal-to-noise ratio. Expression (6.23) shows that at large signals the error is relatively constant, and at small signals it increases due to the noise.

Errors can be also caused by changes in the electronic chain (δ_e) on account of agening of circuit components, fluctuations of supply voltage, etc. In most cases this error can be reduced to $\delta_e \leqslant 0.1\% - 0.5\%$ by stabilization.

g) Total Error

Assuming that the above-mentioned values of different errors to be statistical-
ly independent ($\delta_J = 1\%$, $\delta_\nu = 0.4\%$, $\delta_T(N_i) = 1\%$, $\delta_T(K) = 0.2\%$, $\delta_\omega = 1\%$, $\delta_p(\alpha) = 0.5\%$,
$\delta_p(K_M) = 0.5\%$, $\delta_N = 1\%$, $\delta_e = 0.5\%$), by statistical summing we can see that the
total error of the output signal, δ_U, under standard spectrophone operation
may be about 2%. It is possible to achieve better accuracy by introducing
special systems for stabilizing individual parameters. This however, makes the
OA instrument more complicated and expensive.

In some cases, e.g., in detecting only one type of molecule, it is advisable
to use a two-channel scheme measuring the signal ratio from two spectrophones
in series, having similar characteristics. One of the spectrophones is filled
with the mixture to be analyzed and the other with an etalon mixture containing
the molecuels to be detected in the first spectrophone. If the integration
times are equal, the macroscopic composition of gas mixtures in both spectrophones is identical, the thermal relaxation times τ_T are equal, then the errors
δ_J, δ_ν, δ_T, δ_p and δ_ω in both spectrophones will be correlated. As shown in
[6.3], measuring the signal ratio, even at the level of fluctuations up to
10% - 15%, this technique reduces their effect on the accuracy to 1% - 2%.

6.3.2 Dynamic Errors

Dynamic errors may occur as time variations of certain parameters of OA
instruments. They arise, for example, during laser-frequency scanning, variation of impurity concentration under continuous flow of the mixture through
the spectrophone chamber, as well as during variations of temperature, pressure, electric field, etc. The main source of dynamic error is the limited
detection time of the recording system. For the exponential transient response
of the recording system $h(t) = 1 - \exp(-t/\tau)$, τ being the time constant of the
recording system, the relation between the signals at the input (U_{in}) and the
output (U_{out}) of such a system has the form

$$U_{out}(t) = \int U_{in} \frac{d}{dt} [h(t - t_1)]dt_1 \quad . \tag{6.24}$$

Let the shape of the input signal, $U_{in}(t)$, be a bell-like Gaussian curve.
It is typical, for example, in measuring the Doppler line or the molecular
concentration after the chromatographic column. Let the aim of the measurement
be to achieve maximum spectral (or time) resolution and to reduce it by no
more than 10%. Then, according to [6.22], the half-height duration Δt of the
input signal $U_{in}(t)$ to be measured must be at least three or four times longer

than the time constant of the recording system i.e., $\Delta t \geqslant (3-4)\tau$. The value Δt, in turn determines the maximum permissible rate of frequency scanning $V_\nu : \Delta t = \Delta \nu / V_\nu$, where $\Delta \nu$ is the half-width of the spectral line. For chromatographic measurements Δt determines the gas-flow rate $V_f : \Delta t = V_{eff} / V_f$, V_{eff} being the effective volume of the component to be detected.

Servocontrol of OA systems also give rise to dynamic errors.

6.4 Graduation and Calibration

Quantitative measurements with laser OA instruments are based on a dependence like (6.1) relating the output signal parameters to the medium under analysis. Since these dependences cannot accurately be calculated, the OA instrument scale should be calibrated in terms of the parameter to be detected. Here we deal with the calibration of OA instruments mainly in terms of absorption or concentration.

6.4.1 Graduation by Concentration

Like most of other gas-analytical instruments, the graduation of OA instruments in terms of concentration can be carried out using etalon gas mixtures with a given concentration of the component to be detected in some unabsorbing buffer gas. It is not at all difficult to perform such graduation with rather a small error δ_g (no more than 0.5 - 2%) within relatively high concentrations (no lower than 10 ppm). Yet at lower concentrations, inaccessible to most available instruments, serious problems appear. They are associated with low accuracy of preparation and calibration of the etalon mixtures in the region of microconcentration, since there are no reliable techniques and the effect of gas adsorption on the cell walls becomes significant. This partly explains the fact that in most cases the threshold sensitivity of OA instruments is estimated by extrapolating the signal-to-noise ratio measured at comparatively high concentrations of the component to be detected. In essence, the accuracy of determining the impurity concentration δ_c may be limited by the achieved accuracy for the calibration of the etalon mixtures δ_g, i.e.,

$$\delta_c = (\delta_g^2 + \delta_u^2)^{1/2} \simeq \delta_g \quad \text{for} \quad \delta_u \ll \delta_g \quad . \tag{6.25}$$

In OA instruments, linear in a wide dynamic range $(10^4 - 10^5)$, graduation is reduced, in essence, to determining the proportionality factor between the output signal and the component concentration in the mixture that is constant in this range. This can be realized at one point of the concentration scale. Therefore, calibration may be carried out within the linear section of the high-concentration range. This gives high accuracy. The results of such graduation can be extrapolated to the low-concentration range.

In some cases, when used at the sensitivity threshold, such a technique may produce a large error because of graduation nonlinearity caused by the effect of background signals from radiation absorption in the windows and uncontrollabel impurities in the gas. These impurities can be found both in the buffer gas and in the component under analysis. The contamination can take place on the way to the OA cell chamber or in the chamber as a result of interaction of the buffer gas with the walls. The influence of these factors can be reduced by proper choice of materials for the OA cell and the windows which slightly absorb various gases (for example, stainless steel, teflon, quartz) as well as through careful cleaning of the chamber and the gas.

Operating conditions permitting, it is advisable to use spectrally pure gases like He, Ar, Kr as buffer gases. It is convenient to employ them in estimating the background signals from the windows and walls. If buffer gases differ from one another during measuring and graduation it is necessary to introduce correction factors allowing for the difference in the thermodynamic parameters of gas mixtures and in the parameters of spectral-line broadening γ_B. For example, at equal γ_B the correction factor is equal to the ratio $(\rho C_v)_{grad} / (\rho C_v)_{meas}$ with $\omega \tau_T \gg 1$ and to the ratio K_{grad} / K_{meas} with $\omega \tau_T \ll 1$, being ρ, C_v, K the gas density, thermal capacity and heat conductivity, respectively.

6.4.2 Graduation by Absorption

In contrast to graduation by concentration, graduation by absorption is more universal since with information on the absorption cross-section it can be applied to analysis of different types of molecules.

a) Analysis of Weakly-Absorbing Media

In the simplest case graduation by absorption can be easily obtained from the above-described method in terms of concentration by introducing just a correction factor measured independently, that is the absorption cross section of molecules detected in the etalon mixture. This operation, however, does not solve the problem of graduation of the instrument to the region of very low concentrations. This problem can be partly solved by using molecules with a small absorption cross-section for graduation. The macroconcentrations of such molecules can simulate low absorption in the medium with high accuracy. In this case rather a complex problem of microconcentration calibration is reduced to a simpler problem of precise measurement of the absorption cross section that can be solved successfully by an absorption measurement using multipass cells and comparatively high molecular concentrations.

Particularly, when operating in the IR region it is possible to realize such graduation with a CO_2 laser and etalon mixtures with CO_2 gas, for which we know very accurately the absorption coefficients at laser lines and their dependences on temperature, pressure, and buffer gas type. For example, the absorption coefficient α_{CO_2} in pure CO_2 under normal conditions at the P(20) line of the 10.6 µm band is $\alpha_{CO_2} = (1.56 \pm 0.2) \cdot 10^{-3}$ cm^{-1} [6.23]. The absorption coefficient in CO_2 diluted with a buffer gas $\alpha_{CO_2\text{-buff}}$ at fixed total pressure can be estimated from

$$\alpha_{CO_2\text{-buff}} = \alpha_{CO_2} \cdot C_{CO_2} \left(\frac{\gamma_{CO_2\text{-buff}}}{\gamma_{CO_2\text{-}CO_2}}\right) \quad , \tag{6.26}$$

with $C_{CO_2} \leqslant 10\%$, where $\gamma_{CO_2\text{-}CO_2}$, $\gamma_{CO_2\text{-buff}}$ are the parameters of CO_2-line broadening during collisions like $CO_2\text{-}CO_2$ and CO_2-buffer. The use of mixtures like CO_2 with a buffer allows absorption graduation up to 10^{-8} cm^{-1} accurately to 3 - 5% with the CO_2 concentration no lower than 10 ppm.

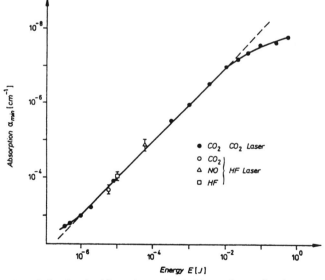

Fig. 6.1. Graduating characteristics of a pulsed spectrophone — dependence of the lowest detectable absorption coefficients α_{min} on energy of HF and CO_2 pulsed lasers [6.3,24]

For illustration, Fig.6.1 gives the results of graduation by a pulsed spectrophone with a pulsed CO_2 laser and the results of measurements of the minimum detectable absorption coefficient α_{min} for some molecules using a pulsed HF laser. The diagram shows α_{min} versus the pulse energy E, expressed by $\alpha_{min} = (S/N)(E\alpha)_{min}/E$, where $(E\alpha)_{min} = 5 \cdot 10^{-10} J \cdot cm^{-1}$ and $S/N = 2$ is the signal-to-noise ratio taken to determine α_{min}. Due to the identity of the gas-mixture macrocomposition (N_2) the results of graduation and measurement agree satisfactorily within the accuracy of 5 to 10%. The nonlinear section in the diagram is due to saturation effects in CO_2 in the high-energy region $E(\Phi_s = 0.15 \, J/cm^2$ [6.25]). In the low-energy region E the "nonlinearity" is caused by a nonuniform distribution of absorbed energy along the spectrophone chamber. The kinetic cooling effect (Chap.7) in the mixture ($CO_2 + N_2$) manifested itself just in a small negative peak at the leading edge of the OA signal and has almost no influence on the accuracy of calibration. Yet according to [6.26], this effect arising from the OA-signal phase variation essentially impedes measuring and calibrating when resonance spectrophones are used.

Graduation by absorption can also be carried out without etalon mixtures. For this purpose the absorption must be determined independently right in the cell, e.g., by comparing the radiation intensity at the cell input and output. However, this can be done only in the region of comparatively high values of the absorption coefficient when $\alpha L \approx 0.1 - 1$. It is convenient in this case to use OA cells with multiple passage of radiation through the chamber. To exclude the influence of nonuniform distribution of absorbed energy along the cell on the accuracy of graduation it is advisable to measure α by the absorption method at multiple passages of the laser beam and to measure the OA signal on a single passage. This technique is basic in calibrating OA instruments to analyze condensed media, especially solids for which it is difficult to prepare etalon samples. In studying weakly absorbing media it is useful during graduation to change to the spectral range where the absorption coefficient is high enough to be measured independently.

These graduation methods are rather universal and with a small modification can be used to analyze substances both in the gas and condensed phases.

b) Analysis of Highly-Absorbing Media

A specific feature of graduation of OA instruments with indirect detection to the region of high absorption coefficients (up to $10^5 \, cm^{-1}$) is that is difficult to determine independently high α by ordinary spectrophotometric techniques. There is a large error of linear extrapolation in the results of graduation from the region of rather small α, where the measurement accuracy of α is high enough, to the region of high α. Therefore particular techniques have been developed to determine the absolute values of α based on using the α dependence of firstly the shape of the OA frequency response [6.27], secondly the OA signal phase [6.28], and thirdly the combined technique of measurement [6.29].

The first (frequency) technique for determining α is based on studying the slope of the OA signal frequency response which varies in two limiting cases: from ω^{-1} (analysis of high-absorption substances) to $\omega^{-3/2}$ (analysis of low-absorption media). The calculation of α is reduced to minimizing the variation of the shape of the experimental frequency response from the theoretical one using (3.20-23) through successive fitting of theoretical results by variation of α. According to [6.27], the accuracy of this technique for determining α in the range up to 10^3 cm^{-1} as the OA signal is measured in the frequency range from 200 Hz to 2 kHz is 10 to 15%. A disadvantage of this technique is that it is necessary to measure the frequency response at each wavelength.

The second (phase) method of detemrining α is based on measuring the OA-signal phase φ related to α [6.28]. For correct measurements the instrumental phase shift φ_0 independent of α should be eliminated. This can be done by phase calibration in measuring the OA signal from substances with α known and determined independently at least at one wavelength. A more direct method to determine φ_0 consists in measuring the OA signal phase at several modulation frequencies. ROARK and PALMER [6.28] considered the accuracy of this method to be 5 to 10%.

The third (combined) method has been tested in measuring rather weak absorption coefficients from 10^{-2} to 10^{-3} cm^{-1} using a solution with KM_nO_4 [6.29]. The values obtained are consistent with the results of measuring α by transmission, i.e., within the limits of 10%.

6.4.3 Graduation by Pressure

Calibration by pressure is specific for analysis of gas media using laser spectrophones. Generally, the dependence of the output signal from the microphone on gas pressure is governed by the microphone parameters, thermodynamic parameters of the gas mixture and the properties of the molecules to be detected. According to (2.40), the microphone sensitivity K_M in the low-pressure region is fully determined by the membrame elasticity, and does not depend on gas pressure since the gas elasticity in the spectrophone chamber is higher. With increasing pressure the gas elasticity is reduced which causes the matching condition (2.41) to be disturbed and the microphone sensitivity will fall off as $K_M \propto P^{-1}$ with $n^* \gg 1$. The dependence of the OA signal on pressure is determined by the time constant of the spectrophone, i.e., $\tau_T \propto P$. Therefore, at $\omega\tau_T \gg 1$ the OA signal P is independent of P and with $\omega\tau_T \ll 1$ $P \propto P$. The properties of the molecules determine the parameter $\alpha(P)$ and the relaxation times τ_{V-T} and τ_{het} (Sect.1.3.1).

For experimental determination of the total influence of these factors on spectrophone sensitivity, except for α, it is necessary to measure the OA signal at the microphone output as a function of the gas-mixture pressure provided that the same laser radiation power is absorbed in the spectrophone chamber. This condition can be realized by analogy with nonselective OA radiation detectors by locating a thin black film with low thermal capacity in the chamber. But this method leads to essential changes in the spectrophone operation at low gas pressures, disregarding the effect of V-T relaxation [6.35]. Therefore, more correct is the calibration of pressure by measuring the OA signal in an absorbing gas with subsequent normalization of the measurement results to a corresponding dependence of the absorption coefficient on pressure. For this purpose it is convenient to use molecules whose absorption coefficient depends little on pressure in the region of collisional broadening of spectral lines. According to (2.28), this holds true if the laser line coincides with the center of the absorption line in a gas. For example, such a situation is typical when a CO_2 laser and CO_2 molecules are used for calibration (Fig.6.2).

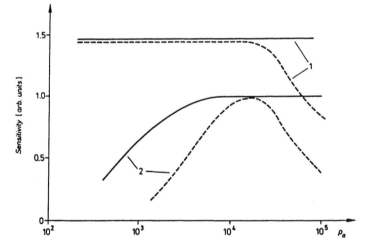

Fig. 6.2. Sensitivity of a microphone (1) and spectrophone (2) as a function of pure CO_2 gas pressure for two ratios of gas and membrane flexibility (2.41). $n^* \ll 1$ (——): a strongly stretched membrane (radius of the chamber r_c = 0.5 cm). $n^* \approx 1$ (---) at P = 2×10^4 Pa, weakly stretched membrane (r_c = 0.25 cm). Radiation source: continuous CO_2 laser; modulation frequency f = 20 Hz [6.3]

It can be seen that the dependence of the spectrophone sensitivity on pressure differs from the identical dependence of the microphone by a decline in the low-pressure region caused by variation of the parameters τ_T and τ_{V-T} with pressure.

The use of highly tense membranes allows the variation of the microphone sensitivity with pressure to be eliminated in the region of relatively high pressure. Yet in this case the microphones sensitivity drops (the figure shows normalized characteristics). It should be noted that the microphone membrane tension T* can easily be estiamted from the pressure for maximum dependence of K_M on P when $n^* \approx 1$ is almost realizable, all other parameters in (2.41) being known. In some cases it is difficult to determine the functional dependence of the absorption coefficient on pressure at a fixed laser frequency. Such a situation takes place, for example, at small absorption cross sections of the molecules when it is difficult to determine α and its dependence on pressure independently from transmission. Then for calibration by pressure it is useful to use the constancy of the integrated intensity of spectral lines, $[\int_\infty \sigma(\nu)d\nu]$, with gas pressure [6.30]. The calibrating technique consists of determining the dependence of $\int_P(\nu)d\nu / P$ on P, which coincides with the dependence of the spectrophone sensitivity on pressure.

6.4.4 Calibration

The reproducibility of an OA instrument is an important metrological parameter affecting measurement accuracy. It is determined during successive inlets of one and the same mixture into the OA cell. In the ideal case the reproducibility of measurements should be commensurate with the measurement accuracy of the OA signal δ_u. But since some parameters of the OA instrument are unstable in time the reproducibility may be worse than this value, and it will determine the resultant measurement error.

The most unstable parameter in the OA instrument is the acoustic detector sensitivity. This is due to frequent action of pressure variation in changing the mixtures to be analyzed. The influence of this factor on the measurement accuracy can be eliminated in several ways.

First, it is possible to improve somewhat the metrological characteristics of acoustic detectors by introducing a system for microphone-sensitivity correction with respect to the reactive and active components of its impedance or by using high-tense metallic membranes. These measures, however, complicate the design of the microphone or reduce its sensitivity.

The second way of improving the reproducibility of measurements consists in regular calibration of the OA cell with etalon mixtures. But this method increases the total duration of the measurement. Besides, there is no guarantee that the detector sensitivity will not change as the mixture to be studied is substituted for the etalon one. A possible way out here is regular correction of the detector sensitivity before or during measurements by forming etalon calibrating signals simulating OA signals. The stability of calibrating signals must be at least one order better than that of the acoustic detectors in use. Calibrating signals can be produced in the following ways.

a) Electrostatic Excitation

Electrostatic excitation can be utilized in calibration of OA signals when a condenser microphone is used. In this case a third electrode is introduced near the microphone membrane and a calibrated-amplitude voltage is applied to it [6.31]. The action of pressure on the membrane is simulated by electrostatic attraction. This method is mostly applied to determine the relative frequency response of a microphone.

b) Pistonphones

Pistonphones are used for absolute calibration of commercial microphones. Here the amplitude of sound pressure P in a closed volume of an OA cell is formed by the vibrating piston located in the cell wall. The sound pressure in the chamber may vary over a wide range according to the dimensions of the chamber, the piston and its vibration amplitude. In practice, the frequency range of the pistonphone is limited at high frequencies of about several kHz by the inertia of the mechanical systems used for piston motion. A modification of this method is the use of a compact and fast-response piezoelectric sound radiator instead of the mechanical piston. Our measurements show that such a source of calibrating signals enables us to reach reproducibility of measurements about 2 - 3% if the instability of microphone sensitivity between two gas inlets is 15% [6.34,38].

c) Thermophones

Thermophones can be employed to calibrate commercial microphones [6.32,33]. Sound pressure is formed due to liberation of heat in the OA cell volume from the electric heater. A thin layer of metal powder 50 - 100 Å thick coating the glass substratum serves as a heater. The amount of liberated heat in it can be easily calculated from the values of voltage and current. In [6.36] a photothermophone was, in addition, coated electrochemically with a thin layer of a metal which simulates a black-body surface because of its nonuniform and patchy structure. This made it possible to eliminate the spectral selectivity of the photothermophone. This device was used to compare the acoustic vibrations formed in the gas as the laser radiation was absorbed on the photothermophone surface and during its electric heating. The accuracy of electric simulation of surface heating by radiation comes to about 3.5%. This device also holds much promise in calibrating OA cells with indirect detection designed to study the surface of a condensed medium.

d) Etalon Plates

An etalon translucent plate periodically introduced into the laser-beam section in the OA cell chamber is used as a source of stable acoustic signal [6.3]. The mechanism of OA signal generation in gases is the same here as for OAS with indirect detection analyzing condensed media. Choosing the material for the plate and its thickness determines the required dynamic range of the calibrating acoustic signals. Since the physical properties of dielectrics are constant, the plates made of them give the most stable source for calibrating signals, and since the absorption coefficient in these plates is rather high the former can be determined independently photometrics. This gives hope for using such plates as a physical equivalent of absorption for aboslute calibration of OA cells without preparing etalon samples.

The absolute sensitivity of acoustic detectors themselves can be determined by such well-known acoustic methods as the compensation, reciprocity, hydrostatic and inertial methods (used basically for hydrophone calibration), calibration in a resonant quarter-wave tube with a Rayleigh disk, etc.

Part 2 Applications

7. Laser Optoacoustic Spectroscopy of Gas Media

Optoacoustic spectroscopy (OAS) is based on measuring the OA signal as a function of wavelength of the monochromatic radiation impinging on the sample. In this chapter we consider mainly the general experimental techniques and potentialities of laser OAS of gaseous media.

7.1 High-Resolution Spectroscopy of Weakly Absorbing Media

Optoacoustic laser spectroscopy of gases is unique in that high sensitivity, i.e., the possibility of operation at very low absorption, and high resolution can be realized simultaneously. Measurements are based on the narrow width of the radiation spectrum of tunable lasers together with a fairly high power level. Before discussing the potentialities of laser OAS specifically we shall consider briefly the technique of OAS with tunable lasers.

7.1.1 Laser Optoacoustic Spectrometers

Progress in the development of tunable narrow-band lasers has opened a wide scope for designing laser OA spectrometers (LOAS) of high resolution. The use of spectrophones to record absorbed power in gases considerably increases in the sensitivity to detect weakly absorbing lines, simplifies the functional scheme and, hence, makes spectral measurement automation much easier.

a) Basic Types of Spectrometers

There are four basic methods of frequency scanning, which can be used to record the spectrum in LOAS: namely linear, stepwise, multichannel and multiplex schemes (Fig.7.1). The choice of a particular scanning method is determined by the research goal and depends on the type of laser, the time of measurement, the sensitivity, etc.

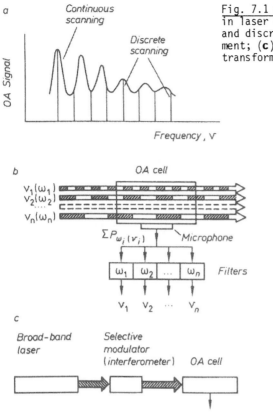

a Continuous scanning

Discrete scanning

OA Signal

Frequency, ν

b OA cell

$\nu_1(\omega_1)$
$\nu_2(\omega_2)$
$\nu_n(\omega_n)$

$\Sigma P_{\omega_i(\nu_i)}$ Microphone

ω_1 ω_2 ... ω_n Filters

ν_1 ν_2 ... ν_n

c Broad-band laser

Selective modulator (interferometer) OA cell

OA-Fourier spectrum

In LOAS with linear scanning continuous frequency tuning of a narrow-band laser is rather slow (Fig.7.1a). The maximum possible scanning rate $V_\nu = d\nu / dt$ is limited by the time of OA signal accumulation, τ_a, and the spectral resolution of the spectrometer $\Delta\nu$: $(V_\nu)_{max} \leqslant \Delta\nu / \tau_a$. For example, with $\Delta\nu = 10^{-2}$ cm^{-1} and $\tau_a = 1$ s, the scanning rate should not exceed $V_\nu = 10^{-2}$ cm^{-1}/s. In linear scanning a rather long time is needed to measure absorption spectra of molecules with high resolution and sensitivity. Specifically, to scan a section of a spectrum just 10 cm^{-1} wide at the scanning rate 10^{-2} cm^{-1}/ s would take about 17 min. Linear scanning can be realized with many types of lasers including dye, semiconductor, spin-flip, high-pressure gas lasers, parametric oscillators, etc.

In stepwise scanning the laser frequency is tuned discretely with a definite step when, after the OA signal is measured at one line, the LOAS is quickly tuned to another line. Then the time losses by transitions from one component of the spectrum to another can be reduced to a minimum and, besides, this also increases measurement reproducibility. Stepwise scanning can be realized with the lasers used in linear scanning. There is, however, a group of low-pressure gas lasers (HF, DF, CO, CO_2, N_2O lasers, etc.) which can only be scanned stepwise because they lase at strictly fixed lines of vibrational-rotational transitions (Fig.7.1a). Stepwise scanning is used mostly for automatic control of the composition of multicomponent gas mixtures.

Multichannel measurements are taken by passing the radiation flux from one or several lasers, with different wavelength and modulated by different sound frequencies, through the OA cell with subsequent discrimination of OA signals at each wavelength using frequency-selective electronic filters (Fig.7.1b) [7.1]. In spite of its complexity, such a method seems to be advantageous for rapid and selective analysis of gas composition, e.g., in monitoring the composition of the atmosphere, in chromatography, etc.

In the multiplex method all the components of the spectrum, not only single discrete lines as in multichannel measurements, can be recorded during the entire measuring time with one OA cell. Laser Fourier OAS examplifies multiplex OAS. Here, the radiation is passed from a wide-band laser through an OA cell. This radiation should first be frequency-modulated in a Michelson interferometer or in some of its modification (Fig.7.1c). In this case it is possible to use wide-band lasers (dye, F-center and semiconductor lasers).

To date several laboratory LOAS and a commercial instrument based on the CO_2 laser have been developed (L-1400 Model "Gilford", USA). The basic parameters of the best LOAS are given in Table 7.1.

The best result have been obtained in the nonoptical region, in the submillimeter range from 0.2 - 2 mm, using coherent electromagnetic tunable radiation of a so-called backward-wave tube (BWT) [7.2]. A radiospectrometer with an acoustic detector (RAD) has already been illustrated in Fig.2.6. The specific feature of RAD is frequency modulation of BWT radiation and the use of a highly sensitive circuit with frequency conversion for OA signal recording by a plane condenser microphone. Thus background signals from the windows and walls of the acoustic detector were minimized and a high threshold sensitivity of about $6 \cdot 10^{-11}$ W $\cdot$ cm^{-1} / $\sqrt{Hz}$ was obtained. This value is just off by a factor of two from the theoretical threshold due to Brownian noise. With such RAD in the submillimeter region results have been obtained for the sensitivity ($\alpha_{min} = 6 \cdot 10^{-9}$ cm^{-1}), the dynamic range (up to 10^6), and the error in absolute frequency (up to $\Delta\nu / \nu = 10^{-8}$).

Table 7.1. Parameters of LOAS

Type of laser	Dye laser [7.9]	He-Ne laser $\lambda=3.3922$ μm [7.8]	Spin-flip CO laser [7.124]	CO_2 laser L-1400 Model	CO_2 pulsed high-pressure laser [7.4]	RAD [7.2]
Spectral range[μm]	0.58-0.61	± 1740 MHz	5.2-5.6	9.2-10.8* (9.2-11.4)**	9.2-10.8	$(0.3-2)\cdot10^3$
Method or device for scanning	intracavity etalon	magnetic field	magnetic field	diffraction grating	etalon with diffraction grating	electronic
Spectral resolution	$7\cdot10^{-4}$cm^{-1}	±15 MHz (10^{-3}cm^{-1})	10^{-1}-10^{-2}cm^{-1}	$1.5-2$ cm^{-1}	$3\cdot10^{-3}$cm^{-1}	$1.5\cdot10^6$Hz
Power source[W]	$20\cdot10^{-3}$	$5\cdot10^{-6}$	10^{-1}-10^{-2}	10^{-1}	10^{-1}(J)	10^{-2}
Minimum detectable absorption cm^{-1}	$3\cdot10^{-8}$	$3\cdot10^{-4}$	$5\cdot10^{-7}$	10^{-7}	10^{-7}	$6\cdot10^{-9}$
Dynamical range	10^4	10^3	10^4	10^4	10^5	10^6
Measuring accuracy of absorption	30-50%			20%	10%-20%	30%
Measuring accuracy of frequency	$7\cdot10^{-3}$ Å	±15 MHz (10^{-3}cm^{-1})		±1 MHz	$2\cdot10^{-3}$cm^{-1}	$\Delta\nu/\nu\approx10^{-8}$

* Parameters of Gilford LOAS

** Laser used on the isotopic mixture $^{12}C^{16}O_2 + {}^{13}C^{16}O_2$ [7.3]

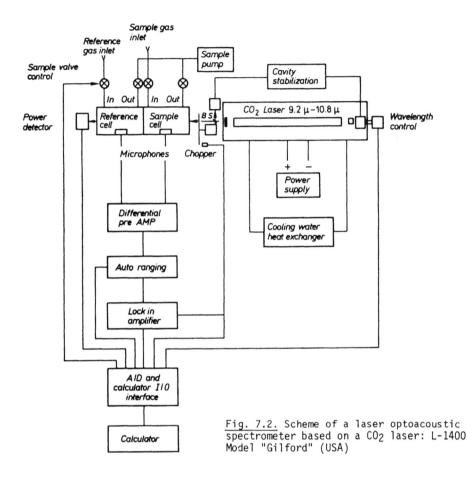

Fig. 7.2. Scheme of a laser optoacoustic spectrometer based on a CO_2 laser: L-1400 Model "Gilford" (USA)

For operation in the IR region a commercial LOAS based on a tunable low-pressure CO_2 laser has been designed. This spectrometer is schematically shown in Fig.7.2, and its parameters are given in Table 7.1. Discrete scanning of the spectrum from 9.2 - 10.8 µm is carried out automatically by turning the intracavity diffraction grating. LOAS operates at about 40 fixed oscillation lines of the CO_2 laser, the distance between them varying from 1.5 to 2 cm^{-1}. During scanning frequency stabilization is realized simultaneously at each line. A specific feature of LOAS is that it comprises a differential scheme with two OA cells which minimizes the background signal and produces difference absorption spectra.

The spectral range of gas lasers can be widened using isotopic variations of the molecules in the active mixture. For example, for a CO_2 laser the following variations can be used: $^{12}C^{16}O_2$, $^{13}C^{16}O_2$, $^{12}C^{18}O_2$ and $^{12}C^{16}O^{18}O$. Thus we can cover a range from 8.8 - 11.4 μm, and a combination of the molecules $^{12}C^{16}O_2 + {}^{13}C^{16}O_2$ allows the total number of generation lines to be increased from 50 or 60 to 130 [7.3]. Spectrometers with discrete tuning are characterized by high spectral reproducibility and simple automation. This makes them very promising for the analysis of molecular impurities in gases.

Optoacoustic spectroscopy in the IR with high resolution and sensitivity at the same time can be realized with spin-flip lasers as well as with pulsed high-pressure gas lasers with continuous frequency tuning; the CO_2 laser in particular [7.4].

Further progress of tunable powerful lasers will promote new models of commercial high-sensitive LOAS with continuous tuning.

b) Spectral Calibration

The spectral calibration of LOAS consits of (a) frequency calibration; (b) determining the instrumental spectral width, i.e., the laser linewidth characterizing the spectral resolution of a spectrometer, and (c) measuring the laser-line frequency instability affecting the accuracy and reproducibility of LOAS.

The general problem of LOAS calibration has been considered in [7.5-6], while below some of its specific features are described. The difficulty in calibrating against the spectrum is explained by the absence of simple methods to measure the frequency in the optical region with a resolution of up to 10^{-3} to 10^{-4} cm^{-1}. For example, calibration gives with common grating monochromators comparatively rough absolute frequency calibration (0.1 - 0.5 cm^{-1}).

The simplest method is absolute frequency calibration with natural frequency standards. The strong single lines of some etalon gases at a low pressure in one OA cell can be used as such standards. To increase the calibration accuracy it is advisable to measure simultaneously the OA signals in the measuring and reference OA cells and then compare the OA spectra of the etalon and sample gases. For example, to identify the lines of a discretely tunable low-pressure CO_2 laser we can use the molecules of SF_6, NH_3, C_2H_4, etc., as reference gases with narrow absorption lines. For submillimeter spectrometers (RAD) SO_2 represents a good reference and for LOAS based on CO_2 laser, the CO_2 molecule itself.

The relative change of laser frequency during tuning is usually controlled with an external Fabry-Perot interferometer, whose dispersion region (the spectral interval between transmission maxima) should constitute just a small part of the laser tuning range. As the frequency is tuned maxima are formed

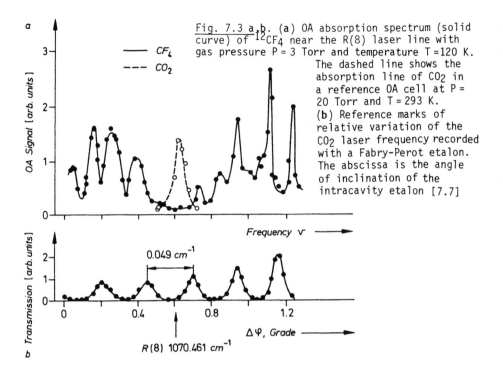

Fig. 7.3 a,b. (a) OA absorption spectrum (solid curve) of $^{12}CF_4$ near the R(8) laser line with gas pressure P = 3 Torr and temperature T = 120 K. The dashed line shows the absorption line of CO_2 in a reference OA cell at P = 20 Torr and T = 293 K. (b) Reference marks of relative variation of the CO_2 laser frequency recorded with a Fabry-Perot etalon. The abscissa is the angle of inclination of the intracavity etalon [7.7]

in interferometer transmission enabling the frequency to be determined by extrapolation. This is illustrated by the OA spectrum of the $^{12}CF_4$ molecule (Fig.7.3). It is measured with a continuously tunable high-pressure CO_2 laser near the R(8) line.

This technique also enables the spectral width of laser radiation and the stability of its frequency to be estimated.

7.1.2 Linear Spectroscopy

a) Spectroscopy with Continuously-Tuned Lasers

Laser OAS has been demonstrated with many molecules and tunable lasers (Table 7.2). It can be used with advantage, for example, to measure the shape, intensity, width and center frequency of single isolated lines as well as the structures of absorption bands due to the Stark and Zeeman effects, Λ doubling and isotopic effects, etc. Such studies have been carried out in the gas-pressure range from 10^5 to 10 Pa, i.e., both with Doppler-broadened lines at pressures below 10^2 to 10^3 Pa and with collisionally broadened lines at higher gas pressures.

Table 7.2. Basic experiments in high-resolution OAS

Laser type	Method of scanning	Spectral range[μm]	Resolution	Molecules under study	References
Ruby	temperature and Fabry-Perot etalon	0.6941-0.6944	0.03 Å	H_2O	7.14
Dye	diffraction grating and Fabry-Perot etalon	0.57-0.31 0.575-0.625 0.58-0.60 0.61-0.65 0.4-0.95	0.5 Å 10^{-3} Å 0.03 Å 0.1 cm^{-1}	S_2O, H_2CS NO_2, J_2 H_2O NH_3,CH_4 C_2H_2, C_3H_6, C_6H_6	7.10 7.117,11 7.9 7.12 7.13
Semiconductor GaAs	pressure and current	0.8-0.9	10^{-2} Å	HF	7.125
F-center	electronic	~3	1 MHz	HCN	7.18
He-Ne	magnetic field	3.39(±1740 MHz)		$^{12}CH_4$, $^{13}CH_4$	7.8,126
Spin-flip	magnetic field	5.2-5.6	0.1 cm^{-1}	H_2O, NO	7.124,15
Semiconductor $Pb_{0.83} \cdot Sn_{0.12}Te$	current, diffraction grating	~10.6		C_2H_4	7.127
High-pressure CO_2	diffraction grating	9.2-10.8	0.05 cm^{-1}	HDS, D_2O $^{15}NH_3$	7.4,16
High-pressure CO_2	Fabry-Perot etalon	9.2-10.8	$3.10^{-3}cm^{-1}$	SF_6,CF_4 C_2H_4	7.17,7

With laser OAS it is very easy to obtain information on the half-width of
a single spectral line and its dependence on the type and pressure of the gas
responsible for the broadening. The high sensitivity increases the accuracy
with which the parameters of line broadening by buffer gases are determined
because low concentrations of the absorbing gas exclude the self-broadening
effect. Such a technique, for example, has been used in [7.8] to study broadening by air and oxygen molecules in the pressure range from 5 to 300 Torr
for CH_4 molecules, using a He-Ne laser with a narrow Zeeman tuning near
$\lambda = 3.39$ μm.

The application of OAS to measure the structure and identify the absorption
bands of a number of molecules has been demonstrated in the visible range
with tunable dye lasers [7.9-13] and the ruby laser [7.14] (see also Table 7.2).

The best results on resolution ($\Delta \nu = 7 \cdot 10^{-4}$ cm^{-1}) have been attained by
ANTIPOV et al. [7.9] who detected 282 absorption lines of H_2O in air in the
spectral range 0.5869 - 0.5936 μm, 40 of them were observed for the first time.
The position of the line centers was defined in two stages. In the first stage
the spectrum of iron was used as a reference spectrum, and in the second stage,
the lines of H_2O were identified in the range indicated. The position of line
centers was determined within $7 \cdot 10^{-4}$ nm. The absorption factors measured in
H_2O were about $0.35 - 4.7 \cdot 10^{-6}$ cm^{-1}.

The use of OAS of high-order vibrational overtones seems to be promising,
as demonstrated in [7.12] in studies of NH_3 and CH_4 molecules. A CW dye laser
was used with an OA cell inside the cavity. The application of hihgly reflecting mirrors produced a high power in the spectrophone chamber which enabled
reliable recording of overtone absorption at 10^{-6} cm^{-1} with a resolution of
0.03 Å. Figure 7.4 shows the spectrum of weak absorption in NH_3 in the spectral
range from 641 to 651 nm [7.12]. The same technique was applied to study the
high overtones of polyatomic molecules, including C_2H_2, C_3H_6 and C_6H_6, using
a resonant OA cell [7.13]. It has been found that as the molecule becomes more
complex the overtone spectrum becomes simpler.

Also OAS is efficient in studying the forbidden transitions in molecules.
In [7.10] forbidden electronic transitions were detected in two sulfur-con-
taining molecules H_2CS and S_2O using a dye laser tunable over the range
310 - 570 nm and 540 - 610 nm. Since these molecules are unstable the measurements were taken at low gas pressures, around 0.3 to 0.5 Torr, under a slow
gas flow through the spectrophone chamber. The sensitivity of the setup allowed
reliable measurement of the oscillator force at a level of 10^{-6}. The optical
path in ordinary absorption measurements would come to 10 km in this case.

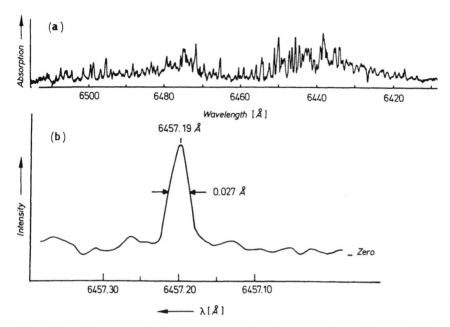

Fig. 7.4. (a) Low-resolution spectrum of weak absorption of NH₃ at 35 Torr. (b) Contour of a single absorption line near 6457.2 Å [7.12]

In various tasks of gas analysis measurements of OA spectra enables more reliable identification of the absorption from the molecules to be detected against the molecular background. This can be illustrated by identifying NO molecules in air at a pressure of 76 Torr near 1886 cm^{-1} with a spin-flip laser [7.15]. In spite of comparatively low resolution of about 0.1 cm^{-1} it was possible to identify the absorption from NO molecules in air with a concentration of 10 ppb.

In the medium IR region the interesting results of high-resolution OAS have been obtained with a continuously tuned high-pressure pulsed CO_2 laser, whose resolution of 0.05 cm^{-1} allowed the OA spectra of the isotopic molecules D_2O, HDS and $^{15}NH_3$ to be studied [7.4,16]. Later the spectra of CF_4, SF_6 and C_2H_4 molecules [7.7] were investigated with a higher resolution of 0.003 cm^{-1}. In particular, it was possible to analyze the structure of the CF_4 absorption spectrum in the band $0 \rightarrow \nu_2 + \nu_4$ near 9.6 μm (Fig.7.3). The measured OA spectra consist of several lines spaced at 0.02 to 0.04 cm^{-1}.

All fundamental transition frequencies to the 25 lowest rotational levels of the ν_3 band of HCN have been measured in [7.18] using OA spectrometer consisting of an F-center laser, an OA absorption cell, and an infrared wavemeter. About 50 lines were determined at a HCN pressure of only 20 mTorr, they had rms deviations of 0.002 cm^{-1} from the calculated values.

Results of high-resolution OAS have been obtained so far only for a limited number of molecules. Therefore, there is a wide field for further research in this direction.

b) Spectroscopy with Discretely Tuned Lasers

When lasers with a fixed frequency or with discrete tuning are used, OAS enables the measuring of the weak absorption coefficient at single points of the spectrum. Information on absorption coefficients (or cross sections) is of particular importance in quantitative analysis of molecular microimpurities in gases as well as in aspects of laser communication and ranging to determine the radiation loss by molecular absorption in the atmosphere.

The absorption coefficients in air and in, so-called, transparency windows usually range from 10^{-5} to 10^{-7} cm^{-1} so that an optical path up to 10 km is required for their systematic measurement. The use of multipass cells necessitates rather large volume gas samples, and measurements under open air conditions are complicated by atmospheric scattering and turbulence. Application of OAS, however, makes local measurement of absorption possible at the required level on a short path (up to 10 to 20 cm) in a small gas volume under investigation (up to several cm^3 at normal conditions), whose composition, pressure and temperature can be varied rather widely.

The basic results of quantitative OAS analysis of weak absorption factors at fixed laser frequencies are collated in Table 7.3. In most of the experiments the OA spectrum was measured by a CO_2 laser, because its oscillation spectrum coincides with the intense absorption bands of many organic and inorganic molecules. In [7.19,20], for example, the OA spectra of about 17 molecular gases have been measured to find the optimum frequencies for their optical pumping since they are used as an active medium in lasers in the far IR spectral region.

In [7.21] the same technique was used to measure the linear absorption coefficients in SF_6, WF_6 and UF_6 in the presence of argon as a buffer at the total pressure of 150 - 300 Torr. The spectrophone was calibrated by determining the coefficient $K = J_{abs} / U_{OA}$ between the electric signals from the microphone U_{OA} and the value of power absorbed in the chamber J_{abs}. At the same time the value J_{abs} was found independently by comparing the radiation at the input and output of the spectrophone filled with the $SF_6 + Ar$ mixture. The absorption factor per unit pressure was calculated from

Table 7.3. Basic gas OAS experiments with discrete-tuned lasers

Lasers	Spectral range[μm]	Absorption level[cm^{-1}]	Molecules under study	References
CO_2	9.2-10.8	10^{-5}-10^{-6}	H_2O	7.24,30,31,32,36
	"	10^{-4}-10^{-5}	$CH_3F,CH_4O,CH_2O,C_2H_3N,CH_3N$, et al	7.19
	"	"	C_2HCl_3, $CClF_3$, CCl_4,CH_2Cl_2,C_3H_4O, $C_6H_{15}N,C_6H_{21}N,H_2O$ et al	7.20
	"	10^{-5}-10^{-6} (cm^{-1}·Torr)	SF_6,WF_6,UF_6	7.21
CO_2	"		toluene; m-xylene, o-xylene et al.	7.22
CO_2	"		acrolein, styrene, vinyl bromide et al.	7.23
$^{12}C^{16}O+$ $^{13}C^{16}O_2$	9.2-11.4	10^{-5}-10^{-6}	C_2H_4, O_3, NH_3	7.3
CO	5.5-6.2	"	H_2O	7.128*,25
DF	3.5-4.1	10^{-8}-10^{-10}	CH_4, N_2O	7.26,35
			H_2O,HDO	7.27,28
DF	3.5-4.1	10^{-7}-10^{-9}	HDO	7.33
DF	"		HCl	7.34
HF	2.7-2.9	10^{-5}-10^{-7}	HF, CO_2, H_2O	7.29

* OTS has been used to study the H_2O molecule (Chap.4)

$$\alpha = -\frac{\ln{(1 - K \cdot U_{OA} / J)}}{PL} \qquad (7.1)$$

where J is the power at the cell output, L is the absorption length, P is the partial pressure of the absorbing gas. It should be noted that at high absorptions the populations of the molecular levels resonant with the laser radiation may change because of increasing gas temperature in the laser-beam section. This results in errors in determining α. For the SF_6 molecule, for example, the absorption cross section increased during laser excitation from the low frequencies of the ν_3 band, while it decreased during excitation from the high-frequency side. High sensitivity of OAS enables measuring the absorption in SF_6 at the wing of the ν_3 band in the 9.4 μm region as well as the absorption in WF_6 at a level of 10^{-4} cm^{-1} $Torr^{-1}$ and in UF_6 at 10^{-6} cm^{-1} $Torr^{-1}$. Figure 7.5 illustrates the OA spectrum of UF_6 at a partial pressure between 10 and 25 Torr and an Ar pressure of 270 Torr.

Information on absorption coefficients is greatly needed in quantitative analysis of molecular impurities in different gases [7.22,23]. To define the optimum analytical lines for detecting the basic atmospheric pollutants in [7.3], their OA spectra were measured with a laser in the isotopic mixture $^{12}C^{16}O_2 + ^{13}C^{16}O_2$. For example, in [7.23] CO_2-laser absorption cross-section data were reported for acrolein, styrene, ethyl acrylate, trichloroethylene,

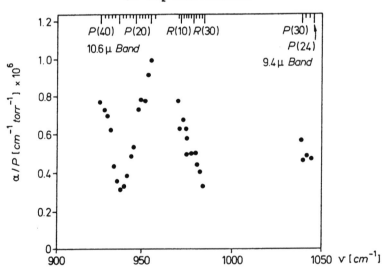

Fig. 7.5. OA spectrum of UF6 molecules at CO2 laser lines [7.21]

vinyl bromide and vinylidene chloride. These data indicate that detection limits at ppb-level is possible for these compounds by the laser OA technique.

A serious problem in measuring in the real atmosphere is the influence of background absorption from the basic gas components of the atmosphere, including H_2O, CH_4, N_2O, etc. Careful measurements of absorption of different radiation lines of CO_2, CO, DF and HF lasers by these components with the use of spectrophones have been carried out in [7.24-34] (Table 7.3). The best sensitivity data have been obtained in [7.26] using a two-chamber longitudinal spectrophone with a differential-pressure indicator. Such a design suppressed the background signals from the BaF_2 windows up to the level of $\alpha_b = 3 \cdot 10^{-9}$ cm^{-1}, so absorption in the CH_4 and N_2O molecules could be measured at an average level of 10^{-8} cm^{-1}, close to their natural atmospheric concentrations, at 17 lines of a CW DF laser in the region of $\lambda = 3.8$ μm. Under pulsed-laser operation it is also possible to measure weak absorption with a satisfactory accuracy by OAS. In [7.35], for example, the DF laser was used to measure the absorption factors in CH_4 at a level of 10^{-7} cm^{-1}. This experiment shows that the deviation of results compared to the results of other works is no higher than 15%.

The great interest is the application of OAS to study weak absorption of certain components of the atmosphere in transparent windows. For example, the absorption coefficients of H_2O at the lines of CO and CO_2 lasers have been measured using both a resonant cylindrical spectrophone [7.24,25] and a differential two-chamber spectrophone of longitudinal geometry with chambers of different length (Fig.5.2e) [7.30].

Water vapor absorption at 161 wavelengths from 9.2 to 11.9 μm of the $^{12}C^{16}O_2$, $^{13}C^{16}O_2$, and $^{14}C^{16}O_2$ lasers was measured using a resonant OA spectrometer in [7.36]. Results were obtained at several precisely determined vapor concentrations in a flow of pure air at a total pressure of 1 atm. The OA spectra of water absorption continuum are obtained for temperatures between 27 and $-10^{\circ}C$ [7.31]. The temperature and water pressure dependencies observed for the continuum suggest that while both collisional broadening and water dimer mechanisms contribute to the continuum, the dimer mechanism is more important in this temperature range. Since the primary cause of the water-vapor continuum absorption within the 8 - 14 μm atmospheric window is controversial accurate absorption measurements at high relative humidity are of great interest. The first OA studies with a CO_2 laser, performed in supersaturated water vapor in an OA diffusion chamber, was presented in [7.32].

The results presented show that OAS can essentially improve the routine technique of measuring weak absorption in multipass cells. When absorption below 10^{-7} to 10^{-8} cm^{-1} was to be determined, this method is more useful than others.

7.1.3 Spectroscopy in Electric and Magnetic Fields

When the gas in the spectrophone is subjected to intense electric and magnetic fields, it is possible to study the Zeeman and Stark spectra of many
molecules. The Stark and Zeeman effects manifest themselves most vividly at
low gas pressures when there is no pressure broadening of lines. Optoacoustic
spectroscopy allows faithful measurements at gas pressures up to 10 - 20 mTorr.
For low-pressure regions we should use combined cells based on a compromise
between OAS and OTS. This method Zeeman and Stark OAS enables us, on the one
hand, to study the Zeeman and Stark effects in weak-absorption media. On the
other hand, the use of Zeeman and Stark modulation right in the OA cell chamber increases the selectivity and the sensitivity of detection of trace amounts
of molecules.

a) The Zeeman Spectroscopy

The first experiments on Zeeman OAS were carried out with the simplest molecule, namely NO. The measurements were made using laser excitation. The CO
laser pumped the transition $V = 0 \rightarrow V = 1$ of the NO molecule. The absorption of
excited states was probed by the tunable spin-flip laser [7.124]. The resolution of 200 MHz obtained in these experiments depended on the frequency instability of the CO laser that was used to pump the spin-flip laser.

Zeeman OAS increases the detection selectivity of molecular impurities
in gases by the magnetic-field tuning of the maximum of the NO absorption
line to the fixed frequency of a CO laser [7.38] as well as by eliminating
the background from BaF$_2$ windows, for example, as the polarization CO-laser
radiation is modulated [7.37]. With linear polarization it is possible by
rotating the $\lambda / 2$ plate to modulate the absorption at a fixed-frequency laser
only when paramagnetic molecules are in a constant magnetic field and to eliminate the isotropic background, particularly from the absorption in the windows. Another way of increasing OAS selectivity is absorption modulation by
a modulated magnetic field. Zeeman OAS can also be applied to the selective
detection of free radicals (H, OH, HO$_2$, Cl$_,$ CLO, etc.) and molecules (NO, NO$_2$)
in different aspects of atmospheric chemistry.

b) Stark Spectroscopy

Stark OAS consists of successive measurements of the OA spectrum of molecules
as a function of the voltage across the Stark electrodes in the spectrophone
chamber. In this case both radiation-power modulation and frequency modulation

can be used. For example, in the submillimeter spectral region the dipole momenta of the PH_3 molecule in the ground and excited states have been measured with a radio-acoustic detector (RAD) operating under frequency modulation with the Stark field strength of up to 1000 V/cm [7.40]. The design and the performance of the OA cell that can be operated in the presence of strong electric fields (larger than 30 kV/cm) and at unusually low pressures (tens of millitors) were described in [7.39,41-43]. By using a Lamb-dip stabilized CO_2 laser and sweeping the static electric field 22 coincidences between NH_3 transitions and CO_2 laser lines were observed in [7.42].

In the OA cell it is possible to carry out Stark modulation of absorption by alternating the electric voltage, with power and frequency fixed. This increases the OAS concentration sensitivity for molecules with a strong Stark effect on account of the decrease in the background OA signals from the windows, walls, as well as the molecules with a weak Stark effect. In [7.44], for example, it has been shown experimentally that in detecting the molecules of NH_2D with a CW $^{12}C^{16}O_2$ laser at the P(20) line of the 10.4 µm band the use of the Stark effect instead of a mechanical obturator for power modulation increases the signal-to-noise ratio by about six times at an optimum gas pressure (about 300 Torr). But under Stark modulation there is an attendant decrease in the absolute value of the OA signal due to the small modulation depth. In [7.44], for example, the OA signal was reduced by 30 times. Therefore, one should thoroughly optimize the parameters of the OA Stark cell, including the choice of an analytical laser line, mutual orientation of the electric field and the plane of polarization, Stark modulation depth and gas pressure. In the submillimeter range OA Stark modulation has been tested to advantage [7.45] with molecules of PH_3 and CH_3OH and a radio-acoustic detector with a cylindrical geometry of the Stark electrodes. It has been illustrated that the use of Stark modulation instead of frequency modulation simplifies the spectrum of these molecules considerably because the signal from the lines with a weak Stark effect are suppressed.

7.1.4 Doppler-Free Spectroscopy

Furthermore, OAS can be used in certain cases in nonlinear laser Doppler-free spectroscopy [7.46]. The spectroscopy of absorption saturation involves OA recording of narrow resonances inside the Doppler absorption line by varying the spectrophone signal as two counter-running wave beams with tunable frequency are passed through the spectrophone chamber. One of them saturates the absorption line and the other one with amplitude modulated, probes its absorption contour (Fig.7.6). By analogy with the experiments on the

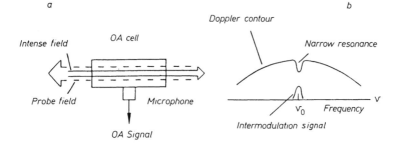

Fig. 7.6 a,b. OA detection of Doppler-free resonances by absorption saturation: (a) measuring scheme; (b) narrow resonances inside the Doppler contour

fluorescence technique of Doppler-free spectroscopy [7.47], modulation of both beams with different frequencies ω_1 and ω_2 is often useful. As a result the useful OA signal proportional to the dip in the Doppler absorption contour is recorded at the total frequency $\omega_1 + \omega_2$ off the zero level (intermodulation OAS) (Fig.7.6b).

Experimentally this technique has been tested in [7.48,49]. In [7.48] the hyperfine structure of the P(93) line of the band 11-0 was measured at the transition B ← X in $^{127}I_2$. The CW single-mode dye laser was used as a radiation source. Its radiation was divided by a semitransparent plate into two beams each 60 mW, modulated with the frequencies $\omega_1 / 2\pi = 757$ Hz and $\omega_2 / 2\pi = 454$ Hz. After focusing, these beams were directed into the spectrophone chamber from opposite sides. Figure 7.7 shows the OA signal measured from a plane condenser microphone at the frequency $\omega_1 + \omega_2$ as a function of frequency. The nonlinear

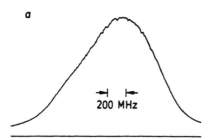

a

200 MHz

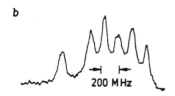

b

200 MHz

Fig. 7.7 a,b. OA spectrum of $^{127}I_2$: (a) Doppler contour of the P(93) line of the 11-0 band of the B ← X transition near 5195 Å. (b) Doppler-free structure of this line. Scanning time: 10 min; integrating time: 1 s; $^{127}I_2$ pressure: 0.2 Torr [7.48] interaction of two waves allows the intra-Doppler hyperfine structure of $^{127}I_2$, about 0.2 Torr.

It should be taken into account, however, that at lower pressures OAS sensitivity drops when the resolution is improved. Specifically, in [7.49] the value of an intermodulation OA signal was studied as a function of CH_3OH pressure with a CO_2 laser at the P(34) line of the 9 μm band. With the power of 0.7 W the sensitivity of the spectrophone made it possible to observe the OA signal up to 20 mTorr of pure CH_3OH. In the same work OAS was used to detect a 10 MHz wide dip at the center of Doppler absorption contour in CO_2 at a pressure of 100 mTorr.

A probing wave in the method of absorption saturation with OA recording of supernarrow resonances can be easily produced when the primary wave reflects from the output window of the spectrophone. Such a scheme, for example, has been used in [7.50] to detect a dip in the Doppler contour of far-IR CH_3OH lasers pumped by CO_2 laser. The OA detection to Doppler-free spectroscopy was used also in [7.41-43,51]. For example, a single-mode widely tunable colour-center laser has been applied to obtain Doppler-free OA spectra of the R(13) and R(16) lines of the ν_3 band of HCN at 120 mTorr pressure with a resolution of 7 MHz [7.51]. By using a Lamb-dip stabilized laser and an OA detection technique the absolute frequencies of a few $^{14}NH_3$ lines were measured with improved accuracy in [7.42].

It is also possible to increase the sensitivity of sub-Doppler spectroscopy in the region of low gas pressures by using OTS for direct recording of vibrationally excited molecules with a pyroelectric detector (Chap.4), e.g., as in [7.52], when a dip 2.5 MHz wide was observed in the Doppler contour of the SF_6 molecule at a pressure of 12 mTorr with a CO_2 laser at the P(16) line. The observation was performed by absorption saturation in the standing-wave field. An OT cell in the form of a cylinder (22 mm in diameter) made of pyroelectric foil (11 μm thick and 6 cm^2 in area) was used. This design made it possible to realize a threshold sensitivity of about $5 \cdot 10^{-8}$ W $\cdot$ cm^{-1} / $\sqrt{Hz}$ at the modulation frequency of 30 Hz.

At the center of a Doppler-broadened absorption line the monochromatic laser wave interacts with those molecules with a zero-order projection of their velocity onto the light-wave direction, i.e., flying in parallel with the wave front. This opens ways for narrowing the spectral lines through eliminating molecules with a large Doppler shift and hence flying at a large angle to the front. This can be realized through forced relaxation at special diaphragms of the molecules not flying in parallel before they transfer energy to other molecules [7.53]. In practice, this can be done with a number of parallel

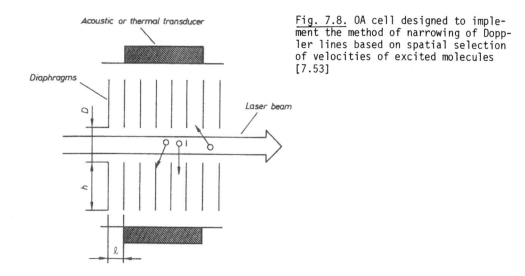

Acoustic or thermal transducer

Diaphragms

Laser beam

Fig. 7.8. OA cell designed to implement the method of narrowing of Doppler lines based on spatial selection of velocities of excited molecules [7.53]

transparent plates placed at right angles to the laser beam in the spectrophone chamber or with several parallel diaphragms placed around of the beam (Fig.7.8). Then the condition for line narrowing is [7.53]

$$\Lambda \geq (h + D) \quad , \quad D \gg \lambda \quad , \quad h/\ell \gg 1 \quad , \tag{7.2}$$

where λ is the wavelength; Λ is the free path of molecules; h is the diaphragm dimensions; ℓ, D denote the distance between the diaphragms in parallel with and transverse to the laser-beam propagation. High efficiency can be obtained at rather high values of Λ, i.e., at low gas pressures. The diaphragm configuration, as shown in Fig.7.8 allows only molecules 1 flying along the diaphragms contributed to the signal, i.e., those with a zero-order projection of their velocity onto the laser beam direction.

7.2 Spectroscopy of Excited Molecular States

According to a Boltzmann distribution at equilibrium the population of high-lying vibrational levels in molecules at normal temperature is small. Therefore the intensity of the absorption bands from excited vibrational states called hot bands is usually small and their systematic study by standard absorption spectroscopy is difficult. The OA method is the most promising in detecting weak absorption from these levels due to its high sensitivity and sufficient accuracy for such measurements.

7.2.1 Thermal Excitation of Vibrational States

In principle, OA detection enables reliable measurement of low absorption from rather high-lying vibrational levels with energies up to $3000 - 5000$ cm^{-1}. In practice, however, it is sometimes difficult to interpret the results correctly due to the potential effect of absorption from the low-lying levels and extraneous microimpurities in the gas. So the measuring technique must be modified to discrimine better between the OA signal and excited molecular states, as has been done, for example, in [7.54] through measuring the dependence of the OA signal on the temperature of the gas mixture. Increase in temperature leads to an increase in the population of excited states. This, in turn, increases the absorption and, besides, discriminates the absorption from separate levels by measuring the slope of the corresponding temperature curves.

Let there be two vibrational states 1 and 2 with energies E_1 and E_2, with $E_1 \ll E_2$. If $\Delta E = E_2 - E_1 \gg K_B T$, which can be fulfilled well for the IR range, the relation $\alpha \propto \exp(-E_1 / K_B T)$ holds true for the absorption factor. From this it is easy to get the following relation:

$$\ln \alpha = - \frac{E_1}{K_B T} + \text{const} \quad . \tag{7.3}$$

Thus, in a plot of $\ln \alpha$ versus $1 / T$ the dependence takes on the form of a straight line. The tangent of its inclination gives the energy of the lower level E_1 participating in absorption.

Measurements of the excited vibrational states of BCl_3 and BF_3 molecules at the lines of a CW CO_2-laser, using the technique described, are given in Fig.7.9. The measurements have been taken in the spectral range of 1072 to 1080 cm^{-1} where, according to estimation, the influence of absorption from low-lying levels is negligible and absorption from excited states can be detected. The spectrophone in use (Chap.5) enables the gas temperature in the working chamber of the spectrophone to be varied from 290 to 600 K. It can be seen from the figure that, as assumed, there are linear sections observed in the dependences which can be related to certain vibrational states. Specifically, in the BCl_3 molecule absorption of a vibraional level with energy $E = 4530 \pm 230$ cm^{-1} has been observed. The presence of several linear sections is explained by successive involvment into absorption of higher-lying levels with a hihger initial energy as the gas becomes heated. The temperature calibration of the spectrophone was performed with CO_2 gas where the structure of vibrational bands and their population as a temperature function in the

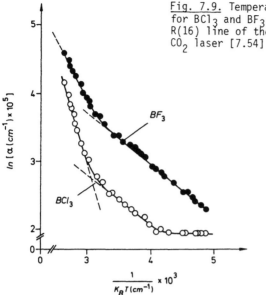

Fig. 7.9. Temperature dependence of OA signal for BCl_3 and BF_3 molecules at 30 Torr at the R(16) line of the transition 02°0-00°1 of a CO_2 laser [7.54]

region of CO_2 laser lines are well known. The calibration shows that the accuracy of level-energy measurement with this technique is about 5% and the error of absolute measurement of the absorption coefficient at 300°C is no higher than 15 to 20%.

It seems useful to widen up the temperature range of spectrophone operation to study the structure of vibrational "quasi-continuum" for many molecules.

7.2.2 Laser Excitation of Vibrational States

Gas irradiation by two IR pulses with different frequencies is more efficient and has a wide scope in investigating absorption at vibrational transitions between excited molecular states. The first pulse excites resonantly the vibrational states, whose thermal population is small. This pulse induces the formation of new absorption bands relating to the transitions between excited vibrational states. These new bands can then be OA detected with tunable laser radiation at the second frequency.

The following three measurement regimes are of most interest: a) pulsed operation of both lasers; b) continuous operation of the first-step laser and continuous or pulsed operation of the second-step laser; c) pulsed operation of the first-step laser and continuous operation of the second step laser.

The advantage of (a) is the possibility of selective excitation of just one or several vibrational states resonant with the laser radiation. It is also

possible to study the excitation kinetics by controlling the time delay between the pulses of both lasers and changing their duration. This technique
was first experimentally tested in [7.55] to investigate the quasi-continuum
of SF_6 molecules with two time-synchronized pulsed CO_2 lasers. The radiation
of the first-step laser with frequency ν_1 was not focused whereas the radiation of the second laser with frequency ν_2 was focused by a lens with focal
distance 40 cm into the central part of a spectrophone with high spatial resolution (Fig.5.6) so that the radiation of both lasers completely overlapped
(the diameters of beams I and II were 3 mm and 0.7 mm, respectively, the central part of the chamber between the diaphragms being about 8 mm). The first-
step laser operated at the P(20) line of the 10.6 μm band in resonance with
the band $0 \rightarrow 1$ of the ν_3 vibration of SF_6. Since the second laser radiation
with frequency ν_2 was tuned within a narrow spectral region, approximately
970 to 1080 cm^{-1}.

The measuring technique consisted in successive determination of absorbed
energy from each beam separately and then with the simultaneous action of the
two pulses on the molecules. Thus the average number of IR photons absorbed
per one molecule in the volume under irradiation from each beam $<n_1>$ and $<n_2>$
was measured, as well as the value $<n_{1+2}>$ from the two beams at the same time.
The value $<n> = <n_{1+2}> - (<n_1> + <n_2>)$ determined the absorption of the second
pulse, which is conditioned by the molecules excited by the first pulse. If
the first pulse is intense enough so that a molecule absorbs several IR photons and can be found in the vibrational quasi continuum, the calculated value
determines the contribution of the vibrational quasi continuum to absorption.
Figure 7.10 thus presents the dependences of $<n_2>$ (Curve 1), $<n_{1+2}>$ (Curve 2)
and $<n>$ (Curve 3) on the frequency of second-step radiation, the other parameters have been fixed.

In the second regime excitation of the molecules and measurement of their
absorption are performed by continuous radiation. Then at spectrophone pressures of about 1 Torr and more the gas mixture is heated uniformly. But, as
a contrast to single-frequency measurement when the gas is heated by the
chamber walls, it is quite simple to realize pure gas heating across the laser
beam to rather high temperatures, about 1000 K. In this case the adsorption
and chemical influence of the chamber walls is minimum. To eliminate molecular
relaxation on the chamber walls at small pressures, the chamber diameter may
be rather large as, for example, in resonant spectrophones. This was realized
in [7.124] to investigate the vibrational transition $1 \rightarrow 2$ in NO molecules using
a tunable spin-flip laser in optical pumping of the molecules at the transition $0 \rightarrow 1$ by CO laser radiation with fixed frequency $\nu = 1917.8611$ cm (Fig.7.11).

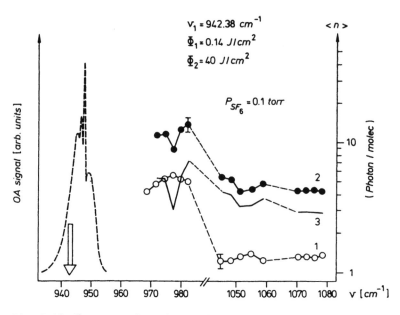

Fig. 7.10. Frequency dependence of absorption in the vibrational quasi continuum of SF_6 at fixed energy fluences of pulses Φ_1 and Φ_2, fixed frequency ν_1
and varying frequency ν_2. The designations for the curves are given in the
text [7.55]

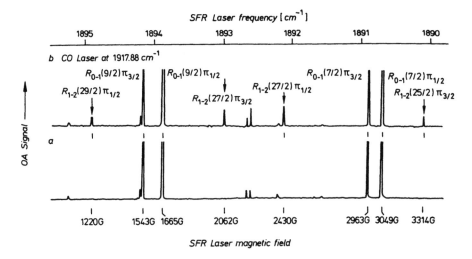

Fig. 7.11. OA signal in NO at 0.5 Torr as a function of spin-flip laser frequency: (**a**) Without CO laser radiation; (**b**) two-frequency excitation [7.124]

Radiation from both lasers was collinearly focused by a lens with focal distance 20 cm into a spectrophone 1 cm long . It can be seen from the figure that as the CO laser is switched on we can observe a number of new lines in the spectrum. Among them it was possible to identify and measure accurately the frequencies of seven lines of the transition $1 \to 2$ in the $^2\Pi_{3/2}$ state and six lines in the $^2\Pi_{1/2}$ state as well as Λ doubling for the $R_{1-2}(27/2)$ $\Pi_{1/2}$ line.

This regime is universal for pure heating of all the molecules by the first-step laser, including those which do not absorb its radiation. For the latter case a chemically inactive admixture with intense absorption bands at the lines of the first-step laser radiation should be added to those molecules whose excited-state spectra are under study.

The third regime (pulsed excitation and continuous measuring of absorption) is in many respects similar to the first regime but, unlike the latter, makes it possible to observe, continuously in time, the kinetics of population and relaxation of excited molecular states.

7.3 Nonlinear Absorption in Molecules

The absorption of IR radiation by a molecule, as the intensity increases, can be divided conventionally into three stages: 1) linear absorption, 2) absorption saturation and 3) multiphoton absorption. In linear absorption the absorbed energy is proportional to the incident energy. Below we consider the efficiency of OAS in measuring nonlinear saturation and multiphoton effects [7.56]. Application of the OAS here makes it possible to investigate nonlinear effects in weakly-absorbing gaseous media and, unlike other methods, it allows direct measurement of absorbed energy in gases.

7.3.1 Saturation Effect

Absorption coefficient of the gas can be reduced because the quantum level populations of the resonant transition participating in absorption are changed. The basic parameter is either the saturation energy $\Phi_s(t_p \ll \tau_{V-T})$ or the saturation intensity $I_s(t_p \gg \tau_{V-T})$ (Sect.2.2).

For experimental determination of the saturation parameters it is necessary to measure the nonlinear dependence of the radiation energy (power) at the output on the energy (power) at the input. For weak transitions, however, when after passing through the medium the energy almost does not change, it is very convenient to measure directly the nonlinear dependence of the absorbed energy itself. This can be done by OAS.

To determine the saturation energy flux Φ_s we may use the following measuring technique [7.57]. The absorption of a rectangular laser pulse with pulse duration $t_p \ll \tau_{V-T}$ in a cell with length L filled with a medium whose initial absorption coefficient is α_0 in a two-level approximation and at homogeneous line broadening is described by [7.58]

$$\ln \frac{\Phi_{in}}{\Phi_{out}} = \alpha_0 L - \frac{1}{2\Phi_s}(\Phi_{in} - \Phi_{out}) \quad , \tag{7.4}$$

where Φ_{in}, Φ_{out} are the pulse energy fluxes at the input and output of the cell, respectively.

When absorption is small, i.e., $\Delta\Phi \ll \Phi_{in}$, Φ_{out} and $\Phi_{in} \leqslant \Phi_s$, the expression can be simplified to

$$\frac{1}{\Phi} + \frac{1}{2\Phi_s} \simeq \frac{\alpha_0 L}{\Delta\Phi} \quad . \tag{7.5}$$

From (7.5) it is seen that in a plot of $1/\Delta\Phi$ versus $1/\Phi$ the dependence has the form of a straight line. When extrapolated to the abscissa it cuts off a length equal to $1/2\Phi_s$, which can be used to estimate Φ_s. Such a technique is valid for laser spectrophones because the OA signal is proportional to the relation $\Delta\Phi/L$. Figure 7.12 presents measurements of Φ_s by a pulsed CO_2 laser in CO_2 mixed with Ar. The measurements were taken with a nonfocused laser beam which allowed accurate measurement of the beam cross section and the satura-

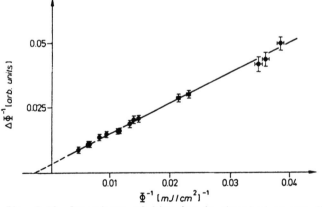

Fig. 7.12. Dependence of $1/\Delta\Phi$ ($\Delta\Phi$ is absorbed energy flux) on $1/\Phi$ (Φ is input energy flux) in the mixture CO_2+Ar at the P(20) line of the $00^01 - 10^00$ band of a pulsed CO_2 laser. Mixture: $P_{CO_2} = 20$ Torr, $P_{Ar} = 1$ atm ($\alpha_0 = 4\times10^{-5}$ cm^{-1}) [7.57]

tion energy (equal to $\Phi_s = 0.15 \pm 0.01$ J/cm^2). In this technique no stringent requirements are placed upon the power level since nonlinearity measurements are carried out essentially at the initial part of the saturation characteristic.

For relaxation of excited molecules or comparatively long laser pulses the measuring value represents the saturation intensity I_s. According to (2.18), this parameter can be determined from the dependence of absorbed power on input power. This technique was used in [7.59,60] to investigate the dependence of absorption in H_2O in air with a resolution of 0.04 cm^{-1} in the spectral range from 694.2 to 694.4 nm on the power of the pulsed temperature-tunable ruby laser. The values of the saturation parameter I_s measured in a maximum of several lines of H_2O at an air pressure of 750 Torr lie within a range from 10^2 to 10^3 W/cm^2, the pulse duration being $t_p = 5 \cdot 10^{-8}$s. If the laser pulse shape is not rectangular and hence complicated here is an error, estimated in [7.59], arising in determining the saturation parameter by the methods above.

7.3.2 Multiphoton Absorption

The interaction of an intense IR field with polyatomic molecules when multiphoton absorption becomes essential is now an important area of nonlinear IR photophysics and photochemistry of molecules. Here we consider briefly only how OAS is applied to study the characteristics of multiphoton absorption without going into the details of this process, which can be found in [7.56,61,62].

The multiphoton character of absorption usually occurs at high energy fluxes of laser radiation. When nonfocused laser beams are used it is sometimes very difficult to achieve a high-energy flux because of optical damage to cell windows or high demands on the energy and divergence of the laser beam. The first factor is particularly important in the IR special region where the intensity threshold of damage of the best optical materials (KBr, ZnSe and NaCl) varies between 10^7 and 10^8 W/cm^2 with pulse duration $t_p = 10^{-7}$s. Therefore, in studies at higher power densities the laser beam should be focused into the center of the cell. This, in turn, calls for selective detection of absorbed energy only from the region of the laser beam's caustic waist, using a pulsed spectrophone with high spatial resolution (Fig.5.6). A spatial-resolution spectrophone enables correct measurements of the most important characteristics of multiphoton processes: the average energy absorbed per molecule <E> in the volume under irradiation or the average number of absorbed quanta <n>(<n> = <E> / hν), and their dependence on the conditions of excitation as well as the spectra of multiphoton absorption.

a) Experimental Accuracy

A disadvantage of OAS is a strong and rather complicated dependence of its sensitivity on gas pressure, particularly in the region of low pressures beginning from 0.5 to 1 Torr. At the same time low pressures are most interesting in studying multiphoton processes to eliminate the effect of V-V and V-T relaxations. So the spectrophone's sensitivity must be calibrated independently, for example, by involving the results of measurements at the same pressures in long test cells (up to 200 cm) with collimated beams of moderate intensity (10^5 to 10^6 W/cm). Since calibration is usually carried out at significant absorption in a low-absorption cell an error may arise due to the variable distribution of absorbed energy along the cell. When the microphone is placed in the center of the cell, the maximum value of the error is [7.63]

$$\delta P = 1 - \frac{2sh(\alpha L / 2)}{\alpha L} \quad , \tag{7.6}$$

where α is the absorption factor in calibration. From (7.6) it follows that the accuracy of OA measurement in the range $0 \leqslant \alpha L \leqslant 1$ is $\delta P \leqslant 5\%$.

The error in determining $<E>$ is due, first of all, to nonuniform energy distribution across the laser beam. To minimize this error, in [7.64] a method was proposed to find the real dependence of $<E>$ on $\Phi = E / A_L$, A_L being the cross section of a laser beam with energy E. The method is based on determining the shape of pulse-energy distribution across the beam. Let the laser beam cross profile be presented as

$$f(r) = \exp[-(r / r_L)^n] \quad . \tag{7.7}$$

The parameters r_L and n are known, and by measuring the value $T = E_{out} / E_{in}$ we can find the real value of energy flux Φ that corresponds to the value of transmission being measured. According to [7.64], for n = 2

$$\Phi = \frac{E}{2\pi r_L^2 \sigma NL} [1 - \exp(-\sigma NL)] \quad , \tag{7.8a}$$

and for n = 4

$$\Phi = \frac{\sqrt{2} \, E}{\pi^{3/2} r_L^2 \sigma NL} [1 - \exp(-\sigma NL)] \quad , \tag{7.8b}$$

where σ is the absorption cross section

$$\sigma = -\ln T / NL \quad . \tag{7.8c}$$

From (7.8) it is possible to find the real dependence $\sigma(\Phi)$ and $<E>(\Phi) = \Phi\sigma(\Phi)$.

After proper processing of the data the basic error in determining the dependence of $<E>$ on Φ is governed mainly by the error in measuring the absolute value of pulse energy E.

In most cases the pulsed OA signal is a train of pulses caused by the acoustic resonant properties of the cell and microphone. When operating with focused beams we should record selectively the first peak of the signal to obtain information on the character of absorption in the region of the beam caustic nearest to the microphone. As a rule, particularly for plane condenser microphones, the value of this peak depends on the cross section and geometry of the laser beam in the spectrophone. This requires that measurements should be carried out at strictly fixed conditions.

b) Intensity Dependence of Multiphoton Absorption

A spectrophone with high spatial resolution was first used [7.65] to measure the dependence of the number of absorbed quanta $<n>$ on the radiation intensity of a TEA CO_2 laser for some simple and complex molecules (Fig.7.13). It has been shown that with an increase in pulse energy with such simple molecules as D_2O and OCS absorption saturation sets in, while for polyatomic molecules, such as C_2H_4 and SF_6, $<n>$ grows without saturation, due to multiphoton absorption.

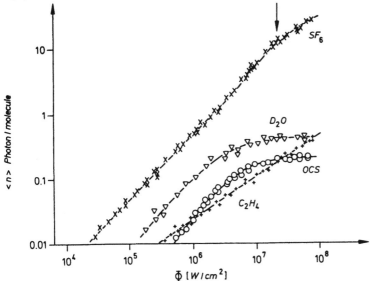

Fig. 7.13. Dependence of $<n>$, the average number of absorbed quanta per molecule, on laser radiation intensity. Experimental conditions (P is gas pressure, ν is CO_2 laser frequency): SF_6, P = 0.25 Torr, ν = 942.38 cm^{-1}; C_2H_4, P = 1 Torr, ν = 954.55 cm^{-1}; D_2O, P = 0.7 Torr, ν = 1079.85 cm^{-1}; OCS, P = 1 Torr, ν = 1045.02 cm^{-1} [7.65]

175

The advantage of OAS in such measurements, as well as in those mentioned above, is its large dynamical range (up to 10^4 - 10^6). It can be easily realized with one microphone on account of changes in its sensitivity, for example, by varying the polarizing voltage within the linear section of amplitude characteristic, by introducing calibrated acoustic attenuators in front of the microphone or with several microphones of different sensitivity.

The multiphoton absorption of molecules at low pressures does not have a clearly defined threshold. The arrow in Fig.7.13 points to a small change in the slope of the intensity dependence (on a log-log scale). This change is caused by dissociation of highly excited molecules and shows that it is possible to define the region of molecular dissociation using this characteristic point. The results of OA measurements of multiphoton absorption may be partially distorted on account of irreversible consumption of absorbed energy by dissociation.

As shown in [7.66], at comparatively high pressures of the buffer gas OAS allows good observation of three stages of IR-radiation pulse absorption by a polyatomic molecule, with increasing density: linear absorption, saturation and multiphoton absorption. The OA detection of the infrared multiphoton excitation of cis - 3.4 - dichlorocyclobutene (DCCB) was used in [7.67]. The dependence of the OA signal on DCCB, on Ar buffer gas pressure and on laser fluence were studied systematically.

Furthermore, OAS makes it possible to study the dependence of multiphoton absorption on the relation between the characteristic relaxation times and the laser pulse duration when pulses of different duration are used as well as on the shape of these pulses [7.68-71].

c) Multiphoton Absorption Spectra

The use of a frequency-tunable pulsed laser allows measuring the OA spectrum of multiphoton absorption of different molecules. Figure 7.14 illustrates the absorption spectrum of SF_6 molecules at the CO_2 laser lines measured by OAS. It can be seen that as the radiation intensity increases the absorption maximum shifts monotonously to the long-wave length side and the absorption band broadens because of vibration anharmonicity.

A cooled-chamber spectrophone enables the OA spectrum of SF_6 at T = 145 K to be measured [7.72]. The measurement were taken with a collimated laser beam using a spectrophone with an electret microphone in its lateral arm. Specifically, it was found that the cooling of SF_6 caused the spectrum to narrow (due to a reduced contribution to hot band absorption) and the width of rotational molecular distribution to decrease. A structure was observed

176

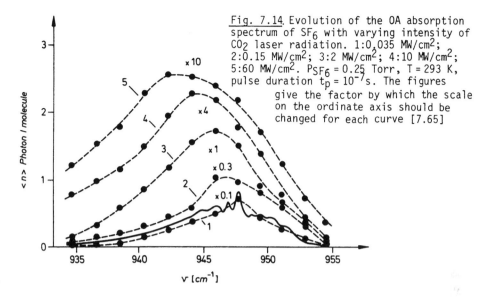

Fig. 7.14. Evolution of the OA absorption spectrum of SF_6 with varying intensity of CO_2 laser radiation. 1:0.035 MW/cm^2; 2:0.15 MW/cm^2; 3:2 MW/cm^2; 4:10 MW/cm^2; 5:60 MW/cm^2. $P_{SF_6} = 0.25$ Torr, T = 293 K, pulse duration $t_p = 10^{-7}$ s. The figures give the factor by which the scale on the ordinate axis should be changed for each curve [7.65]

in the multiphoton absorption spectrum, particularly at the P(20) line, which in [7.73] was explained by the presence of two- and three-photon resonances during anharmonic splitting of the second and third vibrational states of the ν_3 mode.

It is of interest that a sharper structure in OA spectra of multiphoton absorption absent in the linear spectrum was observed in C_2H_4 near the lines of a discretely tunable CO_2 laser P(26), P(14) and P(10) [7.65,74]. With a high-pressure CO_2 laser with continuous tuning the fine structure of the absorption spectrum of C_2H_4 near some lines could be measured with a resolution of $3 \cdot 10^{-3}$ cm^{-1} [7.75].

Two-frequency excitation (Sect.7.2.2) with OA registration of absorbed energy is useful for studying the interaction of molecules with the IR field. Experiments on absorption in SF_6 in a two-frequency IR field using OAS have been carried out in [7.55,76,77]. In [7.77], in particular, it has been shown that the presence of the first step of resonant laser excitation of the ν_3 mode of SF_6 enables the absorbed energy of the second nonresonant IR field coinciding with the band $\nu_2 + \nu_6$ to increase by nearly 20 to 40 times.

By OAS multiphoton absorption can be reliably measured at gas pressures up to 0.05 to 0.1 Torr, below which its sensitivity decreases essentially. But at such pressures the influence of rotational relaxation during the laser pulse is not eliminated completely, which makes it more difficult for us to understand the mechanism of molecular excitation of the vibrational quasi-continuum in the absence of collisions. However, OTS enables the region of lower pressures to be investigated to determine the characteristics of multiphoton absorption of isolated molecules (Chap.4). As shown in [7.78,79], a pyroelectric detector can measure absorption at SF_6 gas pressures up to 0.1 mTorr when the collisional effect between separate molecules may be neglected.

When an OT detector is used to study multiphoton absorption, one basic requirements is that the probability of vibrational relaxation of excited molecules on the detector surface does not depend on the level of vibrational excitation of molecules, which in turn is dependent on laser radiation energy and gas pressure. The validity of this condition with a pyroelectric detector has been tested in [7.79] by measuring the absorbed energy in SF_6 over the pressure range $10^{-1} - 10^{-3}$ Torr and the energy fluence of CO_2 laser pulse from 2 to 110 mJ/cm^2. The OT-detector's linearity over a wide pressure range (Fig.4.2) enables correct measuring of absorbed energy under different conditions of excitation. The spectrum of multiphoton absorption in SF_6 has been measured at 4 mTorr. Results correlate well with those obtained by OAS at higher gas pressures.

Effective applications of OTS still to come may be to investigate multiphoton absorption in molecular beams, the time-of-flight distribution of excited molecules with good temporal resolution (up to 1 μs) as well as to identify and study dissociation products and particularly to determine their kinetic and internal energy.

d) Isotopic Effects

The multiphoton absorption spectra measured by OAS particularly facilitate understanding the process of isotope-selective dissociation of polyatomic molecules in a strong IR field which forms the basis for laser isotope separation [7.61]. With OA spectra of multiphoton absorption in $CH_3{}^{14}NO_2$ and $CH_3{}^{15}NO_2$ molecules measured with a pulsed CO_2 laser an isotope shift in strong IR fields that is not manifested in weak fields has been found, i.e., a multiphoton isotope shift in the vibrational molecular spectrum can be detected [7.80]. Measurement of multiphoton absorption in the isotopic molecular modifications $^{235}UF_6$ and $^{238}UF_6$ at some lines of a CF_4 laser has shown that under these conditions the absorption selectivity [this parameter is determined from the ratio of absorbed energy in two isotopic molecules $J_{abs}(238) / J_{abs}(235)$] exceeds the selectivity under linear absorption [7.81]. Its maximum value obtained at the frequency of 612.2 cm^{-1} equals 2.15 as compared to 1.28 obtained under linear excitation. This effect is caused by selectivity accumulation at successive transitions under multiphon absorption. The volume of information can be increased considerably through simultaneous recording of OA signals and fluorescence of dissociated products as a function of the exciting laser frequency [7.82].

178

7.4 Optoacoustic Raman Spectroscopy

Raman scattering spectra are usually studied by detecting scattered (spontaneous or stimulated) radiation. However, Raman light scattering on molecular vibrations inevitably deposits part of the radiation energy on the molecules due to the difference between the pumping frequency ν_p and the scattered light frequency at the ν_{st} Stokes frequency. This peculiarity can be used to realize an alternative method for studying Raman scattering, based on OA detection of the energy in the molecular gas. This possibility was discussed in [7.83], and recently realized in [7.84-87].

The most natural and efficient method is a combination of OAS and two-frequency excitation of Raman scattering, which is usually used in coherent anti-Stokes Raman spectroscopy [7.88]. However, in OA recording the coherent properties of vibrationally excited molecules are quite inessential since only the energy of vibrationally excited molecules transformed into heat is recorded.

In two-frequency Raman scattering radiation at frequencies ν_p and ν_{st} the difference between which coincides with the vibrational transition frequency, is passed through the gas medium (Fig.7.15). The radiation with a higher frequency ν_p and intensity I_p serves as pumping-induced Raman scattering, whereas the weak radiation with frequency ν_{st} and intensity I_{st} acts as a probe wave at the Stokes frequency. Comparison of the probe-wave intensities before and after the gas cell makes it possible to determine the value of gain g_s and its proportional value of the spontaneous Raman scattering cross section.

In two-frequency excitation of a medium the amplification of the Stokes component can be determined from [7.89]

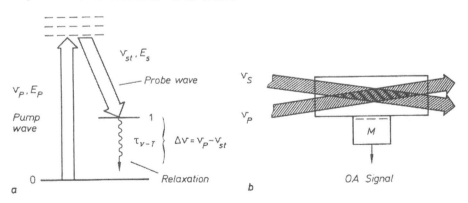

Fig. 7.15 a,b. Method of OA Raman spectroscopy (OARS) with two-frequency excitation of the medium. (**a**) Level scheme of the Raman transition under excitation; (**b**) geometry of gas irradiation

$$I_{st}(z) = I_{st}(0) \exp(g_s z) \quad , \tag{7.9}$$

where $I_{st}(0)$ is the initial intensity of the Stokes-wave propagating in the direction of z, g_s is the gain determined from

$$g_s = \frac{N\Delta}{\pi h \nu_L^2 \tilde{n}_{st} \tilde{n}_p \omega_{st}^3 \Gamma_R} \left(\frac{d\sigma}{d\Omega}\right) I_p \quad , \tag{7.10}$$

where N is the density of molecules; Δ is the density of fractional population difference between two transition levels $\Delta = q_{vib}^{-1}[g_\ell \exp(-E_\ell / K_B T) - g_u \exp(-E_u / K_B T)]$ ($g_{u,\ell}$ and $E_{u,\ell}$ are the degeneracies and energies of the upper and lower levels, respectively; q_{vib} is the vibrational partition function); $\tilde{n}_p$, $\tilde{n}_{st}$ are the refractive indexes on the ν_p and ν_{st} frequencies, respectively; $d\sigma / d\Omega$ is the differential cross-section of Raman scattering; Γ_R is the half-width of the Raman scattering line.

The OA signal amplitude is proportional to the energy of excitation of the medium ΔE_T. With small-gain factors ($g_s L \ll 1$, being the interaction length of laser beams) we have [7.83,86]

$$\Delta E_T / L \simeq \frac{\nu_p - \nu_{st}}{\nu_{st}} E_{st} g_s + E_{st} \alpha_{st} + E_p \alpha_p \quad , \tag{7.11a}$$

$$E_{st} = \int J_{st} dt \simeq J_{st} t_I \quad , \tag{7.11b}$$

where E_p, E_{st} are the energies of pumping and Stokes waves, respectively; α_p, α_{st} are the absorption coefficients in gas at frequencies which determine the amplitude of the background OA signal not related to the Raman scattering effect, t_I is the interaction time of the laser pulses.

From analysis of (7.10,11) it follows that at small α_p and α_{st} the OA signal amplitude is proportional to the intensities of pumping and probe waves (I_p and I_{st}) and their interaction time. To achieve high sensitivity, it is advisable to employ pulsed operation of both lasers when the maximum values of I_p and I_{st} can be obtained. According to [7.86], in conventional measurement the OA signal is about $10^3 - 10^4$ times higher during pulsed laser operation than during continuous operation. Under pulsed measurement the threshold spectrophone sensitivity is $(E\alpha)_{min} = 10^{-9} - 10^{-10}$ J $\cdot$ cm^{-1}. Then from (7.11) under standard conditions ($\nu_p - \nu_{st} / \nu_{st} = 0.1$, $E_{st} = 1$ J), OAS can measure the Raman gain coefficient of up to $10^{-8} - 10^{-9}$ cm^{-1}. The spectral resolution of this technique is determined by the total spectral width of the laser waves at the frequencies ν_p and ν_{st}.

The threshold sensitivity of the OA method of Raman spectroscopy (OARS) can be limited due to background OA signals from resonant absorption of the pumping and Stokes waves at frequencies ν_p and ν_{st}, respectively (for example, due to multiphoton effects or the influence of vibrational molecular overtones). The OA signal of Raman scattering is proportional to the product of the pumping wave intensity ($g_s \propto I_p$) and the Stokes wave intensity. Therefore, if the laser pulses at ν_p and ν_{st} coincide in time, the OA signal is determined by all three members in (7.11a). For two time-spaced pulses the OA signal is generated only from linear absorption, i.e., it is determined by the two last terms in (7.11a). This feature can be used to discriminate against the OA signal of Raman scattering.

The potentialities of OARS have been demonstrated both under CW [7.84] and pulsed [7.85-87] laser operation. In the latter case the radiation source was a high-power Q-switched Nd : YAG laser, whose second harmonic with $\lambda = 532$ nm served as pumping pulse. The third harmonic with $\lambda = 355$ nm was used for optical pumping of a dye laser, its tunable radiation acting as a probe wave. With a repetition frequency of 10 Hz the half-width of the pulses of the pumping and probe waves was 7 ns and 5 ns, respectively. The OA recording was carried out in a spectrophone with an electret microphone having sensitivity of 10 mV / Pa. The spectral resolution (0.5 cm^{-1}) was limited by the dye-laser line width. Figure 7.16 shows the vibrational-rotational spectrum of CH$_4$ measured with this

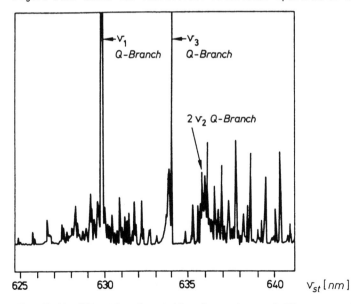

Fig. 7.16. Vibrational-rotational spectrum of CH$_4$ measured by OARS at 50 Torr. Conditions of measurement: $I_p = 1.2$ MW, $I_{st} = 0.5$ MW. The abscissa is the dye-laser frequency (Stokes components) [7.86]

step. The structure observed in the spectrum is due to by the modes ν_1, ν_3 and $2\nu_2$ of CH_4.

The main advantage of OARS compared to other well-known methods of Raman spectroscopy is direct detection of excitation energy. It makes OARS insensitive to fluorescence in the region of ν_p and ν_{st} as well as to various types of light scattering at these frequencies.

Such a combination of the methods of highly sensitive OAS and Raman scattering provides experimentalists with a new rather simple spectroscopic technique useful in studies of the spectra and cross-sections of Raman transition as well as in analytical applications, particularly to gas analysis (Chap.9).

7.5 Measurement of Relaxation Times

Optoacoustic measurement of relaxation times in gases is based on the analysis of the time delay between the radiative excitation of molecules and the appearance of OA signals. Such a delay is caused by the finite rate of transformation of the vibrational energy due to inelastic collisions to the thermal energy of the medium. The relaxation times are determined by measuring some parameters of the OA signal (amplitude, phase, frequency characteristic, etc.) which depend on the value of the time delay. In 1946 GORELIK discussed the possibility of measuring the time of vibrational-translational relaxation in gases (τ_{V-T}) by measuring the dependence of the OA signal on modulation frequency [7.90]. He also pointed out that the phase between the modulated radiation flux and the OA signal contains information on the value of τ_{V-T}. Phase determination of τ_{V-T} was successfully tested in experiments by SLOBODSKAYA [7.91,92]. Later this technique with noncoherent radiation sources was intensely developed and used to measure τ_{V-T} in many molecules [7.93-98].

However, the potentialities of OAS to determine τ_{V-T} are much wider, and lasers provide the best conditions for their realization. Laser radiation makes it possible: 1) to study the processes of relaxation from certain quantum states due to selective excitation of molecules to these states; 2) to measure the time τ_{V-T} in weakly absorbing media over a wide range of pressure and modulation frequencies by virtue of good sensitivity; 3) to realize the pulsed method to determine τ_{V-T}. It should be noted that unlike other known methods [7.5], OA determination of τ_{V-T} is direct, which means that it allows direct study of the transformation processes of the energy absorbed by the molecules to the thermal energy of the medium. It must also be emphasized that it is only the total relaxation time τ_{V-T} that can be simply measured by OAS It is more difficult to determine with it the characteristic times of intra-molecular energy exchange in multilevel systems as well as the differentiated values of τ_{V-T} for each intermediate level participating in the energy exchange. Below we consider in more detail the basic methods to determine τ_{V-T} using OAS.

7.5.1 Pulsed Method. Heating and Cooling of Gas

Excitation of molecules with short laser pulses permits direct measurement of the time of vibrational-translational relaxation τ_{V-T} at the leading edge of pulsed OA signals, following from (2.22). With the radiationless relaxation channel dominant, i.e., with $\tau_{V-T} \ll \tau_{rad}$, τ_{het}, and with the pulse duration complying with the condition $t_p \ll \tau_{V-T}$, this expression in the time interval $0 < t \leqslant 3\tau_{V-T}$ can be presented as

$$P(t) = P_{max}[1 - \exp(-t/\tau_{V-T})] \quad , \tag{7.12a}$$

or on a log scale

$$\ln\left[\frac{P_{max} - P(t)}{P_{max}}\right] = -\frac{t}{\tau_{V-T}} \quad , \tag{7.12b}$$

where P_{max} is the maximum value of pulsed OA-signal amplitude.

To take measurements using this technique it is necessary that some conditions must be fulfilled. First, it is necessary to ensure that $t_p \ll \tau_{V-T}$. Otherwise a long duration of the laser pulse will increase the leading edge of the OA signal which complicates processing the measurements. Second, the response of the microphone in use characterized by the time constant τ_m should be quite fast to measure short τ_{V-T}, i.e., $\tau_m \ll \tau_{V-T}$. The value of τ_m is usually 10^{-5} to 10^{-6} s. As a rule, in practice this condition can be fulfilled by choosing a gas pressure since the value τ_{V-T} varies inversely as the gas pressure. Therefore $\tau_{V-T} = (P\tau_{V-T})/P$, where the value of $(P\tau_{V-T})$ is constant for a specific type of molecule. The choice of low pressure is limited by the relaxation effect of the excited molecules onto the spectrophone walls.

Finally, thermodynamic processes must be dominant in the production of OA signals, whereas the effect of the resonant acoustic properties of the gas chamber as well as the microphone itself must be minimum. Otherwise the super-position of resonant oscillations on the leading edge of the OA signal may substantially complicate processing the measurements. To eliminate this effect it is highly desirable to:

a) irradiate the greater part of spectrophone chamber by laser radiation;

b) place an acoustic damper made, e.g., of rubber or plastic foam on the chamber wall and

c) provide the condition $2\pi f_r \tau_{V-T} \overset{>>}{<} 1$, where f_r denotes the resonant frequencies of the strongest acoustic resonances (usually they are the frequencies of the first longitudinal, radial or azimuthal modes). This can be done by minimizing the spectrophone-chamber volume. This technique has been used in experiments to determine τ_{V-T} [7.99-101]. In [7.99] the value of τ_{V-T} were measured for the vibrational-translational transitions of CO_2, SF_6 and NH_3 using a pulsed CO_2 laser with Q switching. Since the typical values of the vibrational translational relaxation time in the IR range are about $(P\tau_{V-T}) \approx 10^{-5}$ to 10^{-7} s$\cdot$atm the pressures of these gases were reduced at 5 and 20 Torr when $\tau_m << \tau_{V-T}$ (see above). The value of $(P\tau_{V-T})$ obtained for CO_2 equals 10.8 μs$\cdot$atm which within an accuracy of 5% is consistent with the results of measurements taken by different techniques. It has also been shown that an addition of a small amount of H_2O vapor substantially increases the relaxation rate in these molecules. The measurement was taken with a spectrophone, the volumes of its chamber 1 cm^3 and 2000 cm^3. The larger chamber was spherical with mirrored inner walls designed for multiple pass of the radiation reflected from them through the gas.

In [7.99] OAS was used to investigate an interesting effect of kinetic gas cooling. It manifests itself under pulsed operation as a small peak with opposite polarity on the leading edge of the OA signal. For molecular CO_2 gas this effect is caused by laser excitation of vibrational levels with rather a long lifetime within which thermodynamic equilibrium can be established at low-lying energy levels. Here the energy deficit of the depleted lower level participating in absorption is taken from the kinetic energy of molecules. This results in a short-term decrease in gas temperature. More detailed information on the mechanisms of kinetic molecular gas cooling under laser excitation can be found in [7.102]. In [7.103], for example, the feasibility of studying relaxation from high vibrational molecular states with OAS was considered. It was shown that short-time gas cooling is also possible during fast V-V energy exchange between highly excited and low states due to the endothermal character of certain collisional processes. OAS is very suitable to study such relaxation processes because of weak absorption of excited vibrational states.

The pulsed method has been used in [7.100] to measure the total time of collisional deactivation of the O_2 molecules in exciting the vibrational level $V = 1$ in the electronic singlet state $^1\Delta_g$ by the radiation of a pulsed ND-glass laser with $\lambda = 1.06$ μm. Under atmospheric pressure of O_2 the relaxation time

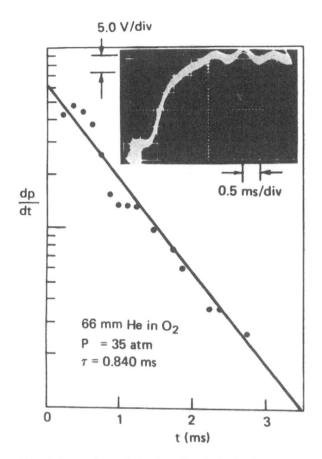

5.0 V/div

0.5 ms/div

$\dfrac{dp}{dt}$

66 mm He in O_2
P = 35 atm
τ = 0.840 ms

0   1   2   3

t (ms)

Fig. 7.17. Value of the OA signal derivative as a time function in O_2. Radiation source is a pulsed Nd:glass laser. At top: OA signal oscillogram [7.100]

is rather long and moreover the absorption at $\lambda = 1.06$ μm is small. So for more optimum conditions to be realized, all measurements were taken at increased pressures of O_2, up to 35 atm; the typical results are presented in Fig.7.17. In this case the value of the OA signal derivative of type (7.12) was set along the vertical axis, so τ_{V-T} can be determined from the slope of the straight line. The oscillogram of the OA signal leading edge presented in Fig.7.17 (inset) has characteristic oscillations due to the resonant properties of the chamber, whose influence was substantially reduced at 35 atm. The relaxation times in pure O_2 thus determined were $(P\tau_{V-T}) = 0.05$ s · atm and in O_2 with helium added $(P\tau_{V-T}) = 0.016$ s · atm.

185

7.5.2 Continuous Method with a Nonresonant Spectrophone

According to (2.27), to determine the time τ_{V-T} of excited molecules we can use the dependence of the phase, the frequency response and the amplitude of the OA signal on this time.

a) Phase Method

The phase method to determine τ_{V-T} was widely used with incoherent radiation sources. Here the main experimental problem is to eliminate the instrumental phase shift caused by the influence of the spectrophone constant τ_T (2.21) by first measuring the instrumental phase shift, then subtracting it from the phase under measurement. Specifically, the instrumental phase shift can be eliminated by measuring the OA signal phase in different spectral ranges, over wide ranges of gas pressures and modulation frequencies, at different concentrations of molecules in the buffer gas as well as by measuring the OA signal from radiation absorption on the spectrophone walls or by adding a fast-relaxing component. To avoid phase distortion with varying modulation frequency we must work in a frequency range where the acoustic resonances of the spectrophone chamber have no effect.

   In practice, the excited molecules can relax through several channels. With successive relaxation through intermediate levels the resulting OA-signal phase φ_Σ can be found from

$$\tan\varphi_\Sigma = \sum_i \tan\varphi_i \quad , \tag{7.13a}$$

where φ_i is the OA-signal phase due to relaxation from the i-th intermediate level. With parallel channels of relaxation through several levels the resulting phase is determined by the vector sum of OA signals formed by relaxation from each level

$$\tan\varphi_\Sigma = \Sigma P_i \sin\varphi_i / \Sigma P_i \cos\varphi_i \quad , \tag{7.13b}$$

where P_i, φ_i are the OA signal amplitude and phase during relaxation from the i-th level. In [7.98], for example, one can find a method for calculating relaxation parameters for parallel channels of relaxation in a three-level scheme typical for many complex molecules when singlet and triplet states are excited in them by visible radiation. Phase determination of τ_{V-T}, including the use of lasers was described in more detail in [7.104-107]. This technique at gas pressures of about 20 Torr allows measuring the relaxation time within 10^{-8} s · atm [7.106].

b) Frequency Method

The frequency method is based on analyzing the dependence of the OA signal on the modulation frequency. According to (2.27), in the frequency range $\omega \ll 1/\tau_T$ the OA-signal amplitude P is independent of frequency, with $1/\tau_T \ll \omega \ll 1/\tau_{V-T}$, the amplitude $P \propto 1/\omega$ and with $\omega \gg 1/\tau_{V-T}$, the value $P \propto 1/\omega^2$. Thus, from the point of inflection, where $\omega\tau_{V-T} \approx 1$, the value of τ_{V-T} can be found. It is also possible to estimate τ_{V-T} by plotting the measured data in coordinates of $P\omega$ versus $[1 + (\omega\tau_{V-T})^2]^{-1/2}$ and by determining the boundary frequency at the 70% level. Such a technique was used with advantage in [7.100] to find the relaxation time of O_2 at pressures up to 35 atm.

c) Amplitude Method

The amplitude method is based on measuring the dependence of the spectrophone sensitivity on pressure in low-pressure regions where collisional relaxation in the gas and deactivation of excited molecules on the spectrophone chamber walls compete. According to (2.27), the dependence of the OA signal on τ_{V-T} manifest itself through the ratio τ_{vib}/τ_{V-T}, where τ_{vib} can be determined from (2.5). The parameters τ_{V-T} and τ_{het} in (2.5) can be presented in the following form: $\tau_{V-T} = (\tau^o_{V-T})/P$, $\tau_{het} \approx \tau_T/\varepsilon_A = \tau^o_T P/\varepsilon_A$, τ^o_{V-T}, τ^o_T being the values of relaxation times at pressure, for example $P = 1$ Torr. Then, the · influence of the radiative relaxation channel being small, the OA signal is proportional to

$$P \propto [1 + P^{-2}(\tau^o_{V-T}\,\varepsilon_A/\tau^o_T)]^{-1} \quad . \tag{7.14}$$

If we find the gas pressure P^* at which the signal value is reduced, say by one half, it is easy to derive from (7.14) an expression to estimate $\tau^o_{V-T} = \tau^o_T(P^*)^2/\varepsilon_A$. This technique was used in [7.108] to measure the accomodation coefficient ε_A for a number of molecules.

Here, to determine τ_{V-T} correctly it is necessary to measure the true dependence (7.14) not affected by the pressure dependence of the sensitivity of the microphone itself and the absorption coefficient of the molecules under investigation. Also, information on the values τ^o_T and ε_A for the spectrophone in use should be available. This technique has been successfully tested in [7.109] to estimate τ_{V-T} for the ν_3 vibration of the CH_4 molecule in the mixtures $CH_3 + H_2$, CH_4 + air and $CH_4 + Ar$ using a He-Ne laser with $\lambda = 3.39$ μm. The spectrophone sensitivity as a function of pressure was determined by the ratio of the OA-signal amplitude to the value of absorption coefficient in a mixture determined independently. The advantage of this method over others is that it is possible to measure rather short relaxation times of $10^{-7} - 10^{-8}$s.

187

The theory of the OA effect in a gas allows, at present, precise calculation of the relative value of the OA signal depending on the ratio between the
relaxation times in the medium under analysis.

7.5.3 Continuous Method with Resonant Spectrophones

Due to high quality of resonant spectrophones ($Q \sim 800\text{-}900$) the OA signal can
be increased substantially as the modulation frequency coincides with one of
the resonant acoustic frequencies of the gas chamber. Due to a bulk generation
of the OA signal the influence of thermal processes near the laser beam is
negligible. Thus such spectrophones can determine τ_{V-T} with the help of the
frequency method at rather high modulation frequencies. Because resonant frequencies are discrete, measurement of the frequency response can be substituted
by measuring the pressure dependence of the OA signal at one of the resonant
frequencies, e.g., at the first radial mode of a cylindrical spectrophone. To
make the processing of the measured results handy the expression for resonant
OA signal (5.3) should be transformed to

$$\left(\frac{1}{\omega P C^*}\right)^2 = (\omega \tau^o_{V-T})^2 \frac{1}{p^2} + 1 \quad , \tag{7.15}$$

where P is the OA-signal amplitude, P is the gas pressure, τ^o_{V-T} is the relaxation time at a fixed pressure (1 atm), and all the terms independent of τ_{V-T}
are reduced to the constant factor C^*. Graphically in the coordinates $(P\omega)^{-2}$
relative to P^{-2} the dependence obtained has the form of a straight line that
when continued on the abscissa determines the value of the pressure P* simply
related time τ_{V-T}

$$\tau^o_{V-T} = \frac{P^*}{\omega} \quad . \tag{7.16}$$

Figure 7.18 shows the results obtained from measuring τ_{V-T} for pure CO_2 by
this method, namely $(P\tau_{V-T}) = 12.4$ $\mu s \cdot$ atm agrees well with results of other
work within an accuracy of 15%.

It is simpler to determine τ_{V-T} from the ratio of two OA signals P_1 and P_2
at one of the resonant frequencies ω_i measured with $\omega_i \tau_{V-T} \gg 1$ and $\omega_i \tau_{V-T} \ll 1$,
respectively, In the simplest case for the spectrophone quality being constant,
the expression for τ_{V-T} is

$$\tau^o_{V-T} = \frac{P}{\omega} \left[\left(\frac{P_2}{P_1}\right)^2 - 1 \right]^{1/2} \quad . \tag{7.17}$$

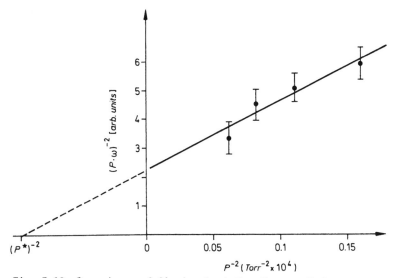

Fig. 7.18. Dependence of OA signal at the first radial mode of a cylindrical spectrophone on CO_2 pressure. Spectrophone parameters: $r_c = 30$ mm, $L = 200$ mm, $f_{001} = 5560$ Hz. Radiation source is a CO_2 laser at the R(16) line of the transition 00^01-02^00 [7.110]

Table 7.4. Comparison of the parameter $P\tau_{V-T}$ [μs · atm] measured in pure CH_4 and $CH_4 + Xe$ by different methods [7.111]

Methods	Resonant OAS [7.111]	Ultrasonic absorption [7.129]	Nonresonant OAS (phase method) [7.130]	Laser fluorescene [7.131]
CH_4	1.8 *	1.7	1.5 - 1.6	1.9
$CH_4 + Xe$	42 **	43	> 40	39

*   Measuring at the first radial mode 001
** Measuring at the 220 mode

Generally for P_2 to be determined the spectrophone must be calibrated using a fast relaxing gas. For some molecules this can be done in a simpler way by adding small amounts of water vapor to the molecules under investigation which will accelerate the relaxation by one to two orders [7.99]. The value of $P\tau_{V-T}$ measured for CO_2 in this way [7.110] at only one frequency (the second radial mode of a cylindrical spectrophone) equals 9.7 μs · atm which agrees rather well with the results obtained by others.

In [7.111] an interesting method has been proposed to determine τ_{V-T} by measuring the resonant frequency as a function of gas pressure with a resonant spectrophone. The characteristic curve closely resembles the ultrasonic dis-

189

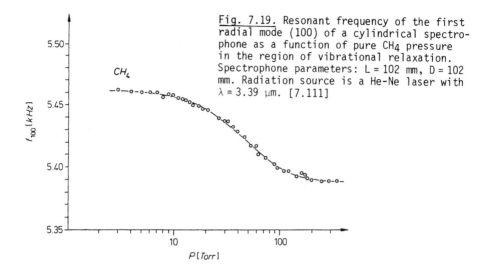

Fig. 7.19. Resonant frequency of the first radial mode (100) of a cylindrical spectrophone as a function of pure CH_4 pressure in the region of vibrational relaxation. Spectrophone parameters: L = 102 mm, D = 102 mm. Radiation source is a He-Ne laser with $\lambda = 3.39$ μm. [7.111]

persion curve (Fig.7.19). In essence, this method compromise between the optical and acoustic methods by measuring the sound velocity in the dispersion area instead of measuring the resonant frequency which is simply related to the sound velocity by (5.4). This enables us to increase substantially the accuracy of τ_{V-T}. The results of measuring τ_{V-T} by this technique in pure CH_4 and in the mixture CH_4 + Xe as well as a comparison of these with those of other methods are given in Table 7.4. We believe that such a technique enables rotational relaxation times to be determined but one should use higher resonant frequencies and operate with lower pressures.

7.6 Chemical Reactions

Using OAS to study chemical reactions is efficient in two respects. First, optical radiation itself can induce a photochemical reaction and simultaneously an OA signal. Studying the OA spectrum under different conditions of measurement (at different gas pressures or exciting radiation energy fluence) enables us to reveal certain peculiarities of photochemical reactions. Second, OAS can be used only to measure the concentration of products (initial, intermediate or final) in various chemical reactions.

7.6.1 Photochemical Reactions

Photochemical reactions manifest themselves especially clearly in the visible and UV spectral ranges where even the energy of a single photon is sufficient to break molecular bonds. The applicability of OAS to measuring the molecular absorption in these ranges was demonstrated in [7.112,113]. In [7.112] the OA effect was revealed in O_2, N_2 and C_2H_6 molecules irradiated by UV radiation from a hydrogen lamp. In the visible and UV ranges the OA measurements are characterized by strong influences of radiative relaxation, photodissociation, photochemical reactions as well as by rather complex relaxation of excited particles through a number of intermediate states. This all complicates the temporal shape of the OA signal.

Photochemical reactions were observed in an OA cell with incoherent sources [7.113,114]. In [7.114], for example, where Cl_2 with H_2 added was excited with UV radiation of a mercury lamp, at first a sharp increase in the OA signal was observed. Then within several hours it dropped gradually to a value smaller than that for pure Cl_2. This phenomenon due with increasing of the OA signal was explained by the formation of an additional OA effect due to the exothermal reaction of Cl_2 with H_2 during the periods of mixture irradiation. This gradual attenuation of the OA signal is conditioned by the irreversibility of this photochemical reaction in which an HCl compound is formed that has no absorption bands in the range of the source in use.

Breaking of molecular bonds under visible irradiation has been revealed in NO_2 and J_2 molecules using OAS [7.113]. Figure 7.20 shows the ordinary absorption and OA spectra of the NO_2 molecule. In (b) there is some attenuation of the OA signal in the short-wave length spectral region which is explained by processes

$$NO_2^* \to NO + O$$
$$(7.18)$$
$$O + NO_2 \to NO + O_2 + 47 \text{ Kcal} \cdot \text{mole}^{-1}$$

at $\lambda < 400$ nm. At the same time at longer wavelengths, when there is no dissociation, only the one process is possible

$$NO_2^* \to NO_2 + 72 \text{ Kcal} \cdot \text{mole}^{-1} \quad . \tag{7.19}$$

Thus, the quantity of heat released with dissociation constitutes 47/72 of the heat released in the absence of dissociation. This manifests itself as a small decrease of the OA signal near 400 nm.

Such experiments with other molecules were discussed in more detail in [7.115]. Clearly the use of visible and UV lasers may substantially increase the excitation selectivity of molecules. Combined with OAS it will probably allow the extremely weak photochemical reactions of molecules in certain quantum states to be investigated.

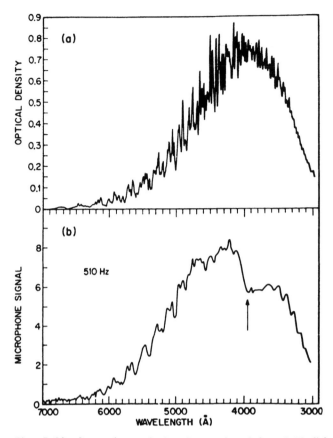

Fig. 7.20. Comparison of the absorption (a) and OA (b) spectra of NO_2 at a pressure of 10 Torr. The photodissociation limit is marked with an arrow [7.113]

High-power IR lasers have made it possible to study the photochemical re-actions of molecules in highly excited vibrational states of the ground electronic state populated due to multiphoton absorption of a great number of IR photons [7.61,62]. Here OAS is convenient both for measuring multiphoton absorption spectra (Sect.7.3.2) and for detecting the energy which inevitably relaxes to heat under multiphoton excitation of molecular gas. Especially useful is the possibility of realizing spatial resolution with a focused laser beam, which was used in [7.80] to study trans-cis-isomerization of $C_2H_2Cl_2$ molecules under vibrational excitation with a powerful pulsed CO_2 laser. In the measurements a combined design of spectrophone was used for simultaneous detection of the energy absorbed in the gas from thermal effects and fluores-cence. The multiphoton absorption measurements of both methods are consistent.

Furthermore, OAS has been used [7.116] to measure multiphoton absorption in cyclopropane to study its photochemical reactions as two different vibrations are excited the C-H group in the 3.22 μm region with a parametric LiNbO$_3$ oscillator and CH$_2$ at 9.5 μm with a CO$_2$ laser. It has been found that the cyclopropane molecule under IR excitation undergoes monomolecular phototransformation through two basic channels: 1) isomerization yielding propylene as the C-H group is excited, and 2) fragmentation yielding many products as the CH$_2$ group is excited.

Due to its high sensitivity acoustic recording may be useful in studying the kinetics of many chemical reactions by recording their attendant temperature fluctuations and hence the gas pressure in an acoustically closed cell with a mixture of reacting components. This possibility was emphasized in [7.29] when observing the noise increase in the spectrophone whose working chamber was filled with HF and N$_2$ molecules at high humidity. In essence, here we are dealing OAS in microcalorimetering chemical reactions followed by small temperature fluctuations $\Delta T = 10^{-5}$ to 10^{-6}°C.

It should be noted that when we study weakly absorbing media the photochemical processes that occur are undesirable because they decrease. This is due to an increase in the spectrophone noise caused by increased thermal gas fluctuations as well as by a decrease in the OA signal since some portion of absorbed power is consumed to break the molecular bonds. According to [7.117] for example, as 2-chloronitropropane was excited by dye-laser radiation in the region from 580 to 610 nm the OA signal recorded was about two orders lower than the calculated one. This is accounted for by photodissociation of these molecules under laser radiation.

7.6.2 Analysis of Reaction Products. Catalytic Reactions

The OA method can be very useful in solving some problems in the research of photochemical transformations in the atmosphere and stratosphere. For example, because of increasing pollution of the atmosphere more interest is being shown in the problem of catalytic destruction of stratospheric ozone (O$_3$) in the presence of the NO and NO$_2$ molecules under solar illumination. The main role is played by the following reactions:

$$NO_2 \xrightarrow{h\nu} NO + O \quad , \tag{7.20a}$$

$$NO_2 + O \to NO + O_2 \quad , \tag{7.20b}$$

$$O_3 + NO \to NO_2 + O_2 \quad . \tag{7.20c}$$

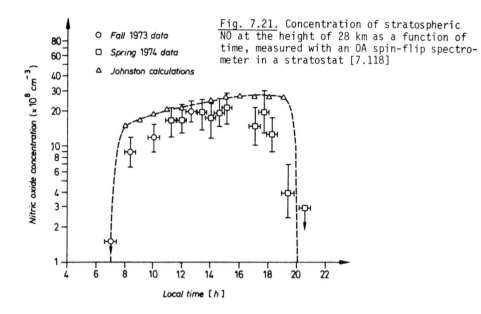

Fig. 7.21. Concentration of stratospheric NO at the height of 28 km as a function of time, measured with an OA spin-flip spectrometer in a stratostat [7.118]

○ *Fall 1973 data*
□ *Spring 1974 data*
△ *Johnston calculations*

Local time [h]

These reactions can be controlled by measuring the kinetics of variation in concentration of any reacting component. Hence, in [7.118] the concentration of NO was measured on a real time scale at the height of 28 km using an OA spin-flip spectrometer in a stratostat. To detect and identify NO in the air the spin-flip laser frequency was tuned to the 1887 cm^{-1} region where two intense absorption lines of NO conditioned by Λ splitting are located. According to the results obtained (Fig.7.21), between sunrise and sunset the concentration of NO increases drastically from the background level to an order which points to its effective formation under UV radiation (its natural concentration is about $1.5 \cdot 10^8$ mol / cm^3). By analogy, we can consider the influence of Freon concentration on the ozone balance in the stratosphere due to formation of Cl/ClO radicals one of which (ClO) can be detected using an isotopic modification of CO_2 laser with $\lambda \approx 11$ μm [7.118].

The potentialities of OAS to control catalytic reactions have been demonstrated [7.118-121]. In [7.118-120] a LOAS based on a discretely tunable CW CO laser was used to study the catalytic reaction in a mixture of NO, CO and H_2 in a reactor above a platinum catalyst at 500 - 800°C. This reaction was performed through detecting its product, that is, the HCN molecules present in the mixture in rather small concentrations. It shows that the yield of HCN decreases to 60 ppm when 5% H_2O is added to the mixture and to 5 ppm and less when SO_2 in a concentration 6 - 60 ppm is added. Yet added O_2 decreases the poisoning action of SO_2 on the catalyst, which is explained by the participa-

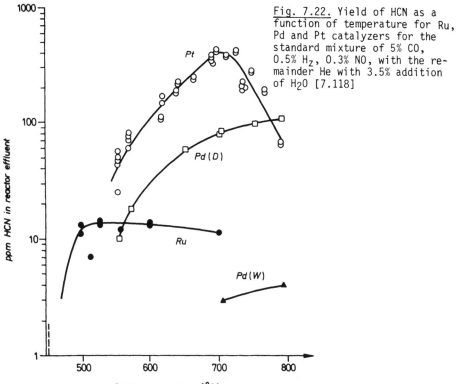

Fig. 7.22. Yield of HCN as a function of temperature for Ru, Pd and Pt catalyzers for the standard mixture of 5% CO, 0.5% H_Z, 0.3% NO, with the remainder He with 3.5% addition of H_2O [7.118]

tion of O_2 in the chemical processes in the surface layer as well as in the change of the catalyst structure. Figure 7.22 shows the yield of HCN measured in these experiments as a function of catalyst temperature for three different types of catalysts. Such a technique allows quick evaluation of small concentrations of other molecules in the process of catalysis (such as NH_3, NO) and their dependence on the type of catalysts, the temperature, and the influence of such admixtures as H_2O, SO_2, O_2. Practically, we can perform continuous automatic control over the course of catalytic reactions and correct the composition of the mixture let into the reactor.

In [7.121] OAS was used to study the kinetics of

$$NH_3 + HNO_3 \rightarrow NH_4NO_3 \quad , \tag{7.21}$$

which is of interest in the formation of atmospheric aerosol and in studies of corrosion processes. The radiation source was a CO laser tuned to 1801.8568 cm^{-1} [the line 13 - 12 P(8)] to detect NH_3 and to 1686.7092 cm^{-1} [the line 17-16 P(12)] to detect HNO_3. A characteristic feature of this mixture is its active interaction with the surface of spectrophone walls so

195

necessitating individual measurement of isotherms for each component. As shown, the molecular concentration decreased by about one order within two hours of measurement. The spectrophone walls were made of stainless steel coated with gold. After determination of individual isotherms the concentration of each component was measured with their simultaneous presence in the chamber with nitrogen as a buffer at about 50 Torr. This made it possible to define the peculiarities and characteristic constants of reaction (7.21) as the component reacted in the gas and on the spectrophone walls.

In [7.122] OAS was used to investigate the addition of ICl to C_2H_4. This reaction is used in effective enrichment of chlorine isotopes. The resulting product C_2H_2ICl can be enriched with ^{37}Cl or ^{35}Cl by 100 times (to 97% enrichment). In addition, OAS was applied to estimate the efficiency of such a reaction from the standpoint of clarifying other potential channels of reaction and deexcitation of ICl under various experimental conditions. The measuring technique consisted of measuring the OA signal in the mixture $JCl + C_2H_4$ with a dye laser before and after irradiation, the radiation line tuned to intense absorption lines.

A similar measuring technique was applied to study the photolysis of nitromethane (CH_3NO_2) irradiated with a mercury lamp from the outlet window of the spectrophone [7.123]. Using a dye laser in the range from 580 to 616 nm it was possible to detect the photolysis products of NO_2 and CH_3NO as well as a delay in the formation of CH_3NO due to an oxidizing reaction as O_2 was added to CH_3NO_2.

8. Laser Optoacoustic Spectroscopy of Condensed Media

Optoacoustic spectroscopy investigate solid, semi conductor and liquid samples which are strongly or weakly absorbing, or which preclude the use of reflectance or transmission spectroscopy due to scattering of the incident radiation. The material presented in this chapter demonstrates the great potentiality of OAS of condensed media.

8.1 Spectroscopy of Weakly Absorbing Media

The use of lasers in combination with OA detection of absorbed energy allows the study of weakly absorbing bands of both liquids and solids.

8.1.1 Liquids

As a rule, in most liquids at room temperature intense continuous absorption bands are observed in the IR, and weak bands only in the visible and near-IR regions. Among the various methods for recording OA signals in liquids, the simplest and most sensitive one is direct detection of acoustic vibrations with a piezoelectric detector in contact with medium; the best results were obtained in pulsed operation. The method and general measurement technique for weak absorption in liquids are similar in many respects to laser OA spectroscopy (LOAS) of gases (Chap.7).

The high sensitivity of LOAS enables to study the high overtone absorption of molecules in the liquid phase [8.1], particularly the overtones of the C-H bond in liquid benzene [8.2,3]. Figure 8.1 presents the OA spectra of three high overtones in liquid benzene measured with a tunable pulsed dye laser. The basic parameters of this laser are the pulse energy of 2 mJ, the pulse duration of 1 μs, the repetition rate of 10 Hz, the spectral half-width of 2 cm^{-1}, and the beam diameter of about 3 mm. The OA cell design is similar to that in Fig.3.1. The OA signal was recorded with a gate integrator that made

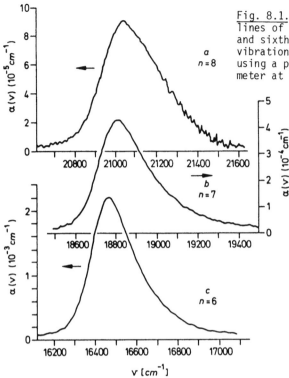

Fig. 8.1. Shapes of the absorption lines of the eighth (**a**), seventh (**b**), and sixth (**c**) overtones of a C-H band vibration in liquid benzene measured using a pulsed dye laser OA spectrometer at T = 25°C [8.3]

it possible to select just the first peak of the oscillating sequence of acoustic pulses (Fig.3.2). The average integration time was about 1 s. It took up to 200 s to record one absorption profile. To determine the absolute value of the absorption coefficient the sensitivity of the OA cell was calibrated when its chamber was filled with pure benzene with a small amount of iodine dissolved in it. In the spectral range under investigation iodine absorbs rather strongly without fluorescence. This allows the absorption coefficient in the medium to be determined independently by comparing the radiation intensity before and after the OA cell.

Figure 8.1 shows that this technique allows reliable studies of the high-overtone absorption profile of benzene with the absorption coefficient ranging from 10^{-5}-10^{-6} cm^{-1}. The overtones $n \leqslant 5$ were measured using the standard spectrophotometric technique, but only OAS made it possible to measure absorption in the 8th harmonic. The results of OAS and other measurements (thermolens, wavelength modulation methods and measurements in long hollow fibers) are in satisfactory agreement,

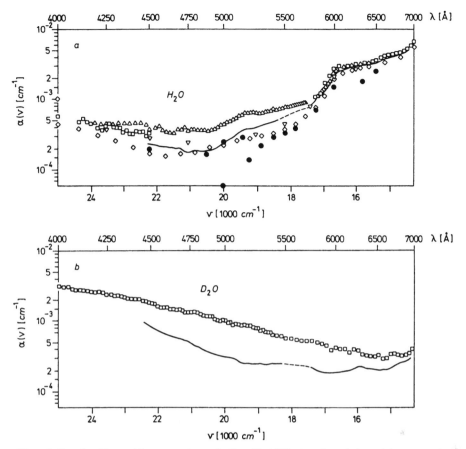

Fig. 8.2 **a,b**. Absorption spectra of the distilled water (**a**) and heavy water (**b**) at T = 21.5°C [8.4]. A solid line shows the results obtained by the OA method. For comparison, results obtained by other methods during recent years are presented (see Ref. in [8.4])

A similar technique can be applied to study weak absorption in water in the visible spectrum, which is essential for underwater laser communication, to determine water purity and dissolved microimpurities, etc. The results of OA measurements of the absorption factor in ordinary and heavy water are presented in Fig.8.2. The OA cell made of stainless steel used in analyzing ordinary water was calibrated with nonfluorescent $KMnO_4$ dissolved in it and for heavy water with $K_2Cr_2O_7$. According to the estimations the measuring accuracy of the absorption coefficient in water at the level of 10^{-3} to 10^{-4} cm^{-1} was ± 10%. As may be seen from the figure, the results of measurements by different techniques are in satisfactory agreement.

In studies of water, the OA signal amplitude was highly dependent in the temperature. This can be explained by the fact that the thermal expansion coefficient β_v, (the OA signal varies proportionally with the value β_v (3.13)), is temperature dependent. For example, as the temperature decreased to 4°C the OA signal is attenuated almost to zero and then below 4°C it again increases, with opposite polarity. To eliminate this phenomena on the measuring accuracy, it is necessary to thermostabilize the OA cell with an accuracy no less than ±1°C or to control the temperature regularly. According to the data obtained, absorption in ordinary and heavy water at different wavelengths is different, in the long-wavelength region the absorption in heavy water is about one order less. This fact due to the difference in the frequencies of OH and OD vibrations, is important when water is used as an optimum solvent or the base in various spectroscopic tasks. The best solvents like water have absorption in the visible region at the level of 10^{-4} to 10^{-5} cm^{-1}. It is advisable to use differential measurement techniques to measure the absorption spectra of impurities dissolved in them at a lower level.

A great advantage of OAS in the analysis of liquid media is its low sensitivity (when the detector itself is properly screened) to scattered radiation. This radiation creates serious problems in other measuring techniques, for example in the absorption method.

8.1.2 Cryogenic Solutions

Cryogenic techniques are widely used in molecular spectroscopy, analytical chemistry and in physical research [8.5]. High transparency of many cryogenic solvents (noble gases, nitrogen, etc.) in the IR range widens the applicability of OAS of liquids to this range. This also enables to carry out studies in low-concentration solutions. At a low temperature, as a rule, the spectral structure of molecules is simplified, for example, the rotational components in the spectrum freeze out. It should be noted that with respect to spectroscopy the properties of various cryogenic systems are still not clearly understood and therefore LOAS is very useful here.

Piezoelectric detectors operate over a wide temperature range, up to 4 K. So it is rather simple to realize cryogenic OAS by inserting a piezoelectric detector into the measuring cell of an optical cryostat. According to the first experimental estimations [8.6-8] such a scheme with a pulsed laser makes it possible to realize the threshold sensitivity at the level $(1 - 2) \cdot 10^{-7}$ J/cm at temperatures up to T = 125 K. Figure 8.3 shows the OA spectrum of C_2D_4 dissolved in liquid krypton at 125 K. The vibration ν_{12} was studied, whose frequency for gas phase equal to 1078 cm^{-1} lies in the regime of the CO_2 laser. In the measurements a pulsed TEA CO_2 laser was used with a pulse duration of

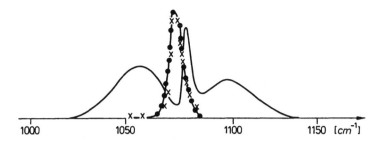

Fig. 8.3. Absorption spectrum of C_2D_4 in the region of ν_{12} vibration. (——):
Spectrum of gaseous $^{12}C_2D_4$; (•): spectrum of C_2D_4 solution in liquid Kr obtained using an ordinary spectrophotometer at $T = 125$ K; x: amplitude of an OA signal in C_2D_4 solution in liquid Kr at $T = 125$ K for discrete lines of a pulsed CO_2 laser at a concentration less than 10^{-3} mol/l. The last two spectra are normalized by the same value [8.7]

$t_p = 100$ ns and a pulse energy of up to 20 mJ. The measuring cell in the cryostat with the solution to be studied is similar to that in common cryogenic spectroscopy. For comparison, Fig.8.3 also shows the spectrum of C_2D_4 produced with a standard spectrophotometer at concentrations of about $2 \cdot 10^{-2}$ mol / l and the temperature range from 165 to 125 K. During OA measurements the absorption coefficient at the frequency of the vibration ν_{12} of C_2D_4 with $T = 125$ K near 1074 cm^{-1} was no higher than $3 \cdot 10^{-3}$ cm^{-1}.

It is possible to carry out OAS in the spectral region from 0.7 to 1.6 μm. For that in [8.9] stimulated Raman scattering in high-pressure hydrogen gas was for down-shifting pulsed tunable dye-laser radiation through first- and second-order Stokes processes. Measurements of the OA absorption spectra of liquid $C_2H_4(T = 113$ K) in the region of 0.7 - 1.6 μm indicate the general applicability of the OA technique to investigate the planetary spectra of several gases.

Thus, these experimental results demonstrate the potentialities of cryogenic LOAS due to the increased sensitivity and the simplicity of instrumental realization.

8.1.3 Solids

Combined with powerful lasers, OAS allows one to study weak bulk and surface absorption in crystals and semiconductors, to evaluate the level of absorbed energy in thin films, to measure the spectra of oxide films in metals, various powders, etc. Depending on the research goal, all the basic methods of OA signal detection can be applied with success (Chap.3), i.e., direct detection with a piezodetector attached to the sample, indirect detection in gas, or the liquid in contact with the sample.

a) Volume Absorption of Crystals

Substantial progress has been made in developing optical materials with rather
low losses. They are of great interest in manufacturing optics for powerful
lasers, fiber optics, etc. It is impossible to investigate such materials and
to define the basic loss mechanisms without developing proper methods for
measuring small absorption coefficients α in solids. The most promising methods
for measuring α were discussed in [8.10]. Essentially OAS can complement the
possibilities of these methods, especially increasing the sensitivity of evaluating α.

The total absorption in a sample, α_{tot}, can be expressed as

$$\alpha_{tot} = \alpha_v \ell + 2\alpha_s \ell_s \quad , \tag{8.1}$$

where ℓ, ℓ_s are the lengths of the sample and the surface layer, respectively.
The value of the product $\varkappa_s = \alpha_s \ell_s$ characterizes the relative fraction of absorbed energy only on the sample surface. As a rule, bulk absorption is determined by fundamental intrinsic properties of the material as well as by
the presence of different bulk absorbing inclusions. Surface absorption depends,
first of all, on the sample's surface roughness, the presence of different
contaminations various atmospheric precipitation, particularly water vapor,
etc. The main problem in such studies is to obtain separate information on
the values of bulk and surface absorption.

Direct recording of OA signals in a sample, which offers real possibilities
for selecting different types of absorption, is the most effective method to
investigate bulk properties (Sect.3.2). To eliminate strong scattered light
this method usually employs a differential technique with two piezodetectors.
One of them is out of acoustic contact with the sample and records only scattered light. The applicability of such a scheme has been demonstrated by
measuring the volume and surface absorption of some optical materials including Si, ZnSe, GaAs, LiF, BaF_2 at the spectral lines of Ar, HF, DF, CO and
CO_2 lasers [8.11]. Specifically, at the HF laser lines the bulk absorption
α_v, in most materials, was 10^{-3} to 10^{-4} cm^{-1} and the surface absorption $\varkappa_s$
was $5 \cdot 10^{-3}$ to $5 \cdot 10^{-4}$. The predominance of surface absorption for some
materials is explained by the presence of various impurities on the surface.
In [8.12],for example, to identify these impurities in SrF_2 and BaF_2 the OA
spectra of surface absorption were measured in the range of CO-laser tuning,
approximately from 5 to 6 μm. HORDVIK and SCHLOSSBERG [8.12] interpreted the
structure measured in the absorption spectra near 1640 and 1720 cm^{-1} as the
contribution made by water vapor and acetone absorption, respectively.

Practice shows that the bulk losses at the level of 10^{-4} to 10^{-5} cm^{-1} can be measured accurately by OAS using lasers with 0.1 - 1 W power. In such measurements it is desirable that the samples of the optical materials to be investigated should have a special form to enable good acoustic matching with the piezoelectric detectors used. But in some cases it seems possible to study absorption directly in available samples, particularly in optical fiber. Results show that such control is possible [8.13]. A considerable OA signal was found in an optical fiber, with 200 μm diameter, with a piezoelectric detector jointed to it, using both a pulsed ruby laser and a cw He-Ne laser. The most important problem here is to exclude scattered radiation. Pulsed laser operation [8.14] which makes it possible to realize the temporal selection of acoustic and light signals is most promising in discriminating signals from scattered radiation. Under continuous operation background signals can be eliminated with the use of the above-described differential scheme with two detectors. In addition, OAS with indirect detection in the gas surrounding the fiber is possible [8.15,16].

b) Absorption in Powders

The difference between powdered substances and solid media is that the former consist of two phases separated by a highly expanded surface. One of the phases is distributed as very small particles in a gas, which is the other phase. When light passes through a fine-grained medium its intensity decreases not only due to light absorption but also due to the effects of diffusive reflection and scattering [8.17].

The thermal properties (heat conduction and heat capacity) of powders and solids may differ essentially, affecting the generation of OA signals. An appreciable effect may be produced by gas convection between grains, especially in coarse-grained systems. With fine grains the main part in heat transfer is played by heat conduction of the intermediate medium, that is air. With an increase in grain dimensions the contribution of heat conduction of the grain itself increases. Thus, the smaller the grain dimension the lower the coefficient of thermal conductivity of the powder since heat conduction of air is much lower than that of solids. Since the thermal diffusion length is reduced, conditions of thermal saturation ($\ell_T \gg \ell_\alpha$) which are undesirable in OA measurements when the OA signal is to be independent of α, do not show up even at low modulation frequencies.

To study powdered substances with OAS we can use different measuring schemes. For highly absorbing powders good results can be obtained with PAS by placing a powder on the substratum in a closed chamber with a microphone and gas. At

low absorption this method may not work due to the low sensitivity. To overcome scattered radiation it is useful to remove the microphone from the chamber
and connect both with a narrow channel of 0.1 to 1.0 mm in diameter. At weak
absorption the substratum must be optically transparent for the laser radiation. It should be noted that OAS has comparatively low sensitivity to scattered radiation. But the OA-signal amplitude itself may depend on the degree
of radiation scattered from the surface layer due to increasing radiation density. For example, at low scattering the ratio between the OA signal from the
sample with scattering and that from the sample without scattering may be
presented as 1+ (scattering section/ absorption section) [8.18].

For weakly scattering samples the OAS measurement is equivalent to a transmission measurement [8.19]. On the other hand, for very strongly scattering
samples the heat conduction in the sample can be neglected, and the OAS measurement is equivalent to a diffuse-reflectance measurement. For intermediate
values of the scattering coefficient, OAS yields information which is not accessible by either transmission or reflection measurements.

For analyzing low-absorption samples indirect detection of OA signals by
inserting the powder into a weakly-absorbing liquid is promising. Here good
acoustic matching of the piezodetector with the medium can be achieved through
an immersion medium. In [8.20] this technique was used to record the absorption
spectra of oxides of rare earth powders (HO_2O_3, Dy_2O_3, Er_2O_3) in the visible
spectrum with a dye laser. The measuring technique consisted of pulsed excitation of these powders in a liquid (ethylene glycol) with a concentration of
about 0.01 gm/cc. The liquid was placed between two quartz plates with a gap
between them of several μm (Fig.3.7b). An area of about 10^{-2} cm^2 was irradiated
with focused radiation. The total mass of the powder under irradiation was be-

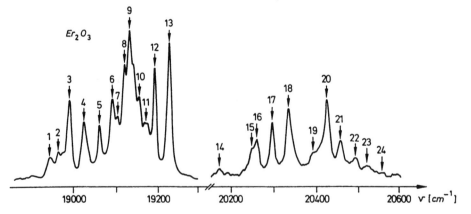

Fig. 8.4. OA spectrum for Er_2O_3 powder at 25°C [8.20]

low 0.1 μg. The acoustic detector was attached to one of the quartz plates at a considerable distance from the irradiation zone. One of the plates had a Γ shape: such a geometry together with time selection of signals minimized the effect of scattered radiation. Figure 8.4 shows the OA absorption spectrum of powdered Er_2O_3 with a spectral resolution of about 10 cm^{-1} measured by the scheme described. The arrows mark the identified transitions of the multiplet $^2H_{11/2} \leftarrow ^4I_{15/2}$. The results of OAS measurement agree well with those by other known techniques.

To eliminate the influence of absorption in the liquid itself, one can first measure the OA spectrum of the pure liquid and thereafter substract it from the measurements with powders.

8.1.4 Multiphoton Absorption

The measuring technique for multiphoton absorption in liquids is almost the same as for gases (Sect.7.3.2). Generally, for multiphoton absorption the OA signal can be presented as [8.21,22]

$$P = K(1 - \eta_L)h\nu\sigma^{(m)} \cdot I^{(m)}NVt_p = K(1 - \eta_L)E\sigma^{(m)}I^{(m-1)}\ell N \quad , \qquad (8.2)$$

where $\sigma^{(m)}$ is the cross section of m-photon absorption, E is the energy of the laser pulse, N is the concentration of molecules absorbing radiation, V, ℓ are the volume and length of excitation area, respectively, and K is the proportionality constant for a specific geometry of the experiment. Definition of K, i.e., calibration of the OA-cell sensitivity can be carried out, besides by the methods described in Chap.6, under full absorption of the incident radiation by the solution in the presence of highly-absorbing substances.

Multiphoton absorption manifests itself in the nonlinear $(P \propto I^m)$ dependence of the OA signal amplitude on excitation intensity. In [8.21] it has been found that when different liquids are irradiated by pulsed ruby-laser radiation (with a duration of 20 ns and a maximum intensity of 400 MW/cm^2) the exponent for kryptocyanine in ethanole (concentration: 10^{-3} mol/ℓ) of carbon disulfide (CS_2), nitrobenzene and ethanole equals 1,2,3-4 and 5, respectively. These exponents, as a rule, agree well with the ratio between the frequencies of the maxima of the liquid's electron-absorption bands and the frequency of the exciting radiation. For example, for carbon disulfide m = 2 between 350 and 360 nm under ruby-laser excitation ($\lambda = 0.69$ μm). For nitrobenzene m = 3 $\div$ 4 and the most intense absorption bands located about 250 nm. For ethanole m = 5, whereas absorption starts around 200 - 210 nm.

Using (8.2) we can estimate the cross section of multiphoton absorption $\sigma^{(m)}$. But in this case it is necessary to calibrate the sensitivity relative to one-photon absorption. To illustrate this let us consider the technique of evaluating $\sigma^{(m)}$ in carbon disulfide. Calibration can be performed with the experimental dependence of the OA-signal amplitude on the energy of exciting pulse (ruby laser) for the cryptocyanine solution. The incident radiation is absorbed completely mainly due to one-photon absorption. Then the radiation energy absorbed in CS_2 is determined from the equality of the OA-signal amplitudes. Since the value calculated for the two-photon absorption cross-section is $\sigma^{(2)} = 10^{-50}$ cm^4/s [8.21], it is consistent with the results of other methods. As in all measurements of multiphoton absorption, the measuring accuracy of $\sigma^{(m)}$ depends, to a great extent, on the measuring accuracy of the spatial-temporal characteristics of the laser beam.

Figure 8.5 shows the spectrum of two-photon absorption of polymethyne dye solution measured with a pulsed dye-laser from the luminescence intensity (curve 1) and by OAS (curve 2). Here one can also see the spectrum of one-photon absorption of a solution measured with a conventional spectrometer (curve 3). Comparison of these curves shows that OAS and the fluorescence method produce similar results. Specifically, the similarity of these spectra

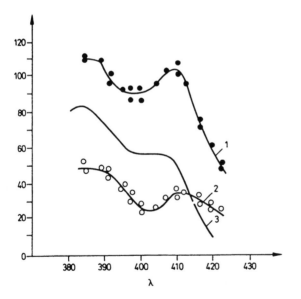

Fig. 8.5. Spectra of two-photon absorption of the polymethyne dye solution: (1): luminescence measurement; (2): OA spectrum; (3): the one-photon absorption spectrum. Wavelengths corresponding to the double frequency of the dye laser are shown along the abscissa axis [8.21]

for polymethyne dye solution shows that in a complex molecule with low symmetry there are no strict selection rules and one- or two-photon transitions take place between the same states.

The pulsed OA technique is also suitable for the investigation of nonlinear absorption in solids [8.23,24]. The extensive two-photon absorption (TPA) study in hexagonal ZnS was reported in [8.23]. The TPA measurements were made over a wide frequency range using 5-ns polarized dye laser. In [8.24] OAS with a 1.06 μm laser has been applied to investigate the third-order multiphoton absorption in selected thallium halides.

8.1.5 Raman Scattering

Like OAS of stimulated Raman scattering in gases (Sect.7.4), we can also realize OA Raman spectroscopy (OARS) of liquids. The possibilities of such a technique were demonstrated in [8.25] using two time- and space-synchronized pulsed dye lasers. One of them forms the pumping wave with frequency ν_p and the radiation of the other acts as a probe wave at the Stokes frequency ν_{st}. The typical values of the pulse energy and the duration are 2 mJ and 10^{-6} s, respectively. The pulses of both lasers were made to coincide in an OA cell whose design was given in Fig.3.1. The threshold sensitivity of such a system allows one to detect weak Raman gains up to 10^{-5} cm^{-1}.

This method has been used to measure the OA spectra of Raman scattering in different liquids including benzene, acetone, 1,1,1-trichloroethane, n-toluene and n-hexane. Figure 8.6 shows the OAR spectrum in benzene for two mutual orientations of polarization planes of the pumping wave and the Stokes component. The polarization of one of the beams (Stokes component) was changed by placing a half-wave plate in its way. The peak in Figure 8.6 near ν_p = 19310 cm^{-1} corresponds to the conditional Raman frequency $\nu_R = \nu_p - \nu_{st}$ = 3059 ± 3 cm^{-1}. The OA-cell sensitivity was calibrated for the Raman gain by comparing the amplitudes of the OA signal during operation of both lasers and the OA signal only from radiation absorption of one laser with λ = 607.1 nm at the 5 th overtone of the C-H vibration in benzene (its absorption factor equals $\alpha = 2.3 \cdot 10^{-3}$ cm^{-1}). The fact that an OAR signal was measured is proved by the proportionality of this signal to the intensities of pump and Stokes waves as well as by its dependence on the collinearity, and spatial and time coincidences of the laser pulses. The depolarization value was determined from the ratio of gains for orthogonal polarizations.

Summarizing OAS allows Raman scattering in samples with strong fluorescence to be studied. Besides, by introducing a time delay between the laser pulses we can measure the relaxation of excited states.

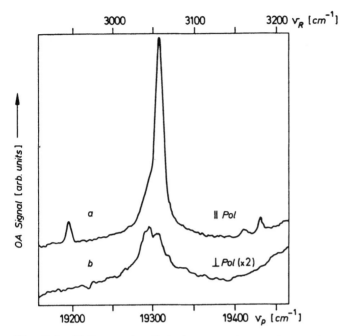

Fig. 8.6. OA Raman signal in benzene at room temperature obtained when the pumping laser frequency had been scanned and Stokes frequency was fixed at $\nu_{st} = 16251$ cm^{-1}. (a) Polarization of both lasers in the same plane; (b) polarization of lasers orthogonal [8.25]

8.2 Surfaces

The output signal of OAS is formed by heat transfer of absorbed energy in a rather thin surface layer of the sample toward the gas. Therefore, this method turns out to be very sensitive to the character of absorption on the sample surface, and this enables it to be used efficiently to study surface states. Depending on the magnitude and character of surface absorption, there are three main areas of application: 1) analysis of highly absorbing substances; 2) analysis of optical coatings, and 3) analysis of highly reflecting materials including metals and semiconductors.

8.2.1 Highly Absorbing Substances

In most cases OA analysis of highly absorbing substances can be realized with conventional nonlaser radiation sources [8.26-28]. The use of lasers is advisable in measuring OA signals with indirect detection at high modulation

208

frequencies (up to several kHz) when the sensitivity of OAS decreases substantially. High modulation frequencies may be required by the necessity of measuring the absorption profile and simultaneously eliminating the effect of thermal saturation, which usually obscures the information from highly absorbing substances.

The possibilities of vibrational OAS of some substances using a CW CO_2 laser were demonstrated in [8.29-31]. In the measurements both an ordinary OA cell and one with an open membrane were used (Fig.5.9). To avoid errors due to scattered light , the walls of the cell were made of polished stainless steel. To reduce the influence of reflection in the windows they were placed at the Brewster angle. In experiments with aqueous solutions it is advisable for such measurements to use a reference channel to exclude the nonstability of the source and to allow for the absorption of water on the windows. Figure 8.7 shows the OA spectra of $(NH_4)_2SO_4$ from aqueous solution and powder measured with a CO_2 laser. The absorption maximum of $(NH_4)_2SO_2$ in aqueous solution (v_3 mode) lies near 10 μm. The small maximum near 10.26 μm for powder was explained in [8.29] as a manifestation of the weak mode v_1 of the SO_4^{2-} ions. A comparatively small value of the OA-signal amplitude in the short-wavelength spectral region is due to the effect of thermal saturation of the sample

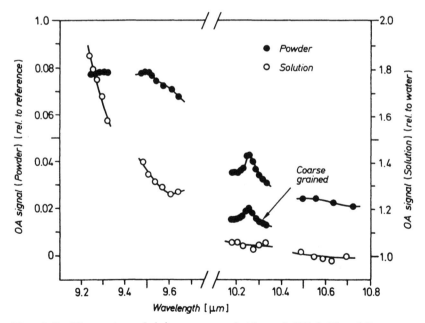

Fig. 8.7. OA spectra of (○) aqueous solution of $(NH_4)_2SO_4$ with a concentration of 58.5 g/kg and (•) powder (diameter of grains ≤ 0.1 mm), obtained using a CO_2 laser with modulation frequency 400 Hz [8.29]

$(\alpha = 2.5 \cdot 10^4 \text{ cm}^{-1})$. The validity of such a conclusion was proved by changing to the modulation frequency of 3 kHz at which a relative increase in OA signal for the ν_1 mode was observed.

A similar technique was successfully used to study $(NH_4)_2SO_2$ in the form of crystals [8.30]. In such measurements one can also use an OA cell with an open membrane demonstrated in [8.31] by measuring the OA spectra with a CO_2 laser. According to the estimations in [8.31], at a radiation power of 0.1 - 0.01 W it is possible to measure OA spectra if the film is several micrometers thick.

8.2.2 Analysis of Coatings

The high sensitivity of OAS with indirect detection can be utilized to measure the surface absorption of optical elements, the thermal and optical properties of various optical coatings (antireflecting, protecting, etc.) and the depth profile of absorption.

a) Signal Generation

The approximate theory of OAS presented in Sect.3.3, is valid for thermally and optically homogeneous samples. However, it can also be applied to thermally and optically inhomogeneous samples. For this purpose we must solve slightly modified equations of thermal diffusion in which some terms (the absorption coefficient and thermodynamical constants) depend on the spatial coordinates in the sample volume. In the case of several layers, for example, with different optical or thermal properties we should use thermal diffusion equations for each layer with appropriate parameters and solve them jointly, taking into account the boundary between different layers.

In [8.32] consideration was given to a particular solution of this problem as applied to a thermally uniform but optically nonuniform sample. The equation found for the temperature on the boundary surface in a solid-gas system has the form

$$T(0,\omega) = \frac{1}{s} \left(\frac{(1 + b) \, \mathcal{H}(s)e^{sl} + (1 - b) \, \mathcal{H}(-s)e^{-sl}}{(1 + b)(1 + g')e^{sl} - (1 - b)(1-g')e^{-sl}} \right) \quad , \tag{8.3}$$

where $s^2 = i\omega / k$. $\mathcal{H}(s)$ is the Laplace transform of the H(x) function related to the absorption coefficient by

$$\alpha(x) = H(x)\left[\frac{1-R}{2K} I - \int_0^x H(y)dy \right]^{-1} \quad , \tag{8.4}$$

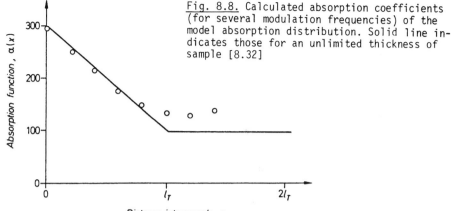

Fig. 8.8. Calculated absorption coefficients (for several modulation frequencies) of the model absorption distribution. Solid line indicates those for an unlimited thickness of sample [8.32]

where R is the reflection coefficient of the sample surface, I is the intensity of incident radiation, the rest of the notations are as in Sect.3.3.

Thus, by measuring the OA signal as a function of frequency, all other parameters known, we can determine $\mathcal{H}(s)$ and with it and the Laplace transform, H(x) and hence $\alpha(x)$. To test this, Fig.8.8 shows the depth profile of absorption at several modulation frequencies in case $x \ll \ell_T$. It can be seen that the model and calculated data agree satisfactorily.

A more complex case of optically and thermally nonuniform samples was considered in [8.33]; some rather cumbersome analytical expressions for OA signals have been obtained. Among other things, it has been shown that the pressence of a nonabsorbing coating on the sample causes the OA signal amplitude to decrease by about $[\exp -(1+i)\ell_c / \ell_T^c]$ times, ℓ_c being the length of the coating, ℓ_T^c the length of thermal diffusion in the coating. We should note that such a coating on the windows of a gas OA cell is an effective means to suppress the background signal from the windows.

b) Experimental Results

The first experimental results showing the applicability of OAS to investigate optical coatings on optically transparent samples were obtained in [8.34-37]. An OA cell with a window as a test sample was used in the measurements. Its coated side faced inside the cell. The average distance between the windows was usually about 0.1-0.5 cm. In some cases the OA cell was cleaned with dry nitrogen to eliminate possible background OA signals caused by radiation absorption in the air filling this cell. With a CW CO_2 laser some coatings on substrata of KRS-5 [8.34], KCl and NaCl [8.36] and ZnSe [8.37] were investigated.

All this was connected with research of optical materials with low optical losses for high-power IR lasers. In [8.34] a $\lambda/4$ layer of calcium fluorite and a two-layer coating (zinc sulfide on thorium fluorite) were studied as coatings for KRS-5. They were applied just on a part of the substratum area which made it easier to obtain separate information on the level of absorption only in the pure substratum and in the coated substratum by scanning the laser beam over the surface. Thus it was found that the ratio of OA signals from the substrata coated with calcium fluorite and without it was almost one order of magnitude at $\lambda = 10.6$ μm. The OA-cell sensitivity makes it possible to determine the relative fraction of absorbed energy in the surface layer at the level of 10^{-4} to 10^{-5} with 10-W radiation power.

For a sample with purely bulk absorption the slope of the frequency response of the OA signal is $\omega^{-3/2}$ instead of ω^{-1} found for a sample with predominant surface absorption, and the phase shift is about $45°$ (Sect.3.3). These conclusions are confirmed by other results [8.36-38]. Specifically, in [8.36] for substrata of KCl and NaCl coated with NaF at $\lambda = 10.6$ μm, the dependence of the OA signal on modulation frequency was close to ω^{-1} over the frequency range from 10 to 100 Hz. The authors [8.36] concluded that the most likely cause of absorption in NaF is the presence of hydroxyl ions.

The OA measurements of weakly-absorbing samples can be distorted by the effects of multiple reflection of light from the sample surfaces, including interference. Given the reflection from the surface of one window, for example, just inside the surfaces of a window a power ratio of [8.36]

$$\frac{J_{exit}}{J_{ent}} = \frac{4\tilde{n}^2}{(\tilde{n}+1)^2} \quad , \tag{8.5}$$

The interference may lead to time fluctuations of the OA signal because of thermal expansion of the sample under passing radiation. In [8.36], for example, within several minutes the OA signal value in a KCl sample varied up to 5 times. This effect could be easily eliminated by a small turn (about $5°$) of the sample about the optical axis.

With two windows in an OA cell the question arises how to increase the accuracy of determining surface absorption in one of them. This problem can partially be solved by measuring two OA signals at two different orientations of the OA cell along the laser beam differing by $180°$. The ratio of the OA signals P_1 and P_2, in a first approximation, as the absorbing layer is placed at the entrance and exit of an OA cell, respectively, equals [8.36]

$$\frac{P_1}{P_2} = \frac{[4/(\tilde{n}+1)^2]\gamma_s + 1}{4/(\tilde{n}+1)^2 + \gamma_s} \quad , \tag{8.6}$$

where γ_s is the ratio of surface absorption factors in the two windows.

The absolute calibration of the OA-cell sensitivity to surface absorption is carried out, as a rule, by the thermophone method (Chap.6), using a window with a metallized layer coating its inner surface. Electric voltage pulses are applied to this coating; the value of the OA signal due to electric heating of this layer is measured. The heat released is determined from the measured current and voltage. This device accurately simulates surface heating by virtue of the similar character of heat transfer between the substratum and the gas both in a thin metallized layer and real surface absorption at a weakly absorbing substratum [8.39]. With the calibration technique described, it has been shown in [8.36] that the level of absorption in the surface film of NaF reaches 6%. With increasing modulation frequency the slope of the frequency characteristic may increase relative to the dependence $P \propto \omega^{-1}$. This is explained by the fact that the thermal diffusion length is decreased to a value comparable with the surface-layer width. As a result, the quasi-volume absorption occurring within the surface layer and the frequency characteristic slope must tend to $\omega^{-3/2}$, according to the data from Sect.3.3.

This question was considered in more detail in [8.37] for different ratios between the optical and thermal parameters of the surface layer and the substratum itself with bulk absorption. The object of research was a sample of ZnSe (0.76 cm thick) with a two-layer antireflecting coating of ThF_4 0.389 μm thick (inner layer) and ZnS 0.319 μm thick (external layer). The experimental results were interpreted using the theoretical models of OA signal generation by ROSENCWEIG and GERSHO (see Chap.3) and BENNETT and FORMAN [8.40] generalized for two-layer nonuniform structures. The results of correlating theoretical and experimental data are presented in Fig.8.9. The experimental dependence of the OA signal on frequency has the form $P \propto \omega^{-n}$ with n = 1.04. The OA-signal phase varies by $20°$ in the frequency range from 1000 to 50 Hz and by just $7°$ between 1000 and 100 Hz. These dependences were obtained using a CW CO_2 laser at 25-W power and $\lambda = 10.6$ μm.

Thus, the results presented demonstrate the effective application of OAS to analyze surface absorption especially in combination with laser calorimetry. Such a technique will make it possible to determine the dominant role of surface or bulk absorption in the formation of optical losses in various materials.

Direct piezoelectric recording of OA signals in the substratum with a coating also enables us to determine a rather low level of absorbed energy in the coating. In accordance with the data from [8.41], for example, with radiation power of up to 1 W, it is then possible to measure the fraction of absorbed energy from 10^{-4} to 10^{-5}. A piezoelectric detector is linked to the back side

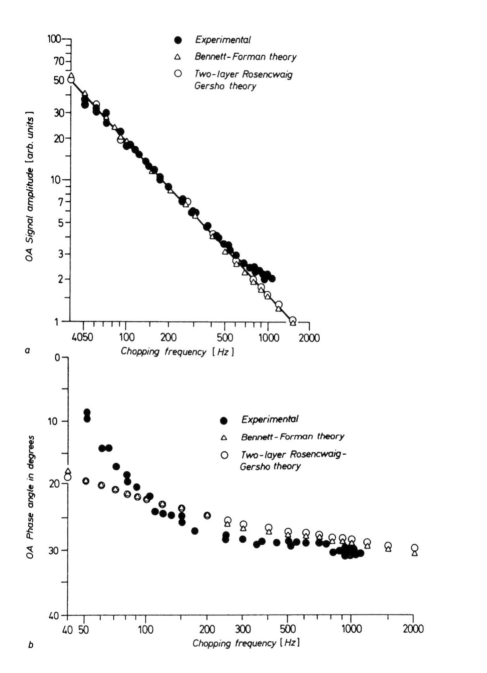

Fig. 8.9 a,b. Comparison of experimental (•) and theoretical (Δo) amplitudes (**a**) and phases (**b**) of OA signal as a function of modulation frequency for the ZnSe sample with two-layer coatings. (Δ): Calculation according to the Bennett-Forman theory for $\varkappa_s/\alpha_v = 0.03$ cm; (o): calculation according to the Rozencwaig-Gersho theory for $\alpha_s = 2.7$ cm^{-1}, $\alpha_v = 0.064$ cm^{-1} [8.37]

Table 8.1. Threshold sensitivity of different techniques to measure absorption in solids [8.10]

Technique	Bulk absorption α_v [cm^{-1}]		Surface absorption x_s	
	Experimental value	Theoretical limit	Experimental value	Theoretical limit
Absorption	$10^{-2} - 10^{-3}$	$10^{-4} - 10^{-5}$	$10^{-3} - 10^{-4}$	10^{-5}
Emission	$10^{-4} - 10^{-5}$	$10^{-6} - 10^{-7}$	$10^{-4} - 10^{-5}$	$10^{-6} - 10^{-7}$
Calorimetric	$10^{-4} - 10^{-5}$	10^{-6}	$10^{-4} - 10^{-5}$	10^{-6}
Optoacoustic (with direct detection)	$10^{-4} - 10^{-5}$	10^{-6}	$10^{-4} - 10^{-5}$	10^{-6}
Interference calorimetry	$10^{-4} - 10^{-5}$	10^{-6}		
Optoacoustic (with indirect detection)		$10^{-3} - 10^{-4}$	$10^{-4} - 10^{-5}$	10^{-7}

of the substratum on one side of the laser beam. A serious problem in such a design, however, is the influence of scattered radiation.

c) Comparison of Different Methods

In conclusion, it is of interest to compare OAS with other methods, reviewed in [8.10]. The sensitivities are listed in Table 8.1. It can be seen that the OA methods are most sensitive in analysis of both bulk and surface absorption. Among their merits are instrumental simplicity as well as a low sensitivity to scattered radiation. The calorimetric method has a similar sensitivity, but it is greatly affected by scattered radiation. Besides, it necessitates a good thermal contact between detector and sample, requires a higher radiation power and is highly inertial. In the absorption method, when measuring low surface absorption the laser beam should be directed along the surface, but that is not always possible. Also, it is highly affected by scattered radiation. In the emission technique corrective measurements are possible only when one absorption type (bulk or surface) prevails, since separation is difficult in this method. Besides, the equipment as a rule, must be cooled to liquid-nitrogen temperature. An additional advantage of OAS is that it allows analyzing small-length samples, for example measuring surface absorption up to 1 μm as well as selecting different types of absorption through measuring the phase of OA signals, frequency responses, etc.

8.2.3 Analysis of Metals

With the high reflection from the sample surface that is characteristic of metals in the IR spectral region, it is difficult to measure absorption by a conventional method taking advantage of the intensity of incident and reflected radiation. This is due to a small relative change in radiation intensity and by the influence of scattering effects. In this case it is very useful to apply OAS in which the energy absorbed directly in the surface layer of the sample is detected. One should keep in mind, however, that by virtue of a high absorption coefficient in metals thermal saturation occurs when operating at high modulation frequencies ($\ell_\alpha \ll \ell_T$). Under these conditions the spectral information is lost, and only the total level of radiation loss by absorption can be determined. But even under these conditions we can use OAS, particularly in a differential scheme with a standard sample, say, to determine and control the losses in a laser mirror, in a diffraction grating, etc.

It is very efficient to use such a technique in determining the absorption in thin dielectric films on the surface of metals. According to [8.42], the variation of reflection from the metal caused by absorption in a thin film has the form

$$\delta R_{\shortparallel} = \frac{16\pi d_f \cos\Theta}{\lambda} \left[\frac{\tilde{n}_f k_f \sin^2\Theta}{\tilde{n}_f^4 \cos^2\Theta} - \frac{\tilde{n}_m f(\tilde{n}_f,\Theta)}{k_m^3 \cos^4\Theta} \right] \quad , \tag{8.7}$$

$$\delta R_{\perp} = \frac{16\pi d_f \cos\Theta}{\lambda} \left[\frac{\tilde{n}_f k_f}{k_m^2} + \frac{\tilde{n}_f^2 \tilde{n}_m}{k_m^3} \right] \quad , \tag{8.8}$$

for the radiation polarized in the plane of incidence ($\shortparallel$) and normal to it ($\perp$), respectively. Here Θ denotes the angle of incidence; $\tilde{n}_f$, k_f and $\tilde{n}_m$, $\tilde{k}_m$ are the real and imaginary parts of the complex refractive incidexes of film (f) and metal substratum (m), respectively; d_f is the film thickness; $f(n_f, \Theta)$ is the function whose value is close to unity. The factor k is determined by the absorption coefficient α : $k(\lambda) = \lambda\alpha(\lambda) / 4\pi$. For most films (10 to 100 Å thick) on well reflecting metals like Au, Ag, Cu and Al give $\delta R_{\shortparallel} \approx 10^{-2}$ and $\delta R_{\perp} \leqslant 10^{-5}$ from (8.7,8). Therefore, the conventional technique even with the use of multiple reflections is little suited for such measurements due to its insufficient sensitivity.

The main advantage of OAS here consists of a high sensitivity in only single reflection from the sample. According to [8.30], the sensitivity of this method makes it possible to determine absorption in 1 / 20 of a natural oxide layer ($d \approx 20$ Å), and at slight modification to the level 10^{-2} to 10^{-3} of a monolayer. This is better than the techniques based on X-rays or electron diffraction. To illustrate the potential of OAS in practice, Fig.8.10 shows the OA spectrum of an oxide layer of Al about 160 Å thick on a substratum (the

216

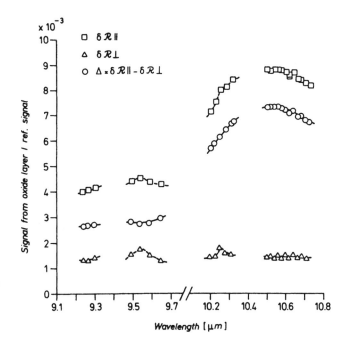

Fig. 8.10. OA spectrum of the oxide film on Al obtained by anodization for the different polarizations of radiation [8.30]

layer produced by anodizing) measured with a CO_2 laser (power: 40 mW, angle of incidence: $\theta = 60°$, f = 30.5 Hz); the cell windows were made of BaF_2. For polarization in the plane of incidence (∥) there is a well-pronounced absorption maximum near $\lambda = 10.5$ μm ($\nu = 947.9$ cm^{-1}) identified as a longitudinal mode of the vibration Al-O.

The optoacoustic method and the attenuated-total-reflection (ATR) method have their pros and cons for the investigation of a surface layer. The ATR-OA method combines their pros and eliminates their cons [8.43-46]. In this case a sample film or thin plate is put onto a prism base and is irradiated by an amplitude-modulated wave so as to obtain total-internal reflection at the sample-gas or substrate-sample interface. The energy absorbed by the sample is recorded by the OA method. The main advantages of this method in comparison with the traditional ATR method are: 1) the higher sensitivity; 2) the qualitative measurement due to the absence of problems due to the sample-prism contact; 3) the absence of losses due to scattering, reflection, etc; 4) the possibility to measure the absorption profile varying the modulation frequency change. Consequently, the ATR-OA method is unique for studying surface phenomena.

Such a technique can be used for operation in vacuum, in studies of chemical and catalytic reactions on a surface, and to investigate the vibrational spectra of thin films at different temperatures, etc.

8.3 Circular Dichroism

Circular dichroism (CD) arises in optically active media or in media placed in a constant magnetic field [8.47]. The CD effect manifests itself in the ability of matter to absorb left and right polarized radiation in different ways. The parameters of CD are the absorption difference for right and left circularly polarized light $\Delta\alpha = \alpha_L - \alpha_R$ and the average (isotropic) absorption $\bar{\alpha} = (\alpha_L + \alpha_R) / 2$. The sensitivity and spectral resolution of conventional dichrometers are insufficient to solve many important problems. OAS combined with a laser source offers a possible solution.

8.3.1 Experimental Schemes

Information on CD spectra can be obtained by measuring the OA signal as a function of the wavelength of adequately modulated radiation. Possible types of modulation are illustrated in Fig.8.11.

It was difficult to carry out the very first pre-laser measurement of CD because in almost all cases a secondary effect was used: the transformation of plane-polarized light to elliptically polarized light as a light beam traverses an optically active sample. The use of lasers simplifies substantially the measuring of CD spectra since lasers provide plane-polarized radiation of rather high power. The respective OA technique comprises successive or simultaneous measurement of the OA-signal amplitude for right and left circularly polarized light and determines their difference.

As for the instrumental realization, successive measurement of signals is simplest as intensity-modulated radiation is passed through a sample first for right and then for left circular polarization (Fig.8.11a). Discrete switching of polarization is usually performed by turning a quarter-wave plate by an angle of +45° so that in one position the plane-polarized light changes to right circularly polarized light, and in the other position it changes to left circularly polarized light. A disadvantage of this method is low measurement accuracy, because for most known optically active substances the difference between the absorption coefficients $\Delta\alpha = \alpha_L - \alpha_R$ is very small. Usually it does not exceed several percent of the absolute value of the absorption coefficients, i.e., $\Delta\alpha \ll \alpha_L, \alpha_R$.

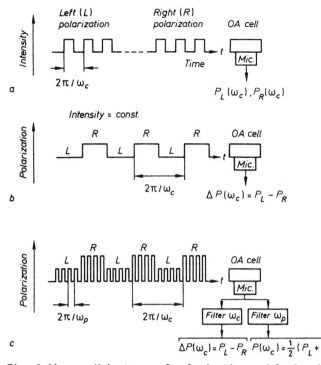

Fig. 8.11 **a-c.** Main types of polarization modulation in an OA dichrometer: (**a**) sequential modulation, (**b**) simultaneous modulation and (**c**) two-frequency modulation [8.48]

From the standpoint of minimizing the errors in $\Delta\alpha$, the scheme in which a light beam of constant intensity passes through the substance under investigation is prefered (Fig.8.11b). The beam polarization is modulated with frequency ω_c so that alternately left or right circularly polarized light appears. Such modulation can be realized, for example, with an electrooptical modulator operating on the Pockels effect, with plane-polarized light directed onto it, or with a photoelastic modulator. When the absorption coefficients for left and right polarization differ, a differential OA signal appears at the frequency ω_c, whose value is related to the parameter $\Delta\alpha$. To exclude the spurious amplitude modulation in this scheme, stringent requirements are placed upon the stability of the laser intensity for the different polarizations. The magnitude of the differential OA signal varies about the average which corresponds to the average absorption of the sample, $\bar{\alpha}$.

The third scheme (Fig.8.11c) in which, in addition, the intensity is frequency (ω_p) modulated with the $\omega_p \gg \omega_c$, is more suitable for the simultaneous measurement of $\Delta\alpha$ and $\bar{\alpha}$ [8.48]. Filters discrimine OA signals at ω_p and ω_c.

As in the previous scheme, ω_p is simply related to the parameter $\Delta\alpha$, which ω_p carries information on the value of $\bar{\alpha}$. Therefore this scheme makes it possible to determine the ratio $g_{CD} = \Delta\alpha / \bar{\alpha}$, called the CD dissymmetry factor.

The change $\Delta\alpha$ between different wavelengths may be positive or negative. The OA signal determines only the magnitude of $\Delta\alpha$, its sign can be defined by measuring the OA-signal phase.

An important requirement in measuring CD is the isotropy of an optical scheme. This poses a problem in conventional schemes mainly due to the anisotropy in sensitivity of different radiation detectors. For OA circular dichroism account must be taken only of spurious phase anisotropy in the entrance window of the OA cell. This can be easily reduced to minimum if the window is strictly perpendicular to the optic axis.

In some cases when studying CD very strict requirements are placed on the absence of an external magnetic field. This requirement can be relaxed by using more involved measuring schemes [8.49].

8.3.2 Sensitivity of CD Method

An OA CD spectrometer can be realized both through direct recording of OA signals in different media including gas, liquid and solids, and through indirect recording of OA signals in condensed media. For weakly absorbing media ($\alpha L \ll 1$) the value measured for the difference of two OA signals $\Delta P = P_L - P_R$ is directly connected with the value $\Delta\alpha = \alpha_L - \alpha_R$. For highly-absorbing media this interconnection is more complicated. But the latter is more typical in practice, since the CD effect manifests itself in the region of absorption [8.50,51].

The sensitivity limit for measuring $\Delta\alpha = \alpha_L - \alpha_R$ depends on the OA-cell sensitivity and the laser power as well as on the accuracy of determining the absolute value of α. The threshold sensitivities of different OA cells and measuring schemes are given in Table 6.1 (Sect.6.1.2). Accordingly, the best results are obtained with direct recording of the OA signals in the sample. For example, if the laser power (energy) is up to 1 W (J), this method can be used to measure the absorption coefficients in gases and liquids at the level 10^{-8} to 10^{-9} cm^{-1}. The same threshold can be obtained for the parameter $\Delta\alpha$ as well. Therefore, with an absorption coefficient in the media under investigation around 10^{-2} cm^{-1}, OAS allows the parameter $g_{CD} = \Delta\alpha / \bar{\alpha}$ to be measured at the level 10^{-6} to 10^{-7}.

Thus, an important advantage of OAS is a high signal-to-noise ratio which makes measuring rather small values $\Delta\alpha$ possible. In this case, however, stringent requirements are placed upon the suppression of OA-signal fluctua-

tions, which result from fluctuations of laser parameters, parasitic modulation of intensity as well as by the imperfection of the devices for polarization modulation of the laser light. It should be emphasized that it is possible
to determine small variations of $\Delta\alpha$ only at a relatively high power level,
necessitating the use of laser sources in such measurements. According to the
results in [8.51], to measure the parameter g_{CD} at a level of 10^{-4} at
$\omega / 2\pi = 50$ Hz and $\alpha = 10^3$ cm^{-1} under typical CD measurement conditions in condensed media it is necessary that the radiation intensity be at least $1 W / cm^2$
at the OA-cell input.

   In [8.52] an OA spectrometer was applied to measure the spectrum of magnetic CD in the monocrystal NdMoO$_4$ (2 mm thick). A dye laser excited the
transition in the Nd^{3+} ion with strong absorption. At room temperature the
spectrum of magnetic CD related to the Zeeman splitting of ground and excited
states. With a time constant of 1 s, a magnetic field intensity of 0.7 T and
a polarization-modulation frequency of 3 kHz, the noise level of the scheme
corresponded to the least detectable value of $g = \Delta\alpha / \bar{\alpha}$, that was about 10^{-3}.
The same setup was used to record the spectrum of linear dichroism of NdMoO$_4$.
In the experiment the ratio ℓ_T^S / ℓ_α was about 0.4 at 3 kHz, corresponding to
optically and thermally thick samples. In the frequency range from 300 to
3000 Hz the frequency dependence of the OA signal was close to $\omega^{-3/2}$ agreeing
well with theory.

8.3.3 Comparison of Different Methods

Besides using OAS (OACD), CD can be measured by the absorption method (through
transmission) (TCD), the luminescence method (LCD), and through Raman-scattered
radiation (RCD). It is of interest to compare these methods.

   In the most widely used method TCD, the transmission of left and right circularly polarized radiation is measured. This method can be applied only to
optically thin uniform and nonscattering samples. The advantage of OAS over
TCD is the large dynamic range and the possibility of analysis of both
weakly and highly absorbing objects, the influence of radiation scattering
is minimum.

   The measuring principles of OACD and LCD are similar; in LCD the absorption
of differently polarized radiation is measured by detecting luminescence. This
method is sensitive and selective enough, but its realization cells for such a
relaxation channel that the fluorescent signal should be clearly related to
the absorption level of radiation of different polarizations. Thus OACD and
LCD complement each other.

In conclusion, we must note that the OA method described for studying CD can be applied to analyze a wide range of optically active materials, including crystals, polymers, biological samples like membranes, etc. In particular, OAS is useful for analyzing samples solved or suspended in highly scattering media.

8.4 Photoactive Media

Some part of absorbed energy is transformed to heat and produces an OA signal. Another part may be spent on photoactive processes which cause the OA signal to decrease. Then the magnitude of the OA signal can be presented as [8.53]

$$P = KJx_f\eta_{nr} = KJx_f\left(1 - \sum \frac{\eta_{ph}\Delta E_{ph}}{N_A h\nu} - \frac{\eta_L\nu_L}{\nu} - \eta_{pv}\right) \quad , \tag{8.9}$$

where J is the radiation power at the OA cell input; x_f is the fraction of energy absorbed in the sample; K is the proportionality factor depending on the thermal properties of the sample, the geometry of the OA cell, the modulation frequency, etc.; the factors η_{nr}, η_{ph}, η_L and η_{pv} define the quantum yields of nonradiative relaxation, photochemical reactions, luminescence and photovoltaic effects, respectively; ΔE_{ph} is the change of internal energy (by a mole) connected with the formation of products in photochemical reactions; N_A is the Avogadro number; ν, ν_L are the frequency of the exciting radiation and the "average" frequency of luminescence, respectively. In the absence of photoactive processes the OA spectrum fully equivalent to the ordinary absorption spectrum.

The situation changes in the presence of photoactive processes. As applied to gases, for example, Fig.7.20 shows that the consumption of some part of absorbed energy on dissociation of NO molecules distorts the OA spectrum relative to the absorption spectra. This manifest itself as a small decrease of the OA signal in the short-wavelength region. The presence of a radiative channel of relaxation can even cause some absorption bands of the molecules to disappear completely in the OA spectrum. In [8.54], for example, it has been shown that for benzene vapor at 24 Torr in the spectral region above 245 nm the absorption, luminescence and OA spectra are almost identical, i.e., the quantum yields of luminescence and nonradiative transitions vary slightly with wavelength. An vice versa, in the spectral region shorter than 245 nm the quantum luminescence yield drops drastically, manifested in the disappearance of one band in the luminescence spectrum but not in the absorption and OA spectra.

When studying the OA spectra of photoactive media at different conditions (temperature, pressure, composition, concentration, etc.) we can obtain information on the character and degree of different photoactive processes from some distortions in the OA signal. When such processes show up simultaneously it becomes more difficult, of course, to perform a differential study. The simplest way to solve this problem is when only one of the photoactive processes is dominant. Below we shall consider such specific cases in more detail.

8.4.1 Quantum Yield of Luminescence

The quantum yield of luminescence η_L is important for evaluation of the laser efficiency and studying the relaxation mechanism in various systems. Usually the quantum yield of luminescence is measured from the relative luminescence intensity of the substance under investigation and the substance with known quantum yield. In such measurements we should introduce a correction for the spectral sensitivity of the detecting system, the position and the form of the luminescence spectra of the sample and the standard, etc. All these corrections complicate the measurement procedure. These shortcomings are not characteristic of the calorimetric method for measuring the luminescence yield [8.55], in which the heating due to light absorbed in fluorescent and nonfluorescent solutions is compared. This method, however, requires rather long-term and complicated measurements.

It is simpler and faster to measure the absorbed energy transformed to heat with OAS. Whose advantages here are its high sensitivity and also its indifference to the spectral composition of exciting radiation and luminescence. It is of importance only that the laser energy should leave the OA cell and not participate in the generation of the OA signal due to, say, its absorption in the walls or entering the acoustic detector. To meet this requirement it is advisable that the walls of the OA cell should be made of material transparent to luminescent radiation and the detector should be somewhat removed from the sample and screened from scattered radiation. When operating in the visible region we can obtain good results with OA cells made of quartz or plexiglass. It should be noted that since it is possible to work with weakly absorbing media we can neglect the effect of luminescence reabsorption in the sample.

a) Combined Technique

In the most general case η_L can be found from (cf. list of symbols)

$$\eta_L = \frac{E_L}{E_{abs}} = 1 - \frac{E_T}{E_{abs}} \quad . \tag{8.10}$$

223

With no other photoactive processes present

$$E_{abs} = E_L + E_T \quad . \tag{8.11}$$

Thus, according to (8.11,12), to determine η_L it is sufficient to measure independently any two pairs of parameters: E_{abs} and E_L, E_{abs} and E_T or E_L and E_T. From the discussion, preference should be given to the parameters E_{abs} and E_T. This technique is applicable to case of high absorption where E_{abs} can be determined accurately and independently by the conventional absorption method.

The most accurate value of η_L can probably be obtained by simultaneous measurement of E_{abs}, E_T and E_L. Here we should mention [8.54] which treated gas media. In [8.54] the three parameters were measured in a gas (benzene vapor at 24 Torr and $T = 24°C$), using the absorption, luminescence and OA methods. The value of η_L for benzene vapor is $0.19 + 0.02$ near $\lambda = 253$ nm, which is in good agreement with measurements performed by other methods.

This technique cannot be applied to analyze either highly or weakly absorbing media since then it is difficult to determine accurately and independently the value of E_{abs} by the conventional absorption method.

b) Method of Etalon Solutions

In the method of etalon solutions η_L can be determined by comparing two OA signals from the solution under investigation and the etalon solution with the quantum yield η_L^{et} known. The identity of the thermal properties of the solutions is the most important requirement, since it is essential in establishing the same conditions for the generation of OA signals in these solutions. The last requirement can be easily met due to the high sensitivity of OAS that allows operating with low-concentration solutions. Then the influence of the thermal properties of the dissolved substances on the solvent properties is minimum. It is most desirable to use standard solutions with a zero quantum yield, which produces the simplest measuring technique.

The possibilities of such a technique were demonstrated in [8.56] by determining η_L for the solution of rhodamine GG in water with a CW argon laser with $\lambda = 488$ nm and 0.5 W power. To determine the power absorbed and transformed to heat used OAS with a piezodetector placed in an OA cell 20 cm long and 1 cm in diameter. The measurements were carried out at the resonant frequency of 2350 Hz. To avoid considerable variations in OA-cell sensitivity that affects the accuracy of η_L (mainly on account of uncontrollable air bubbles in the cell), the OA-cell sensitivity was corrected by normalizing the OA signals to the acoustic signal initiated in the cell from the nearby acoustic radiator. The solution of $K_2Cr_2O_7$ in a concentration of $0.4 \cdot 10^{-4}$ M, in which no luminescence and photochemical processes are observed in the range from 400 to 500 nm, was used for absolute calibration. The concentration of rhodamine GG was $3 \cdot 10^{-7}$ M. The quantum yield in it was $\eta_L = 0.96\pm0.02$, which was in satisfactory agreement with the published results.

In the absence of calibrating solutions with zero quantum yield it is possible to use solutions with the quantum yield of fluorescence, η_L^{et} to be known. Then it is necessary to compare the amplitudes of OA signals at similar values of absorbed energy. In [8.57], for example, the conditions of complete absorption of the exciting radiation in the cell in the absence of absorption saturation were provided by appropriately choosing the concentration of the solutions and the intensity of exciting radiation. The value of η_L^x was found from

$$\eta_L^x = \frac{\nu}{\nu_L^x} - \frac{P_x}{P_{et}} \frac{\nu - \nu_L^{et} \eta_L^{et}}{\nu_L^x} , \qquad (8.12)$$

where ν_L^x and ν_L^{et} are the average fluorescence frequencies of the solution under investigation and the etalon solution, respectively; ν is the frequency of the exciting radiation; P_x/P_{et} is the amplitude ratio for OA signals in the solution under investigation and the standard solution. In [8.57] this technique was used to measure the yields for four solutions (rhodamine B in ethanol, etc.) with the second harmonic of a ruby laser.

c) Spectral Method

The necessity of calibration using etalon solutions with known quantum yield may no longer arise in analyzing solutions in which we know a priori the variations in relaxation from excited levels in the process of frequency-tuning the exciting radiation. It is known from [8.58,59] that in dye solutions as well as in certain solid samples the transitions from the upper excited singlet states "i" with energy E_i to the lower excited state "1" with energy E_1 are nonradiative as a rule, whereas the transitions from the lower excited state to the ground state are either radiative or nonradiative. We can therefore determine η_L by comparing OA signals at different frequencies. The advantage of this technique is that it allows to measure OA signals in one and the same medium, and this eliminates errors due to differences in thermal properties of different samples.

The technique with indirect detection described in [8.58] was used to determine the quantum yield in $[(CH_3)_2N-C_6H_4]_3-C^+Cl^{-1}$ dissolved in water. The upper excited states were pumped with short-wavelength radiation from a xenon lamp with $\lambda = 306$ nm. The second spectral range at $\lambda = 617$ nm was chosen because it was necessary to pump only the lower excited state with energy E_1. The value of η_L obtained was 0.02 ± 0.01. A similar technique was used in [8.60] to study the character of luminescence in powdered ruby at different concentrations of Cr in Al_2O_3. For this purpose, the OA signals were compared by exciting two different absorption bands in the ions Cr^{+3}.

The basic difficulty in finding the quantum yield of luminescence from the highly excited electronic states of molecules in liquid solutions is that it usually has a small value ($\eta_L \leqslant 10^{-4}$). A substantial measurement error is here introduced by luminescence of the impurities in the solvent and the basic substance, which absorbs the UV exciting radiation.

There is one more technique to determine η_L that does not necessitate absolute calibration of the measuring system with an etalon. It consists of measuring the OA signal as a function of concentration of the luminescent ions in solids. It is mainly based on the effect of concentration quenching of luminescence due to increasing interaction between separate ions. The efficiency of such a technique was demonstrated in [8.61] by analyzing Nd^+ ions in glass using an argon laser.

High sensitivity makes OAS suitable to the degree of deactivation of excited molecules to the character of quenching the molecular luminescence as a result of different external and internal factors, e.g., temperature or pressure variations, or increase in relative concentration of the luminescent component.

8.4.2 Photochemical Reactions

The fraction η_{ph} of absorbed energy can be consumed in photochemical processes. Then the OA spectrum differs from the absorption spectrum. From the relative decrease in OA signal it is possible to determine the parameter $\eta_{ph}\Delta E$. This parameter may be determined in many respects similarly to those described above. For example, to determine $\eta_{ph}(\nu)\Delta E_{ph}$ generally we should measure the absorption spectrum, and calibrate the OA-cell sensitivity. The measurements can be performed using only OAS, say, by comparing two OA signals from the sample, first, when photochemical processes take place, and then when they are absent, as done in [8.62] by studying the photoreactions in chloroplast.

An important problem is how to obtain information on intermediate photochemical reactions. The way of solution lies in an additional measurement of the OA-signal phase and its dependence on frequency. The technique applied to photochemical processes in biological media was described in [8.53]. The technique becomes more complicated when several types of processes, say, luminescence and photochemical reactions take place simultaneously.

8.4.3 Measurement of Photovoltaic Energy

When photovoltaic processes dominate, OAS makes it possible to determine the efficiency of energy transformation to photovoltaic processes. The efficiency η_{PV} is determined by the relationship between relative OA signal and photovoltaic energy.

In the case of a pure photovoltaic process occurs, (8.9) can be written for the specific resistance load R_L as

$$P(R_L) = K[1 - \eta_{PV}(R_L)] \quad . \tag{8.13}$$

To solve (8.13) for $\eta_{PV}(R_L)$ requires a knowledge of K, the energy absorbed by the sample, which is sometimes difficult to obtain. A photovoltaic device, however, has zero energy-conversion efficiency under the open-circuit (OC) condition. Thus $K = P(OC)$, and

$$\eta_{PV}(R_L) = \frac{P(OC) - P(R_L)}{P(OC)} \quad . \tag{8.14}$$

The results for n^+ / p Si solar cells (40 Ωcm) are presented in Fig.8.12. The sample was exposed to visible radiation with $\lambda \geqslant 420$ nm. The figure shows that the power-output and OA-signal dependences are well correlated or rather re-

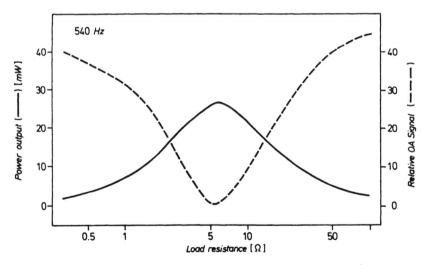

Fig. 8.12. Electrical output power (——) and OA signal (---) as functions of load resistance for a Si sample. The modulation frequency was 540 Hz, and the radiation source was a 450-W Xenon lamp [8.64]

present "mirror" images of each other. When radiation of 0.16 W/cm^2 strikes the sample the power in the circuit goes to 0.028 W/cm^2. This corresponds to 17% efficiency of transformation of absorbed energy to the photovoltaic processes. Some useful information can be obtained from measuring the OA spectrum of the sample in an open electric circuit and under a load close to optimum. In [8.64] the values obtained for the parameter η_{PV} from the ratio of OA signals in the spectral range from 420 to 1080 nm were about 0.2 - 0.3.

It is also possible to determine the generation efficiency of photocarriers with the help of OAS by measuring the degree of decrease in the OA signal as a function of the voltage applied across the sample: η_{PV} is estimated from (8.9) through the ratio of two OA signals in the presence and absence of a voltage. Specificially, in [8.63] this technique was used to find the value of η_{PV} for thin organic films (1 μm) applied in electrography. The characteristic feature of such a metering circuit is direct piezoelectric recording of an OA signal in the films applied to the substratum together with excitation of the carriers by the modulated radiation of a Kr^+ laser with $\lambda = 676$ and 799 nm. The modulation frequency of about 40 kHz corresponds to a mechanical resonance in the design of sample fixing, which causes the OA signal to increase by up to two orders of magnitude. Such conditions of excitation with a low-power laser compared to pulsed excitation is particularly convenient to study samples with a low threshold of optical destruction. The value η_{PV} determined by this technique varies between 0.04 and 0.01 over the voltage range from 450 to 200 V. The merit of OAS is its simplicity and that there is no need to measure the absolute radiation intensity and the number of free charge carriers in the sample.

The combination of picosecond spectroscopy with the OA detection was realized in [8.65,66]. The idea is to registrate the integrated OA signal formed by two picosecond pulses. The time delay between these pulses was varied. Similar techniques allow to probe fast relaxation processes with time resolution, limited only by the optical pulse widths even though the OA detection mechanism is much slower.

9. Laser Optoacoustic Analytical Spectroscopy

One of the most valuable and important applications of LOAS is the analysis
of trace quantities of atoms and molecules. The efficiency of OAS in solving
such tasks is understood by its high concentration sensitivity (to 0.1 - 1 ppb),
relative simplicity, linearity over a wide dynamic range (10^4 - 10^5), relatively fast response (tens of seconds), comparatively small volume of sample
(several cm^3 for gases under normal conditions), an accuracy sufficient for
many applications (to 5 - 10%). Finally it has the added advantage that can
be fully automated.

9.1 Analysis of Molecular Traces in Gases

The most effective method in detecting molecular traces in different gas media,
including air is OAS. Interest in this problem is stimulated greatly by increasing demands to control atmospheric pollution [9.1,2].

9.1.1 Laser Optoacoustic Gas Analyzers

Basically, there is almost no difference between the schemes of laser optoacoustic spectrometers (LOAS) and laser optoacoustic gas analyzers (LOAGA).
For example, all the LOAS described in Sect.7.1.1 can operate as gas analyzers.
The use of frequency-tunable lasers in these instruments considerably improves
the OA spectral selectivity and makes it possible to identify single microcomponents in gas mixtures by high-resolution spectra. The parameter to be
measured in LOAGA is the absorption coefficient while in LOAS it is the micro-
impurity concentration. This results in a number of specific features both
in the schemes of LOAGA and in the technique of measurement and calibration
of these instruments. Firstly, LOAGA can be divided into three basic types:
working, etalon, and unique instruments.

a) Working LOAGA

Working LOAGA are constructed for highly sensitive local control, with a
measured accuracy (5 to 20%) of just one or several types of molecules at
different points or in mobile units. These must be rather inexpensive, simple
and reliable instruments which allow automatic operation. Such LOAGA must
constitute the least number of elements: a laser, a modulator, an optical
system and a measuring OA detector. To eliminate the fluctuations in laser
power one should either stabilize the power at a given level or introduce
a reference channel with an additional radiation detector; the latter is much
simpler.

In working LOAGA it is possible to use simple laser designs operating at
fixed frequencies or tuned over a narrow spectral range. Especially convenient
in such tasks are gas lasers with discrete tuning to single vibrational-rotational lines of various molecules (CO_2, CO, HF, etc.). As a rule, to increase rapidity of analysis it is necessary that a gas-flow OA cell should
be used. In detecting microimpurities at an ultimate level, for example in
background monitoring of the atmosphere, the OA cell can be placed inside the
laser cavity.

b) Etalon LOAGA

Etalon LOAGA has an increased measuring accuracy (up to 0.5 - 2%) so that etalon gas mixtures can be calibrated. Such instruments are usually constructed
for operation under stationary conditions. The measuring accuracy can be increased by using special circuits and the stabilization of basic parameters
(laser and OA detector), and furthermore by employing appropriate technique
of LOAGA graduation and calibration. Figure 9.1 illustrate one such LOAGA
intended for a He-Ne laser at $\lambda = 3.39$ μm to measure the concentration of CH_4
in different gases in the concentration range from 10^{-4} to 1%, accurately to
2 - 4% [9.3,4]. The specific feature of this scheme is the use of two OA detectors located in series along the optic axis: a reference OA detector with
a definite concentration of CH_4 (no higher than 0.1%) and a measuring detector. Such a scheme essentially reduces (up to an order) the influence of the
following factors on the measuring accuracy: fluctuations of laser power and
frequency, modulation-frequency instability and slow variations in ambient
temperature. To improve reproducibility, LOAGA contemplates microphone sensitivity, calibrates directly during measurement by forming an acoustic signal in the OA cell chamber from the internal piezoelectric emitter.
Potential changes in microphone sensitivity are included by normalizing the

230

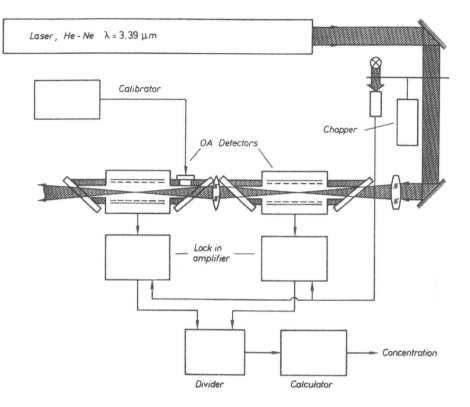

Fig. 9.1. Scheme of a laser optoacoustic gas analyzer (LOAGA) with increased accuracy of measurement [9.3]

OA signal from absorption in the medium to the signal from the internal piezo-element. Such calibration enables reproducibility of measurements up to 2%.

The graduating characteristic of the LOAGA is presented in Fig.9.2. It was produced by measuring the OA signal in mixtures of CH_4 and air, with successive dilution of the original mixture having a high concentration of CH_4 in air. The nonlinearity in the region of high concentration of CH_4 is explained by a nonuniform distribution of the absorbed energy along the OA detector. The graduation characteristic of the instrument in the region of high CH_4 concentrations becomes less nonlinear when the reference OA cell is placed before the measuring one. The bend in the region of low concentrations is conditioned by the presence of CH_4 in the air used to dilute the mixture. The microphone sensitivity can also be calibrated by thermophones, using a plate with the metallic coat heated by current. For an elongated OA detector it is more convenient to use a thin wire located along the chamber wall [9.5].

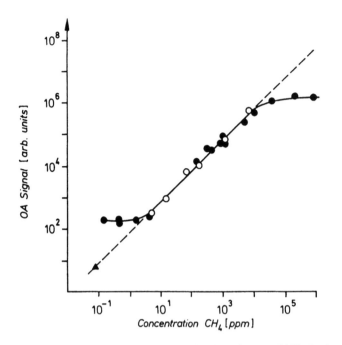

Fig. 9.2. Graduating characteristic of the LOAGA designed for analysis of methane in air. (•): Calibration by mixtures prepared under laboratory conditions; (o): calibration by mixtures tested independently with a chromatograph accurate to 2 - 4%; (Δ): estimate for the level of background signals from the windows [9.3]

c) Unique LOAGA

Unique LOAGA is designed for selective analysis of microimpurities in multi-component mixtures as part of a stationary universal multichannel gas analytic system. In such systems we may use complicated nonstandard equipment, for example, lasers with a large tuning range, particularly IR lasers on HF, DF, N_2O, CO, CO_2 and their isotopic modifications or high-pressure lasers with fine frequency tuning. To widen the scope of LOAGA one can combine them with other methods and instruments. Using an OA detector as a chromatographic detector, for example, enables the analytical selectivity to be increased essentially by producing a two-dimensional OA spectrochromatogram (Sect.9.5). The stationary work condition of such LOAGA allows the parameters of the gas mixtures in the OA detector to be varied over a wide range both by pressure and temperature, which considerably increases the informative capacity of the measurements.

9.1.2 Concentration Sensitivity in Various Experiments

Table 9.1 contains the most important results of applying OAS to detecting microimpurities in gases. Of most analytical interest is the IR spectral region, approximately from 2 to 20 μm, since many gaseous impurities to be detected have strong vibrational-rotational absorption lines in this region.

High concentration sensitivity of OAS was first demonstrated by KREUZER in [9.11] when he detected methane in nitrogen using a He-Ne laser tunable over a small range by a magnetic field with $\lambda = 3.39$ μm, with the laser power 20 mW. According to his results, the sensitivity of a nonresonant OA detector with a cylindrical condensor microphone allows concentrations of methane of up to 0.01 ppm to be detected.

Sensitivity of the same order was obtained in [9.16] in detecting NO in air with a tunable spin-flip laser of 15 mW power and pumped by a CO laser. In this case measuring OA spectra with rather high resolution (about 0.1 cm^{-1}) allows the absorption lines of NO to be identified with assurance against the background of the absorption bands of H_2O (Fig.9.3). Such a technique, in particular, has allowed estimating the content of NO in the air sampled near a highway and directly from a car exhaust; it equals 2 and 50 ppm, respectively.

This technique has successfully detected NO molecules in a concentration of up to 10^8 cm^{-3} at 28 km altitude using a stratostat-borne LOAS [9.17]. The measurements were carried out by tuning the spin-flip laser frequency to two

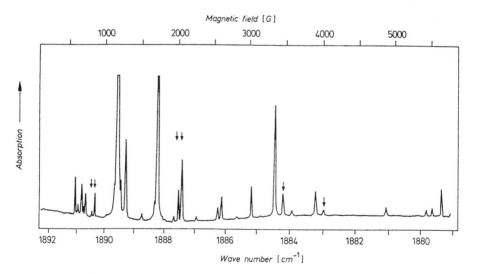

Fig. 9.3. Spectrum of part of the absorption band of the H_2O molecule near 1886 cm^{-1} with NO admixed in a concentration of 20 ppm at a total pressure of 76 Torr. The lines of NO are marked with arrows [9.26]

Table 9.1. Results of experiments to detect molecules with LOAS

Laser	Operating conditions* (AM)	Type of OA detector**: resonant(R) and nonresonant(NR)	Threshold sensitivity [(Jα)·W·cm⁻¹]	Threshold absorption coefficient α_{min} [cm⁻¹]	Molecules detected	Threshold of detection C_{min} [ppb] for S/N = 1	References
Dye laser λ=0.29-0.31 μm	AM	R	$4 \cdot 10^{-10}$	$7.5 \cdot 10^{-9}$	SO_2 in N_2	0.12	[9.6]
Ar laser λ=0.48 μm	AM	R			NO_2 in air	20	[9.7]
Dye laser λ=0.48-0.62 μm	PP	NR	$2.4 \cdot 10^{-9}$		ditto	40	[9.8]
Dye laser λ=0.57-0.62 μm	AM	R	$2.5 \cdot 10^{-8}$	10^{-7}	ditto	10	[9.9]
HF laser λ=2.7-2.9 μm	MP	NR	10^{-9}(J·cm⁻¹)	10^{-8}	HF, NO, CO_2 in N_2	10^3	[9.10]
He-Ne laser λ=3.39 μm	AM	NR	10^{-9}	$1.6 \cdot 10^{-4}$	CH_4 in N_2	200	[9.11]
" " "	AM	R	10^{-8}	10^{-6}	CH_4,n-butane in N_2	10^3	[9.12-14]
CO laser λ=4.75 μm	AM	NR, R	$5 \cdot 10^{-9}$, $5 \cdot 10^{-10}$	10^{-6}, 10^{-7}	CO in N_2	10^3, 10^2	[9.15]
spin-flip laser λ=5.5 μm	AM	NR	10^{-9}	$5 \cdot 10^{-7}$	NO in air	0.1-10	[9.16-18]
CO laser λ=5.2-6 μm	AM	NR	$4 \cdot 10^{-9}$	$5 \cdot 10^{-7}$	NH_3, NO, NO_2 etc.	0.1 for NO_2	[9.19]
CO_2 laser λ=9.2-10.8 μm	AM	NR	ditto	ditto	C_2H_4, benzene,etc.	0.2 for C_2H_4	

Laser	Operating conditions* (AM)	Type of OA detector**: resonant*(R) and nonresonant(NR)	Threshold sensitivity $[(J\alpha)$ $\cdot W \cdot cm^{-1}]$	Threshold absorption coefficient α_{min} $[cm^{-1}]$	Molecules detected	Threshold of detection C_{min} [ppb] for $S/N = 1$	References
CO laser $\lambda = 6\mu m$ and					Explosive vapor N6,	$2.4 \ 10^3$ (for EGDN)	[9.20,21]
CO_2laser $\lambda = 9 - 11\ \mu m$	AM	NR	10^{-8}	10^{-7}	EGDN, DNT	$0.28 \ 10^3$ (for N6)	
CO_2 laser	AM	R	$1.5 \cdot 10^{-9}$	$1.5 \cdot 10^{-9}$	SF_6 in N_2	0.01	[9.22]
" "	AM	R	10^{-8}	$9 \cdot 10^{-8}$	NH_3 in N_2	2.8	[9.23]
	AM	NR	$6 \cdot 10^{-10}$	10^{-8}	NH_3 in N_2	0.3	[9.25]
CO_2 wave-guide laser	AM	R	10^{-8}		C_2H_4 in N_2	20	[9.24]
CO_2 high-pressure laser	MP	NR	10^{-9} $(J \cdot cm^{-1})$	10^{-8}	HDS in H_2S	100	[9.25]

* AM: amplitude modulation; MP: monopulse regime, PP: pulsed-periodic regime

** The parameters of resonant OA detectors are presented in Table 5.3

adjacent absorption bands of NO at about 1887 cm (Fig.9.3). Such a technique makes it possible to estimate the daily variations of NO and H_2O, which is important for studying stratospheric ozone destruction. In this experiment the equipment was calibrated automatically during measurements.

Discretely tunable gas lasers on HF, DF, CO, CO_2, N_2O and isotopic modifications of these molecules are rather simple and handy in service. The possibility of applying these lasers is based on the natural coincidence of some vibrational-rotational laser lines with the absorption bands of several molecules. To increase sensitivity, it may be useful in some cases to tune the absorption line to the fixed line of laser radiation via the Zeeman shift as the gas is acted upon by a magnetic field [9.27]. Some results with CO and CO_2 lasers in OA analysis of microimpurities were given in [9.28].

Furthermore, OAS holds promise for detecting complex polyatomic molecules and particularly explosive vapors like nitroglycerine (NG), ethylene-glycol-dinitrate (EGDN), dinitrotoluene (DNT) and trinitrotoluene (TNT). The intense absorption bands of these compounds due to the NO_2 group are concentrated near 6, 9 and 11 - 12 μm. This allows lasers based on CO, $^{12}C^{16}O_2$ and $^{13}C^{16}O_2$ to detect then [9.20,21]. The lowest detectable concentrations it was assumed that four laser lines would suffice to identify each compound. Absorption for these compounds varies from 0.4 to 7 atm^{-1} cm^{-1}. It is difficult to detect these compounds in air because of the background absorption of such atmospheric components as NO, NO_2, CH_4, and H_2O. The spectral region from 9 to 12 μm is mostly free from absorption bands of these molecules. Another difficulty in analying explosive vapors is their strong adsorption by the walls of the measuring cell. Therefore the cell must be made of chemically inert materials like stainless steel and teflon, and it is also desirable to heat the OA cell.

Most of the works listed have been carried out with the simplest OA cell in which the sensitivity is limited mainly by the background OA signals from the windows ($\alpha_{min} = 10^{-6} - 10^{-7}$ cm^{-1}). Such measurements can also be taken with resonant cells. But unlike nonresonant systems, they generally have lower accuracy. The mean accuracy in OA detection of trace amounts of molecules varies from 5 to 20%.

As concerns the high sensitivity, CW laser operation with modulation is most convenient although in certain cases good results can be obtained under pulsed operation, too [9.10,29]. In [9.10], for example, a pulsed HF laser OA detected HF and NO in a concentration of up to 1 ppm.

According to the data from Table 9.1, most air pollutants including NO, NO_2, CO, SO_2, NH_3, C_2H_4, etc., can be detected using OAS, their concentrations approximating the background content of these microcomponents in the atmosphere, that is from 10 ppb to 1 ppm. This offers considerable scope for local monitoring of the most important atmospheric pollutants with LOAGA.

The OA detection of molecules with highly isolated electron transitions (SO_2, NO_2 and NO) can be realized in the visual range as well with dye lasers. Moreover, OAS may be applied to nonpolar molecules, such as H_2, N_2, O_2, as well as to vapors of some metals.

The high sensitivity of OARS (Sect.7.4) means that it is effective not only in studying weak Raman transitions but also in detecting small concentrations of certain molecules. According to the data given in [9.30], at rather moderate values of pulse energy of the pumping laser and the Stokes wave OARS detects CH_4 and CO_2 molecules in a concentration of up to several ppm.

However, OARS is inferior to conventional OAS of resonant absorption in threshold sensitivity by 2 to 3 orders. This is because of a small cross section of Raman scattering compared to that of resonant absorption in the IR region. Besides, OARS is technically complicated, and since the OA signal depends on many parameters, particularly on the degree of spatial and time coincidence of both laser beams, it is difficult to achieve a high measuring accuracy with this method.

The advantage of OARS is its universality, since one measuring setup in the visual spectral range allows, in principle, many molecules to be detected with a sensitivity sufficient for some applications. In conventional resonant OAS it is still quite difficult to do this because of the lack of sufficiently powerful IR lasers which would cover the entire IR spectral range. Good transparency of the gases in the visual spectral range which forms the basis for gas mixtures (O_2, N_2, Ar, etc.) minimizes the background OA signal from resonant absorption of the pump and Stokes waves.

9.2 Analysis of Impurities in Condensed Media

Generally speaking, OAS is less effective in analysis of impurities in condensed media than in gases. This is due to a comparatively high level of background absorption in the solid sample. Nevertheless, there are some tasks which LOAS can solve better than conventional methods.

9.2.1 Impurities in Liquids (Including the Cryogenic Case)

The best threshold sensitivity of OAS can be achieved by direct detection of OA signals from a liquid using a piezoelectric detector in contact with it.

Table 9.2. OAS concentration sensitivity in analysis of impurities in liquids

Detected component	Solvent	Laser	Absorption coefficient[$mol^{-1}cm^{-1}$]	Threshold of detection for S/N=1 PPT*	ng/ml	Ref
β-carotene	chloroform	Ar laser λ=488 nm and 514.5 nm J=0.7 W	$1.02 \cdot 10^5$ $3.7 \cdot 10^4$	12	0.08	[9.31]
cadmium	chloroform	Ar laser λ=514.5 nm J=0.5 W	$7.9 \cdot 10^4$	7	0.01	[9.32]
KMnO$_4$	water	Ar laser λ=514.5 nm	$1.8 \cdot 10^3$		3	[9.33]
benzene	CCl$_4$	pulsed dye laser λ=607.1 nm		0.1%		[9.34]
vitamin A	H$_2$O	pulsed N$_2$ laser λ=0.33 µm		1	$2 \cdot 10^{-3}$	[9.70]
hematoporphyrin	ethanol				0.3	[9.36]
protoporphyrin	ethanol				0.09	[9.36]
chlorophyll b	ethanol				0.3	[9.36]
U(IV)	H$_2$O	dye laser λ=0.414 µm			$8 \cdot 10^{-7}$ M/L	[9.37]
U(VI)	H$_2$O	dye laser λ=0.66 µm			10^{-6} M/L	[9.37]

* 1 ppt = 1 part of admixture per 10^{12} parts of solvent

The results of such a recording scheme to analyze microimpurities in various liquids are collated in Table 9.2. The cell designs described in Sect.3.1.2 can be used in such measurements. In [9.32,35], for example, OA cells were used in the form of hollow piezoceramic cylinders with optically transparent windows on their ends. These cells then act as a measuring chamber with a liquid and a sensitive pressure detector. Usually OA cells are placed in an acoustically and vibrationally isolated chamber to reduce the influence of vibrations and external acoustic interference.

The highest absolute sensitivity of OAS for liquids can be achieved under pulsed operation. In analyzing the trace amounts of elements the choice of laser operating conditions is not so critical, because the threshold sensitivity is limited not by the noise of the electronics but by background absorption in the solvent, which is minimum in the visual range and usually lies between 10^{-4} and 10^{-5} cm^{-1}.

The concentration sensitivity of OAS usually increases as the influence of background absorption in the solvent reduces. Therefore differential or compensating measuring circuits as well as frequency modulation of laser radiation should be used. In [9.31], for example, discretely frequency was modulated through successive commutation of an Ar laser at the 488 and 514 nm. This decreases the background signal in chloroform from $9 \cdot 10^{-4}$ to $2.2 \cdot 10^{-5}$ cm^{-1}, which corresponds to the detection threshold of β-carotene of about $9 \cdot 10^{10}$ mol/cm^3 or 12 ppt. Sensitivity of the same order was obtained in detecting the extraction of Cd in chloroform using amplitude-modulated radiation of an Ar laser [9.32]. Output-signal linearity has been demonstrated within three orders, the concentration of Cd varying from 0.05 to 50 ng/ml.

Thus, the concentration sensitivity of OAS when analyzing nonfluorescent elements with high absorption in the visual range turns out to be adequate and this makes it promising to be applied to detect trace amounts of impurities in different liquids.

In Sect.8.1.2 we discussed the application of OAS to spectroscopy of cryogenic liquids. According to the first experimental estimates [9.38,39], the threshold sensitivity of a cryogenic OA cell with a piezoelectric detector under pulsed CO_2-laser operation was about 10^{-7} $J \cdot cm^{-1}$. The absorption coefficient for many molecules in the gas phase between 10 and 100 $cm^{-1} \cdot atm^{-1}$ increases by three orders during gas condensation. Therefore, even with a rather moderate pulse energy of up to 10 mJ the concentration sensitivity of OA cryogenic spectroscopy is very high and comes to 0.1 - 1 ppb.

So it is quite possible to cool and dissolve the gas mixture to be analyzed in cryogenic liquids. Compared to the analysis of the same impurities in the gas phase, the main advantage of cryogenic OAS is higher sensitivity (from one to two orders) with better selectivity, since on cooling the vibrational spectrum is narrowed and the absorption coefficient increases to a maximum.

9.2.2 Impurities in Solids

Direct detection of OA signals is sufficiently sensitive to measure absorption in solids at a level from 10^{-5} to 10^{-6} cm^{-1} with up to 1 W laser power (Sect. 8.1.3). Such high sensitivity with a small sample length (0.5 to 1 cm) enables this technique to analyze small concentrations of impurities in solids. This problem is especially pertinent to solid-state electronics in the production of pure semiconductors. The optimum range for analysis of, say, a Si crystal is the spectral range from 1.1 to 2.5 μm in which the natural absorption in Si is small enough. It was shown in [9.40] that with the threshold of absorption measurement only 10^{-3} cm^{-1}, OAS detects impurities like Au, Zn, S and

Pt at a level of 10^{13} cm^{-3}. The technique described may also be very useful in determining the purities in different IR crystals (ZnSe, NaCl, BaF$_2$, etc.) by measuring the respective absorption coefficients of powerful laser lines.

9.2.3 Impurities on Surfaces

High sensitivity of OAS to surface absorption enables it to be applied to analyze small quantities of substances in different phases (in the form of powder, film, solution, etc.) on the sample surface. This is useful for studying oxidation, corrosion, catalysis, etc. In taking such measurements it is very important to choose an optimum spectral range in which the component to be detected gives the maximum possible OA signal at a minimum level of background OA signal from radiation absorption in the bulk. So the sample must be either slightly absorbing or highly reflecting.

The analytical possibilities of the OA method were demonstrated in [9.41] by detecting powdery arsenaso III mixed with Al$_2$O$_3$ powdery. The radiation source was a He-Ne laser at the $\lambda = 632.8$ nm in which powdery Al$_2$O$_3$ absorbs radiation very slightly. The minimum quantity of the absorbing component in the mixture thus detected was $1.8 \cdot 10^{-5}$ g.

Rather good threshold sensitivity can also be obtained at a comparatively high level of background OA signal. This can be illustrated by [9.42] where ammonia sulfate $(NH_4)_2SO_2$ in aqueous solution in concentrations from 20 g/kg (with one-channel scheme) to 0.1 g/kg (with differential scheme) was detected with a CO$_2$ laser, of up to 0.1 W power. The differential signal was linear as the ammonia sulfate concentration varied within two orders (Fig.9.4). The detection threshold in the differential circuit was limited mainly by intensity fluctuations at $\lambda_1 = 9.25$ μm and $\lambda_2 = 10.7$ μm. Detection of ammonia sulfate is important because it can be found in aerosols in the atmospheric and plays an important part in the global thermal balance.

Quantitative and qualitative analysis in thin-layer chromatography may become an important area of application. The quantitative composition of separated substances can be studied directly on the chromatographic plate by OAS. This was demonstrated even with nonlaser sources in the visual spectral range by studying spots of benziliden-acetone with different concentrations [9.43] or through quantitative determination of fluorescein in silica on a glass substratum [9.44]. Usually the differential technique at two wavelengths is used in such measurements. The detection threshold obtained for fluorescein was 20 ng. There is no doubt that a more powerful tunable laser will enable us to reduce this threshold and improve the technique of background-signal subtraction from the plate itself.

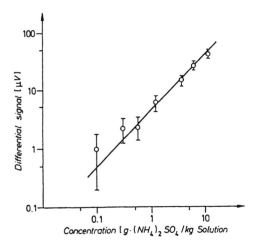

Fig. 9.4. Differential OA signal
as a function of the concentration
of ammonia sulphate in water with
modulation of laser radiation at
the wave-lengths P(24) (9.25 μm)
and P(28) (10.675 μm) [9.42]

9.3 Isotopic Analysis of Molecules

An important area of application may be the analysis of molecules with different
atomic isotopes (C, N, O, H, B, etc.), which is essential for many tasks in
biology, medicine, agrochemistry, geology, etc. Hence the isotopes to be studied
must be in compounds convenient for OA analysis, e.g., CO_2, H_2O, NH_3, BCl_3,
etc.

There are two trends in OA isotopic analysis of molecules. First, measurement of the concentration of specific isotopic molecules in natural isotopic
mixtures; and second, precise measurement of isotopic ratios and their small
variations. The advantage of OAS here lies in the possibility of detecting
molecules with rare natural isotopes (with small concentrations up to 10^{-6}%)
in comparatively small volumes of gas samples (10^{-2} to 10^{-1} cm^{-3}) at normal
conditions: the OAS method is simple, efficient, and rapid. Besides, it is
characterized by nondestructive control and the possibility of fast replacement of measuring cells in analyzing chemically aggressive gases. The accuracy currently obtained (3 to 5%) suffices for many tasks. Besides, it is
still possible to improve it considerably.

Better results can be obtained by analyzing molecules in the gas phase in
the IR range, because many such molecules have intense vibrational-rotational
lines and a comparatively large isotope shift. This enables us to use well-
developed methods of IR spectroscopy with accessible and handy lasers for
their detection. But in a number of cases cryogenic OAS or OARS may prove
more useful. Some practical possibilities of OAS isotopic analysis will be
considered in more detail below.

241

9.3.1 Isotopically Selective Detection of Molecules

Compared to the problem of detecting molecular traces in different media, the detection of isotope molecules is somewhat simpler since in most cases we have to analyze rather simple mixtures of two or several absorbing components. With a large isotope shift the most simple task is selective detection of different isotopes. For example, the vibrational spectra of H_2S and HDS molecules with a width of about 200 cm^{-1} in the ν_2 band are overlapped a little because of the large isotope shift, as deuterium is substituted for hydrogen ($\Delta\nu_{is} = 157$ cm^{-1}). This factor makes it possible to detect HDS molecules in their natural isotopic mixture with H_2S (content of HDS is $4.5 \cdot 10^{-2}$%) when the CO_2 laser is tuned to the long-wavelength edge of the absorption band of H_2S where the absorption of H_2S is minimum [9.25] (Fig.9.5). In some cases it is advisable to dilute the mixture under analysis with a buffer gas to increase the spectrophone sensitivity. The optimum relative pressure is determined from the equilibrium condition of the elasticity of gas and sensitive microphone membrane. For HDS molecules the optimum relative pressure of N_2 equals about 70 torr.

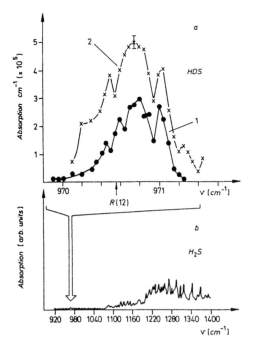

Fig. 9.5. (a) Absorption line of HDS in a natural isotopic mixture with H_2S measured with an OA spectrometer with a pulsed fine-tunable high-pressure CO_2 laser near the line R(12). (1): pure H_2S at a pressure P = 2.5 Torr; (2): mixture $H_2S + N_2$, P_{H_2S} = 2.5 Torr, P_{N_2} = 75 Torr. (b) Absorption spectrum of H_2S measured with a conventional absorption spectrometer [9.25]

Some molecules in the IR range feature rather a wide vibrational-rotational spectrum at a relatively small isotope shift. Then, to detect their isotopic modifications, we should use narrow rotational-vibrational lines and narrow-band tunable lasers. A high-pressure CO_2 laser enables selective detection of the $^{15}NH_3$ molecules in a natural concentration of about 0.3% since it can be tuned to the isolated line sP(1, 0) with $\nu = 943.07$ cm^{-1} near which there are no strong lines from $^{14}NH_3$ [9.25].

In some specific cases OA isotopic analysis can also be performed with discretely tunable gas lasers. According to [9.45,46], for example such practically important molecules as $H_2^{18}O$, $HD^{16}O$ and $^{15}N^{16}O$ can be detected with a CO laser and D_2O and $^{12}C^{13}CH_4$ with a CO_2 laser. A DF laser detects HDO molecules in their natural concentration in H_2O by [9.47]. Here the isotopic analysis of HDO should be performed at the laser line with $\nu = 2772.45$ cm, where the absorption factor in HDO is maximum (0.686 $atm^{-1} \cdot cm^{-1}$), while in H_2O it is negligible.

The accuracy obtained in detecting HDO at atmospheric pressure using a single-channel nonresonant spectrophone was 10%. To identify the absorption in the molecules under analysis, measurements were taken for the OA absorption spectra of H_2O and for the enriched mixtures with HDO and D_2O.

Through successive tuning of the laser to the absorption line of the corresponding isotopes and by comparing the resulting signals one can determine not only the concentration of each isotope but also the ratio of these concentrations, i.e., the isotopic ratio. However, with better accuracy and speed this problem can be solved using some special schemes, considered below.

9.3.2 Measurement of Isotopic Ratios and Their Variations

Figure 9.6 shows a scheme for measuring the absolute value of the isotopic ratio and its small variations [9.25,48], based on measuring with periodic comparison the OA signals in two spectrophones through which radiation is passed at two frequencies coinciding with the intense absorption bands of the corresponding isotopes. This scheme measures with CO_2 lasers the isotopic ratio of C in $^{12}CO_2$ and $^{13}CO_2$ as well as the isotopic ratio of B in $^{10}BCl_3$ and $^{11}BCl_3$ and their isotopic modifications.

The beams of both lasers are alternately directed into two spectrophones by a mirror modulator with the commutation frequency ω_0. The first spectrophone is filled with a mixture with known etalon isotopic ratio, and the second with a mixture of unknown isotopic ratio. The first spectrophone acts as reference, whose error signal controls the power of one of the lasers through the servoloop and a servoelement, an electrooptical modulator, for

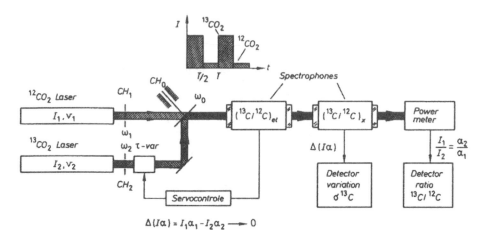

$$\Delta(J\alpha) = I_1\alpha_1 - I_2\alpha_2 \longrightarrow 0$$

<u>Fig. 9.6.</u> Scheme for OA measurement of isotopic ratios in molecules [9.25,48]

example. As a result, the power levels of the lasers are stabilized in a ratio inversely proportional to the ratio of absorption coefficients and hence to the concentration ratio of the corresponding isotopes. Measuring the ratios of these powers gives information on the absolute value of the isotopic ratio $^{12}C/^{13}C$. To increase the measuring accuracy of $^{12}C/^{13}C$ (up to 0.5 - 1%) the power of each laser should be recorded by the same photodetector through additional modulation of each radiation by the frequencies ω_1, $\omega_2 \gg \omega_0$, with subsequent extraction of the signal from each laser beam at the photodetector output using selective amplifiers.

In measuring small relative variations of the isotopic ratio between the two mixtures we must measure the error signal from the second spectrophone since it carries information on the variation δC. This scheme allows very small values of δC to be detected due to stabilization of the laser power ratio as well as elimination of the influence of laser, frequency fluctuations and variations in microphone sensitivity on the measuring accuracy. Threshold sensitivity can be estimated from (cf. list of notations)

$$\delta C_{min} \approx \frac{1}{J}\left[(J\alpha)_{min}^2 + \Delta(\overline{J\alpha})_{min}^2\right]^{1/2} \quad , \tag{9.1}$$

where $[\Delta(\overline{J\alpha})_{min}^2]^{1/2}$ is the differential threshold sensitivity of the servo-system; $(J\alpha)_{min}$ is the threshold sensitivity of the spectrophone. In the ideal case the condition

$$\left[\Delta(\overline{J\alpha})_{min}^2\right]^{1/2} \leqslant (J\alpha)_{min} \ll J\alpha \tag{9.2}$$

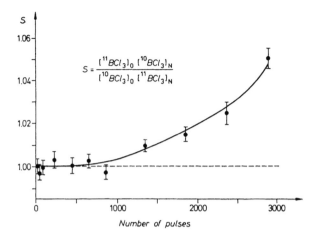

Fig. 9.7. OA measurement of separation factor S of the ^{10}B isotope versus the number of laser pulses as $^{11}BCl_3$ is excited [9.48]

should be met. For the typical measuring conditions $[(J\alpha)_{min} = 10^{-8}$ W $\cdot$ cm^{-1}, $J\alpha = 10^{-2}$ W $\cdot$ cm$^{-1}]$, when (9.2) is realized, from (9.1) we get $\delta C_{min} \approx 10^{-4}\%$. Thus, the ultimate sensitivity of the scheme is high enough. In practice, the detection threshold δC_{min} is determined by technical fluctuations in laser power and frequency. To illustrate the potential of the scheme, Fig.9.7 presents the results of its application to determine the enrichment of ^{10}B isotopes during their separation by isotope-selective photodissociation of BCl_3 molecules by a high-power pulsed CO_2 laser. Two CO_2 lasers at the lines R(36) (985.49 cm^{-1}) and P(18) (945.98 cm^{-1}) of the band $00^{\circ}1 - 10^{\circ}C$ coinciding with the absorption bands of $^{10}BCl_3$ and $^{11}BCl_3$ were used. With their help it was possible to measure the separation factor of ^{10}B as a function of the number of laser pulses of excitation of $^{11}BCl_3$.

In isotopic OA analysis it is most advisable to use nonresonant spectro-phones since resonant ones have a shortcoming, i.e., a comparatively large volume of the chamber and a strong dependence of the OA signal on composition temperature and modulation frequency which causes the measuring accuracy to decrease to 10 - 20%. Laser OA analyses of isotopic molecules using IR absorption spectra are not universal at present and do not have such a high measuring accuracy as mass-spectrometric analysis. But for special tasks their advantages are simplicity, high sensitivity, short measurement time and the possibility of automatic control in mobile setups.

9.4 Analysis of Aerosols

Quantitative and qualitative analyses of aerosols in various media are important for atmospheric optics, particularly to control air pollution or to estimate the loss of laser radiation in long optical paths, in chemistry, and biophysics to study cloudy or colloidal solutions, etc. Most existing methods of analyzing the quantitative composition of aerosols are indirect and produce significant measurement errors. Analysis using the extinction coefficient (scattering and absorption) and scattered radiation suffers from the main problem of correctly separating scattering from absorption effects in aerosols as well as allowing for the influence of aerosol dimensions on the level of scattered radiation. These shortcomings are not typical of OAS which directly measures radiation absorption in aerosols by detecting the OA signals arising in heat transfer from the radiation-heated aerosol to the medium. The main advantage of OAS here is that it enables measuring weak absorption with low concentrations of aerosols at natural conditions, the influence of scattered radiation being negligible. This is due to a small transfer of absorbed power from the walls to the gas (about 10^{-3}) as well as by the possibility of reducing the influence of scattered radiation, for example, by decreasing the cell length. With OAS absorption by aerosols in both gases and liquids can be measured.

9.4.1 In the Gas Phase

The formation of OA signals in gases from suspended aerosols is complicated. Depending on the measuring conditions, the following effects can contribute to the generation of OA signals: a) heat transfer from aerosols to the gas through thermal conductivity; b) piston action of the thin gas layer in contact with an aerosol particle, or c) purely mechanical expansion of heated particles.

Radiation absorption by aerosols depends essentially on their dimensions. For example, Rayleigh scattering is dominant when $r_p \ll \lambda$ is valid, r_p being the characteristic dimension of particles. Here the absorption, in a first approximation, is proportional to the total mass of the particles and does not depend on their dimensions. This makes OAS differ from other methods, specifically, from electron microscopy and the method utilizing scattered radiation.

The characteristic time of heat transfer from aerosols to the gas is essential for the kinetic of the OA signal. In case of uniform spherical particles with radius r_p the time of cooling after the uniform heating is [9.49]

$$\tau_p = \frac{\rho_p C_p r_p^2}{3K_g} \ ,$$

(9.3)

where ρ_p, C_p are the density and thermal capacity of a particle, and K_g is the gas thermal conductivity. For different particles in air (chalk, sand, clay, water, etc.) with an average size $\bar{r}_p = 0.25$ μm, the time τ_p lies in the range $(1-5) \cdot 10^{-6}$s.

The process of heat transfer is most effective when $\omega\tau_p \ll 1$ and when the entire volume of a particle takes part in the production of the OA signal. When this condition is not valid, the efficiency of heat transfer in the particle-gas system reduces, which is equivalent to a decrease in the OA signal. Then the OA signal depends on the dimensions of the particles. This, in particular, enables one to detect the absorption from particles of the required dimensions by choosing a specific modulation frequency. For example, with a resonant spectrophone at 4000 Hz, the OA signal is formed mainly by particles of no more than 3 μm dimension in [9.50].

With $r_p \ll \tau_T$, τ_T being the thermal constant of the spectrophone, the instrumental constants for molecular and aerosol absorption coincide. Thus the spectrophone, designed for analysis of aerosols can be calibrated by etalon molecular absorption in gas. Otherwise, the calibration must be carried out using etalon aerosol mixtures.

Since $\tau_p \propto r_p^2$ we can estimate the average size r_p of the particles by measuring the parameter τ_p obtainable by measuring the leading edge of a pulsed OA signal or the shape of frequency resonance of the OA signal by analogy with the determination of the time of vibrational-translational relaxation in gases. The main problem in measuring weak absorption by aerosols in natural media, particularly in air, is eliminating the influence of attendant molecular absorption in the gas. With one spectrophone in use, this can be done by comparing the OA signals during successive filling of the spectrophone at first with a gas and aerosols and then only with the pure gas, the aerosols being removed by special filters. A more convenient scheme which eliminated the molecular background practically on a real-time scale is differential measuring with two spectrophone placed in series along the optical axis. The mixture with aerosols flows through one of these spectrophones, then through the filters which remove the aerosol, to the second spectrophone.

The potentials of the differential scheme were demonstrated in [9.51] by measuring the absorption by quartz dust in air at the CO_2 laser lines. The measurements were performed with resonant spectrophones operating at the longitudinal modes of a small-diameter internal resonant chamber (Fig.5.3c). One

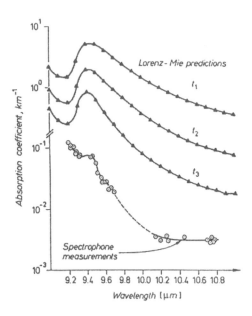

Fig. 9.8. Absorption in quartz dust at the CO_2 laser lines estimated theoretically by an approximation of Lorentz-Mie theory and experimentally by the OA method. Average radius of the particles is 0.015 µm. Average concentrations are 35 mg/m^3 ($t_1 = 11$ min), 10 mg/m^3 ($t_2 = 27$ min) and 4 mg/m^3 ($t_3 = 47$ min) at the different times. The experimental dependence corresponds to the interval between t_2 and t_3 [9.51]

Labels in figure: Lorenz-Mie predictions; t_1; t_2; t_3; Spectrophone measurements; Absorption coefficient, km^{-1}; Wavelength [µm]

spectrophone was filled with pure air, and the mixture with quartz dust continuously flowed through the second one which had no external windows. The special filters controlled the concentration of particles with time. The concentration and their size distribution were determined independently with an electron microscope and a counter which operated on the scattering effect. Measurement of OA signals from two spectrophones reduced the influence of molecular absorption in the air mainly by H_2O and CO_2 to 10^{-7} cm^{-1}.

The results for absorption in quartz dust using such a scheme are given in Fig.9.8. It may be seen that the theoretical results are in satisfactory agreement with experiment. Some discrepancy was attributed in [9.51] to the deviation of the particle form from sphericity, the effects of double refraction and uncontrollable inclusions in quartz.

A similar differential scheme, with nonresonant spectrophones though, was used in [9.52] to measure absorption in acetylene and diesel smoke at the lines of Ar and CO_2 lasers. Quantitative determination of the absorption coefficient α_A in aerosols was based on

$$ P = K \frac{\alpha_A}{\alpha_E} \frac{1}{L}\left[1 - \exp(-\alpha_E L)\right] J \quad , \tag{9.4} $$

where L is the cell length, α_E is the extinction coefficient, and K is the instrumental constant. The mass coefficient of absorption, $\varkappa_M$, which can be

found from $\varkappa_M = \alpha_A / M_A$, M_A being the mass concentration of aerosols, was estimated. In quantitative measurements the spectrophone sensitivity was calibrated
by molecular absorption in gas (by NO_2 for the Ar laser and by trichloroethylene for the CO_2 laser) and the coefficient of extinction per pass was determined independently. The mass coefficients of absorption measured in acetylene smoke tuned out to be 8.3 ± 0.9 and 0.76 ± 0.1 m^2/g at $\lambda = 0.5145$ and
10.6 μm, respectively. Within the limits of accuracy the OA signal was proportional to the mass concentration of aerosols up to 200 mg/m^3. Extrapolating these results to the region of small concentrations, with a residiual
molecular absorption coefficient of 10^{-5} cm^{-1} in the gas, gives the detection
limit of aerosols at $\lambda = 0.5145$ μm the value of $(M_A)_{min} \approx 10^{-6}$ g/m^3.

In OAS of aerosols it may be difficult to allow for the influence of particle
sedimentation on some part of the spectrophone. But, according to the conclusions
from [9.52], when the flow rate reaches 1 ℓ / min, this effect may be neglected
within the limits of error, ± 5%. The potential influence of this effect can
also be reduced by using a spectrophone with a large ratio of its volume to
the wall surface, specifically a resonant cylindrical spectrophone with a radial or azimuthal mode. In [9.50], for example, such a spectrophone with a
rather high threshold sensitivity $(J\alpha)_{min} = 2 \cdot 10^{-9}$ $W \cdot cm^{-1}$ at 4000 Hz was used
to measure absorption in cigarette smoke at the line of an Ar laser. The ratio
α_A / α_{SC} measured, α_{SC} being the scattering coefficient, was equal to 7. The
level of the background signal from scattering corresponded to absorption of
about 10^{-7} cm^{-1}.

9.4.2 In Solid-Phase Sedimentations

Another measurement technique first collects atmospheric dust and aerosols by
filters, and then places the particles on the surface of a slightly absorbing
plate for OAS analysis. The analysis sensitivity may increase due to preliminary accumulation of particles and elimination of the effects connected with
particle sedimentation on the cell walls during their analysis in gas phase.
But here it becomes difficult to perform quantitative determination of the
absorption coefficient, because the measuring conditions deviate from the real ones and there is uncontrollable heat distribution between the substratum
and the dust. In essence, this problem is similar to studying powder with OAS.

To overcome the background signal from radiation absorption in the windows
and substratum we should use a differential measuring scheme with two OA cells.
In [9.53], for example, a two-chamber longitudinal cell with one common window
was used. One of the chambers had a NaCl plate with dust on its surface, and

the other had a plate with a clean surface to compensate for the background from radiation absorption by the material of the plates. Spectrally pure nitrogen was let into the chambers to exclude molecular background. The absorption spectrum of atmospheric dust at the CO_2 laser lines with average particle size of several μm is similar to the dependence presented above in Fig.9.8.

In [9.54] the measuring technique described is applied to other particles, specifically to quartz dust and acetylene black, with laser radiation at three wavelengths 1.06, 9.4 and 10.6 μm. The results of calibrating the OA cell by three independent methods — with an etalon-absorption gas $(CO_2 + N_2)$, a high-absorption sample like acetylene black, and gas heating from a nichrome wire — coincide within 25%, and the quantitative results are in satisfactory agreement with the measurements from the different techniques.

In [9.55] OAS was used to measure the absorption spectrum of asbestos in the range of HF-laser tuning from 3800 to 3450 cm, where a strong absorption band of hydroxyl group doublet lies (Fig.9.9). Asbestos is a widespread pollutant which can be found in water media as chrysotile fibers with about 250 Å diameter and in the air as fibers with a diameter of several μm. For studying asbestos in air a differential scheme with two cells was used. In one cell asbestos was applied to the output-window surface. As seen from Fig.9.9, the characteristic doublet manifests itself well in the OA spectrum.

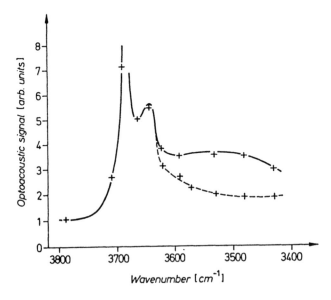

Fig. 9.9. OA spectra of asbestos at the HF laser lines. (——): spectrum of natural asbestos; (---): spectrum of degassed asbestos [9.55]

The absorption in the short-wave length region is explained by the influence of water on the surface of the fibers. The detection threshold of asbestos is estimated to be at a level of 0.1 ng/cm^3, which corresponds approximately to its natural content in drinking water in many cities.

9.4.3 In Liquids

Analysis of aerosols in liquids is essential in studying biochemical solutions, for controlling the degree of purity of different liquids, etc. The potential of OAS was demonstrated in [9.56] in determining $BaSO_4$ particles quantitatively in colloidal aqueous solution. The measuring technique comprised an Ar laser and an OA cell with a piezoelectric detector in the form of a hollow cylinder. Measurement of the spectrum of $BaSO_4$ showed its weak spectral dependence, and in particular, a slight decrease of absorption with increasing wavelengths. The linearity of the relationship between the OA signal and the concentration of SO_4^{-2} remained constant as the concentration varied from $2 \cdot 10^{-8}$ to $4 \cdot 10^{-5}$ g/ml, i.e., within three orders. With the background absorption level, in the solvent, of 10^{-5} cm^{-1}, the concentration threshold was about 5 ng/ml for S/N = 1. This value is nearly two orders lower than that detectable with a tur-bidimeter, with a 50 mm long cell.

The colloidal tellurium was determined by means of laser OAS in [9.57]. The concentration of hydrochloric acid, the amount of stannous chloride and the standing time were investigated in order to obtain the most suitable conditions for the reduction reaction. The detection limit and the reproducibility were 2.5 ppb and 3.9%, respectively.

9.5 Selectivity of Optoacoustic Analysis

Very often in estimating the potential of OAS in different cases we should take into account not only its threshold sensitivity and accuracy but also such an important parameter as selectivity. This parameter characterizes the efficiency of OAS for identification of the molecules under study, and its ability to discriminate useful signals from the background. The background in OA cells may be caused by radiation absorption in foreign molecules as well as at elements of the measuring cells. Some methods to decrease the influence of such elements were considered in Sect.5.2. Therefore here we consider, in more detail, an interesting case of selective OAS of molecules in the presence of interfering components. It should be noted, however, that the methods to increase selectivity considered below can also be used in suppressing instrumental background. Especially important is the problem of increasing selectivity in the analysis of multicomponent gas mixtures, e.g., in the control of atmospheric pollution or in biomedical applications.

At present there is no general definition of the selectivity of different methods of gas analysis by which they could be compared. Here the following relation is used as a quantity measure of selectivity:

$$S_x = \frac{C_B}{C_x} \quad , \tag{9.5}$$

where C_x denotes the concentration of the molecules, which can be detected in the presence of background molecules with concentration C_B. In our case the equality of OA signals from background molecules and those to be detected is taken to be the detection threshold. The simplest way of estimating S_x is for one background component. In the presence of several background components the value of S_x also depends on their relative concentrations.

It is convenient to estimate the efficiency of the method to increase selectivity to be considered below, using the concept of the selectivity-increase factor F. This factor can be determined from the value of the increase in the signal-to-background ratio by the use methods of selecting the OA signals formed by the molecules under detection. In practice an increase in selectivity has sense only when the concentration threshold of OAS by the noise of the receiving-recording system is realized.

9.5.1 Enhancement of Selectivity for Binary Mixtures

In binary mixtures it is necessary to detect one type of molecule surrounded by other molecules. This is the most simple situation.

a) Spectral Selection

Generally spectral selection of molecules is a complicated problem that can be solved in different ways for three types of molecules: 1) simplest two- or three-atom molecules and radicals; 2) molecules with rather a small number of atoms and with a well-resolved vibrational-rotational structure; 3) complex molecules without sharp electron and vibrational resonances. The highest detection selectivity is obtainable when simple molecules in simple molecular surroundings are to be detected.

b) Wavelengths Modulation of Radiation

Frequency modulation proves highly effective in detecting molecules with sharp absorption lines against a background of nonselective or slightly selective absorption from an undesirable component. In laser OA instruments two main types of frequency modulation are possible: discrete and harmonic.

Discrete frequency modulation is more easily realizable, for example, with two independent lasers. In practice the first type of modulation can increase selectivity of detection up to two orders, i.e., F = 30 - 100.

Harmonic frequency modulation in which the OA signal is proportional to the derivative of the absorption line contour may be used to suppress the background signal from undesirable components. The recommendations for the optimization of ratio between the OA signals caused by radiation absorption in the background molecules and those under analysis can be found in [9.28].

c) Multifrequency excitation of molecules

Having a priori information on the structure of energy levels which form absorption in the background molecules and in those to be detected, we can increase selectivity through multi-frequency, or simply, two-frequency excitation of the coupled transitions in the molecules to be detected, with energies $h\nu_1$ and $h(\nu_1 + \nu_2)$ having a common sublevel (Fig.9.10). The respective inten-

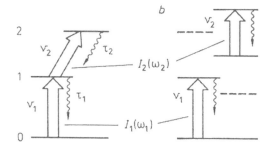

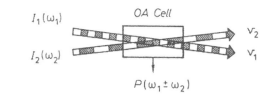

Fig. 9.10 a-c. Two-frequency excitation of the molecules to be detected, at coupled transitions: (a, b) energy structures of the molecules to be detected and the background ones, respectively; (c) irradiation scheme [9.25]

sities are modulated with the audio frequencies ω_1 and ω_2. Because of inevitable background absorption in other molecules whose absorption lines coincide with the ν_1 and ν_2 frequencies, OA signals arise at the audio frequencies ω_1 and ω_2 which do not differ from the OA signals from the molecules to be detected. Yet due to selective two-step excitation of these molecules OA signals with frequencies $\omega_1 \pm \omega_2$ carry information on the concentration of the molecules to be detected only. An increase in selectivity of such a method can be explained by the small probability of coincidence of the laser frequencies ν_1 and ν_2 with the bound transitions of two different types of molecules.

A decrease of the spectral width and the collisional width of the spectral absorption line to the level of the Doppler width permits the ultimate spectral resolution in linear spectroscopy and maximum detection selectivity. Direct measurement of Doppler broadening makes it possible to determine the mass of molecules and hence roughly to identify the type of these molecules.

d) Nonspectral Methods of Selection

The main indication of molecular selection for nonspectral methods of selection may be differences in some parameters of these molecules, such as the saturation parameter, the relaxation time, and their dependence on gas temperature and pressure, differences in multiphoton absorption spectra, different behavior in electromagnetic fields, etc. For simplicity, we consider all possible methods of selection by analyzing a binary mixture of absorbing molecules A and B.

e) Selection Using Saturation Parameters

This type of selection is based on the difference in the saturation parameters for the molecules A and B. Selective detection of molecules A in the presence of background absorption by molecules $B(I_S^B \geqslant I_S^A)$ can be achieved by recording the OA signals on the second harmonic of the modulation frequency. The method is based on nonlinear absorption in the molecules A, which gives rise to the second harmonic during modulation of the laser intensity, whereas the absorption in the molecules B is still linear and does not produce OA signals in the second harmonic [9.11].

f) Selection with Multiphoton Absorption

This method of selection can be applied in an analysis of polyatomic molecules in the presence of background absorption from simple molecules (Sect.7.3.2).

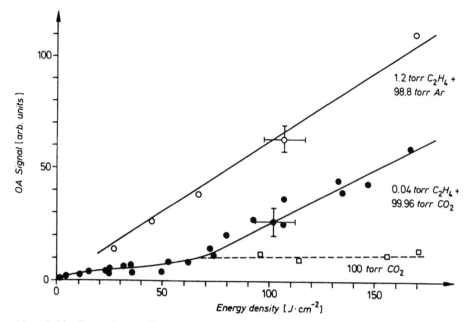

Fig. 9.11. Dependence of the OA signal on CO_2 laser pulse energy fluence at the line $P_9(16)$ for the molecules CO_2 and C_2H_4 and the mixture $C_2H_4 + CO_2$ [9.29]

As the radiation intensity increases, energy absorbed by polyatomic molecules continuously increases on account of multiphoton effects, whereas in simple molecules only the saturation effect shows up. The essence of the method is illustrated in Fig.9.11 which also shows the dependences at the CO_2 laser line for the C_2H_4 and CO_2 molecules as well as for their mixture $C_2H_4 + CO_2$ [9.29]. It may be seen that at high radiation intensities it is possible to increase the detection selectivity and sensitivity for the C_2H_4 molecules in the presence of CO_2 by nearly 7 times. Analysis of the slope of the dependence of the OA signal on radiation intensity allows us to identify the type of molecules with dominant absorption at a chosen intensity.

It is possible to identify molecules using nonlinear and multiphoton absorption spectra, [9.58] for example, the characteristic narrow resonances in the spectra of some molecules [9.59].

g) Selection Using Relaxation Times

This method is based on selection and identification of molecules having different times of vibrational-translational relaxation τ_{V-T} in complex mixtures. It has been shown in Sect.7.5 that a number of techniques determine the value

of τ_{V-T} using OAS. The same techniques can also be used to select and identify molecules by the τ_{V-T} parameter. Under continuous operation, for example,
in analysis of a binary mixture of molecules A and B with the relaxation times
$\tau_{V-T}^A \gg \tau_{V-T}^B$ at a modulation frequency satisfying the condition $\omega \tau_{V-T}^B \geqslant 1$, we
can increase the selectivity factor with respect to the molecules B to the
value $F = \tau_{V-T}^A / \tau_{V-T}^B$. In [9.60] with a CO_2 laser and a resonant cylindrical
spectrophone with a resonant frequency of about 4000 Hz and with a small
amount of aqueous vapor added to CO_2 at 100 Torr, the OA signal is increased
by about one order of magnitude because H_2O influences acceleration of the
relaxation process in CO_2.

h) Temperature Variations

The technical possibilities for modified spectrophone designs (Sect.5.5) enable
measurements over a wide range of gas pressures and temperatures, which in
some cases prove useful in increasing the selectivity of OA analysis. For example, with increasing temperature the selectivity towards molecules which
absorb from excited states increases, and with decreasing temperature the selectivity increases for molecules which absorb from ground states. By discriminating temperature dependences of OA signals by the techniques described
in Sect.7.2, it is possible to identify the type of molecules with dominant
absorption.

With a decrease in temperature the molecular absorption bands become nar-
rower and, as a result, the selectivity increases due to decreased probability
of absorption of the bands from different molecules. This technique is especially effective when the gas is cooled down to its transformation to the liquid
state until the rotational structure in absorption disappears completely. The
selectivity-increase factor may be as high as $F = 10^2 - 10^3$.

Selection methods may also be based on revealing the differences in the
behavior of molecules in electric or magnetic fields, i.e., on the Stark and
Zeeman effects. The benefits of these methods are indicated in experiments on
Stark and Zeeman OAS (Sect.7.1.3). Absorption modulation in alternating fields
seems to be optimum here.

9.5.2 Analysis of Multicomponent Mixtures with Etalon Spectra

It is usually difficult to increase OAS selectivity when analyzing simultaneously several molecular components with overlapping absorption bands especially when these bands are wide and not specific. Then we can apply two approaches.
First, it is possible to measure the OA spectrum over a wide range and to de-

termine the composition of a mixture with the help of the known OA spectra of separate molecular components. Second, we can separate the mixture, e.g., by chromatography, and then detect individual components by their known spectra. In this and following sections we consider both approaches briefly, using a few examples from this field.

Let us assume that the basic gas, not absorbing the laser radiation, contains a mixture of N absorbing gases whose concentrations $C_j(j = 1 \ldots N)$ can be determined by measuring the OA signals at different frequencies $\nu_i(i = 1 \ldots M)$. At each frequency ν_i the output signal from the spectrophone ν_i at low concentrations (absorption), when $\alpha L \ll 1$, is determined by the following linear relation:

$$U_i = K_i J_i \sum_{j=1}^{N} \varkappa_{ij} C_j \quad , \quad i = 1 \ldots\ldots M \quad , \quad j = 1 \ldots\ldots N \quad , \quad (9.6)$$

where J_i is the laser power; K_i is the spectrophone sensitivity; $\varkappa_{ij}$ is the absorption (concentration) coefficient from the j-th component at the i-th laser frequency (dimension in terms of $cm^{-1} \cdot atm^{-1}$ or $cm^{-1} \cdot ppm^{-1}$). By assuming that the absorption factor $\varkappa_{ij}$ from separate components have a low concentration in the mixture, they are assumed additive and so (9.6) is fulfilled. The coefficients of spectral line broadening, times of V-T relaxation, heat conductivity and other parameters which affect the values of K_i and $\varkappa_{ij}$ depend on the properties of the basic gas in the mixture, which is assumed to be unchangeable. The value of the absorption coefficient in the medium at a given line α_i $[cm^{-1}]$ can be determined from U_i and J_i measured with an OA spectrometer, and K_i from calibration. When α_i is determined at all wavelengths, we get a system of linear concentration equations which has the matrix form

$$(\varkappa_{ij})_{NM} \times (C_j)_N = (\alpha_i)_M \quad . \quad (9.7)$$

In (9.7) there is a sequence of matrices of standard absorption coefficient, concentrations and absorption coefficients for the medium.

The system (9.7) with N = M has a solution, and only one, when the determinant of the matrix of the etalon coefficients is not equal to zero, i.e., $\det(\varkappa_{ij})_{NM} \neq 0$. From (9.7) we have

$$(C_j)_N = (\varkappa_{ij})_{NM}^{-1} \cdot (\alpha_i)_M \quad , \quad (9.8)$$

where $(\varkappa_{ij})_{NM}^{-1}$ is the inverse matrix.

The problem of increasing selectivity here is reduced to increasing the determination accuracy of comparatively small values of C_j. The accuracy depends mainly on the spectral specificity of the terms in the matrix $(\varkappa_{ij})_{NM}^{-1}$ and the accuracy of their presetting as well as on the error of determining α_i. In the limit the ultimate selectivity can be obtained when the accuracy is restricted by the internal electronic noise of the OA spectrometer and the matrix is diagonal in the form $(\varkappa_{ij})$, i.e., at each frequency only one component should absorb. In practice, a decrease in selectivity is due to a low spectral specificity of the polyatomic molecular spectra, a limited tuning range of the lasers in use, as well as a considerable error in determining α_i (5 - 10%). Besides, the feasibility of the measuring equipment usually does not allow presetting the etalon coefficient with an accuracy better than 3 to 5%.

At fixed conditions of measurement in the spectrophone chamber (e.g., in a rapid local analysis of air) one way to increase selectivity is to increase the set of laser frequencies so that $M > N$. Generally, the criterion of optimum choice of analytical lines may be the minimum error in determining analytical results. It should be noted that the accuracy of determining the concentrations of separate components in the mixture can become better or worse, as the number of analytical lines increases. The latter case may require a great number of low-current measurements taken at the wings of absorption bands of separate components.

Another possible way of increasing the selectivity of the OA analysis of multicomponent mixtures is to obtain concentration equations of the type (9.6) at different conditions of measurement, such as pressure, temperature, modulation frequencies, values of applied electric or magnetic fields, etc. When the matrix terms $(\varkappa_{ij})_{NM}$ depend substantially on the conditions of measurement listed above, the method may prove to be more effective and technically simpler than that of increasing the number of analytical laser lines.

All this is illustrated by the results obtained from estimating the accuracy to determine the concentration of ethylene in the real atmosphere in the presence of absorption from CO_2 and H_2O molecules and rather nonselective and relatively small background from other microcomponents at four lines of the CO_2 laser, $R_{10}(20)$, $P_{10}(14)$, $P_{10}(16)$ and $P_{10}(26)$ [9.61]. For example, with the measuring accuracy of 1% for the OA signal the threshold of detecting C_2H_4 in its natural atmospheric concentration increase to 50 ppb, while in an ideal case it is equal to 0.3 ppb.

9.5.3 Optoacoustic Spectroscopy Combined with Chromatography

All the above methods, unfortunately, cannot solve properly the problem of selective detection of complex polyatomic molecules. This is due to the large

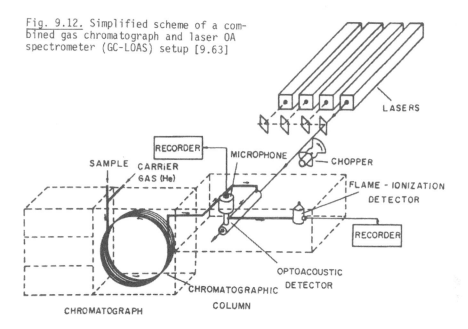

Fig. 9.12. Simplified scheme of a combined gas chromatograph and laser OA
spectrometer (GC-LOAS) setup [9.63]

width of the IR absorption bands which also very often overlap. Combined with
other analytical methods and particularly with chromatography OAS may solve
this problem.

The measuring technique in a combined system such as a gas chromatograph
(GC) and a LOAS consists of preliminary. separation of a multicomponent mixture
into fractions in the chromatographic column. From the column outlet these
fractions enter the laser OA detector chamber at succesive times (Fig.9.12).
The OA detector operates as a selective chromatographic detector in which
separate compounds are identified by two features: 1) the time of their escape
from the column (the so-called time of retention) as in ordinary detectors
of gas chromatographs, and 2) OA absorption spectra.

In gas-chromatographic analysis with OA detection one may deal with the
following typical situations:

1) Some chromatographic peaks are resolved well. In this case an OA detector increases the reliability of identifying separate compounds. This is particularly useful when their number in the sample is large. Such a situation
is characteristic, for example, of an analysis of different types of hydrocarbons, particularly aromatic hydrocarbons against the background of saturated ones.

2) Some chromatographic peaks are poorly resolved. Then the resolution of
the GC-LOAS system can be increased by choosing spectral ranges of laser operation in which the radiation is absorbed mainly by one component.

259

3) Some chromatographic peaks are not resolved. Then the analysis of multi-component mixtures with an OA detector is reduced due to preliminary separation in the chromatograph to a simpler task of analyzing binary or triple mixtures. This last task is accomplished by measuring OA signals at different wavelengths.

Moreover, in detecting some gases with strong absorption bands in the IR (SF_6, NH_3, etc.) the OA detector sensitivity may exceed the sensitivity of ordinary chromatographic detectors.

The GC-LOAS system can provide two basic types of measurement: a) with continuous gas flow and b) cyclic gas flow. Under cyclic conditions the separate chromatographic peaks are utilized in the cell for OA spectrum measurement under static conditions.

In continuous operation the compounds separated at the time of their escape from the column successively enter the OA detector chamber in a continuous flow with the carrier-gas (usually He or N_2). Under these conditions the following special requirements are placed upon the OA detector: 1) effective operation in continuous gas flow at a rate of up to 50 ml/min; 2) operation in the temperature range up 350 - 400 K to overcome condensation of compounds on the detector elements (windows, walls or microphone); 3) minimization of the chamber volume to eliminate chromatographic peak broadening; 4) a small time constant and fast tuning over the spectrum, essential for efficient analysis of separate peaks in a period of 2 - 20 s. Some of these requirements are conflicting and their fulfillment in practice requires a compromise.

The potential of the combined GC-LOAS system was first studied in [9.62] and later in [9.63]. The laser OA detectors used in both works were similar in design. Their chambers were made of a stainless-steel cylinder 15 cm long, and 0.4 cm in diameter having a volume of about 2.3 cm^3 [9.63]. The gas flow at the exit of the chromatographic column was injected into the detector chamber and then let out through small-diameter tubes to get rid of the pressure drop and to prevent acoustic noise from coming into the chamber. The detector was connected to the heater of the chromatograph and could be heated up to 400°C. To maintain its working capacity the sensitive microphone was removed far from the heated chamber and was connected with the latter via a small-diameter tube cooled with flowing water. To eliminate the parasitic "memory" effect which may be caused by remains of the gas probe in the microphone volume, the tube connecting the microphone to the chamber was subjected to counterflowing gas. The gas was mixed with the sample after passing through the detector chamber, which prevented decrease of the OA signal due to sample dilution. The threshold sensitivity of such an OA detector, limited mainly, by the background from the ZnSe windows was about 10^{-7} cm^{-1} with respect to the absorption coefficient [9.62,63,64].

The specific feature of the GC-LOAS measuring system in [9.63] was that the gas flow behind the OA detector entered a flame-ionization detector conventional in gas chromatography (Fig.9.12). This allowed proper comparison of the potentialities of both detectors on a real-time scale. Tunable $^{12}C^{16}O_2$, $^{13}C^{16}O_2$, $^{12}C^{18}O_2$ lasers at $\lambda = 9$ to 11.4 μm and a He-Ne laser with $\lambda = 3.39$ μm were used as radiation sources.

To illustrate the potential of the combined GC-LOAS system, Fig.9.13 shows results for primary, secondary and tertiary isomers of butanol. These isomers are resolved rather well behind the chromatographic column, proven by the measurement using a flame-ionization detector (Fig.9.13a). However, we cannot identify separate isomers well enough only by the time of retention from the column. At the same time, measuring OA signals at different wavelengths (Fig.9.13b) allow more reliable identification of these isomers by their characteristic IR absorption spectra.

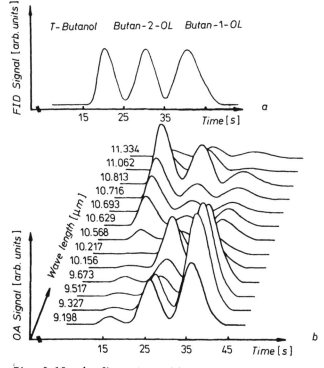

Fig. 9.13 **a,b**. Chromatographic measurement of butanol isomers: (**a**) with an ordinary flame-ionization detector; (**b**) a two-dimensional OA chromatogram produced with a laser OA detector [9.63]

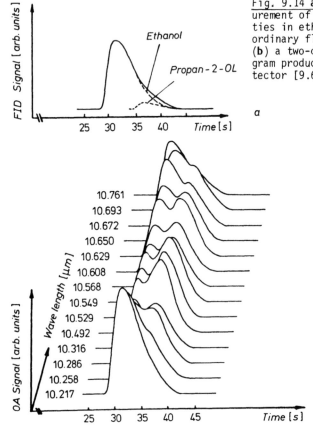

Fig. 9.14 **a,b.** Chromatographic measurement of isopropyl alcohol impurities in ethyl alcohol: (**a**) with an ordinary flame-ionization detector; (**b**) a two-dimensional OA chromatogram produced with a laser OA detector [9.63]

By choosing a proper spectral range it is possible to obtain a high sensitivity of the OA detector only to a certain compound in the mixture. This increases the resolution of the chromatograph and the concentration sensitivity of the system. Figure 9.14, for example, demonstrates the possibility of reliable detection of isopropyl alcohol impurities in ethyl alcohol as the CO_2 laser is tuned to the line ($\nu \approx 950$ cm^{-1}) at which the impurity has maximum absorption and, vice versa, the absorption in ethyl alcohol is minimum. This enables us to precisely identify these compounds at sufficient resolution with an OA detector (Fig.9.14b), whereas in recording with a flame-ionization detector the peak from only ethyl alcohol dominates (Fig.9.14a). Great potential of the GC-LOAS system have been demonstrated in [9.63] by quantitative determination of meta- and para-xylols using IR absorption spectra (these compounds could not be separated by our gas chromatograph), identification of different types of hydrocarbons, cis- and transisomers of hepten-3, etc.

The experiments described are characterized by a long duration of measurements due to successive observation of all the chromatogram peaks at each laser line. It is possible to increase the response by using lasers with fast wavelength tuning. In particular cases, for example, in detecting one component in a complex mixture, the multiplex spectrometer may be useful (Fig.7.1). In this scheme the radiation modulated with different frequencies at different wavelengths are passed through the OA detector chamber [9.65].

Fast spectral identification of a substance during its passage through the OA-detector chamber (1 to 10s) can be achieved by fast tuning via the scheme described in [9.66]. The potentialities of such a system are demonstrated by selective analysis of complex mixtures of alcohols and hydrocarbons.

The observation time of one peak and the spectrum tuning rate can be matched due to programming temperatures in the column with the aim to "extend" the chromatographic peak in time.

The sensitivity and the selectivity of the laser OA detector with other promising chromatographic detectors are shown in Table 9.3. The OA detector is second in selectivity to some, the mass spectrometer in particular. This results mainly from a limited tuning range of the CO_2 laser used. Selectivity of the GS-LOAS system can be furthermore increased by employing lasers with a large range of frequency tuning as well as using methods to increase OAS selectivity. The combined GC-LOAS system can complement gas chromatography.

The use of laser OA detectors in fluid chromatography holds much promise, too [9.67-70]. The main advantage then of such detectors over wide-spread

Table 9.3. Threshold sensitivity and selectivity of detectors in in gas chromatographs [9.62]

Detector type	Threshold sensitivity* pg/S	Selectivity	
By specific heat conductivity	100		
Flame ionization	2		
Nitrogen (N)-phosphorus (P)	0.05	70000 (P)	35000 (N)
Mass spectrometer	0.05	1000	
Laser OA		10	
by ethylene	0.2		
by 2-pentanone	26		

* Flow rate: 30 ml/min and S/N = 1

UV detectors lies in their much higher sensitivity with simultaneous miniaturization of the detector volume. In [9.67], for example, with an Ar laser
with $\lambda = 488$ nm used as a radiation source, the OA detector sensitivity was
25 times higher than the sensitivity of the standard UV detector with a deuterium lamp at $\lambda = 254$ nm. The parameters of the OA detector were a length of
11 mm, an internal diameter of 1.5 mm, and volume of 20 µl; the windows were
made of quartz and the chamber of stainless steel. The OA signals were recorded with a piezoelectric detector joined to a platinum foil sheet 0.1 mm thick
and forming part of the detector chamber wall. The measurements were carried
out at a frequency of about 4 kHz with the best signal-to-noise ratio. The
noise was mainly due to pulsations in the fluid flow caused by the pump. The
average rate of solvent flow (methanol) caused by the pump was 1 ml/min. The
laser OA detector were demonstrated by measuring three molecular isomers of
chloro-4-(dimethylamino) azobenzene in a concentration of 10^{-5} M. The dynamic
range within the linear part was 10^{4}. This pioneering study also indicates
that a combination of LOAS and fluid chromatography may have considerable
promise.

10. Laser Optoacoustic Microspectroscopy

Spectra of microscopic objects play an important role, for instance, in material, science and the technology of optical and electronic devices. Here, OAS complements the potentials of conventional methods of microscopy (optical acoustic, electronic, etc.) and in some cases it enables us to obtain quite new information about the microstructure of various media. In this chapter, we consider the principles of laser OA microspectroscopy and the results of the first such experiments.

10.1 Principles of Optoacoustic Microspectroscopy

Figure 10.1 compares simplified schemes of two types of scanning microscopes: standard and optoacoustic. In the standard technique the microscopic nonuniformities in a sample are detected by recording the transmitted or scattered radiation. In the OA microscope the power absorbed in the sample is recorded directly by an acoustic detector. Radiation focusing combined with spatial

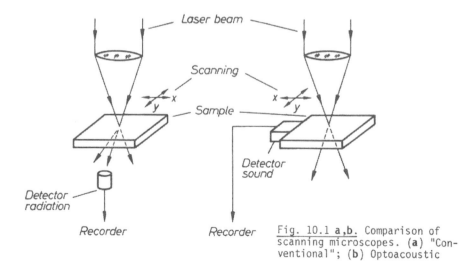

Fig. 10.1 a,b. Comparison of scanning microscopes. (a) "Conventional"; (b) Optoacoustic

265

scanning of laser radiation enable us to obtain information on the bulk micro-structure of the sample under investigations.

The principle of laser OA microscopes is based on three main physical processes responsible for the formation of the OA effect in different media: radiation absorption → the generation and propagation of heat waves → the production and propagation of acoustic waves. The term "heat wave" is used here, according to [10.1], to make the analysis more convenient since the diffusive propagation of heat from a periodical source may be formally treated as the propagation of a wave in a medium with high losses, its length λ_T being comparable with the length of thermal diffusion in this medium, i.e., $\lambda_T \simeq 2\pi \ell_T$, and $\ell_T = \sqrt{2k/\omega}$, k being the heat diffusion coefficient. In principle, all successive processes of transformation of absorbed energy to acoustic vibrations provide information on the microstructure of a medium, owing to nonuniformities in optical, thermal or acoustic properties.

A necessary condition for high spatial resolution of the OA microscope is to optimize the volume of radiation-particle interaction, in strong radiation focusing. High spatial coherency of the laser radiation leads to focusing into extremely small volumes which, in the ultimate limit, are comparable with a diffraction volume of the order of λ^3, λ being the wavelength. In the OA microscope, however, this condition does not suffice since the resolution is also influenced by the opto-thermo-acoustic parameters of the medium and the conditions of measurement (modulation frequency, method of OA signal selection, etc.). Below we consider this problem in more detail for some particular cases.

In accordance with two main schemes to record absorbed energy in condensed media, an OA microscope may indirectly detect acoustic vibrations by a microphone in the gas in contact with the sample, or directly record acoustic vibrations in the sample by a piezoelectric detector joined to it. When each type of nonuniformity is dominant there may be three limiting cases of microscopic analysis, which can be realized by OAS: optical, thermal, and acoustic (Fig.10.2). Such a classification is rather conventional but it is useful for understanding of the main characteristics of OA microspectroscopy (OAMS).

10.1.1 Optically Nonuniform Samples. "Optical" OA Microscopy

With optically nonuniform samples it is assumed that a sample is, however, thermally and acoustically uniform. Optical uniformity here means local variations of the absorption coefficient in the sample volume. The presence of optical nonuniformities leads to a nonuniform distribution of absorbed energy in the volume, which finally manifests itself as variations in the OA signal when the beam is scanned across the sample (Fig.10.2a). For the main schemes

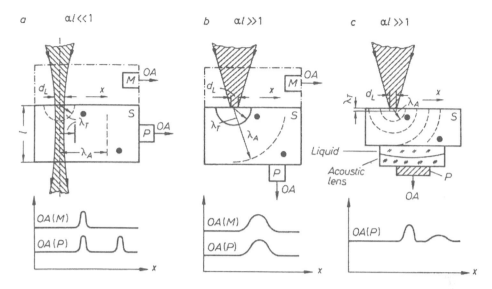

Fig. 10.2 **a-c.** Types of OA microscopy: (**a**) optical, (**b**) thermal, (**c**) acoustic (M is a microphone, P is a piezodetector). Bottom: variation of signals as the beam is scanned along the coordinate in the presence of two nonuniformities shown as dark circles

of OA microscopy (OAM) the transversal spatial resolution is determined by the beam diameter d_L.

Longitudinal spatial resolution depends on the form of the sample, the type of measuring scheme and the signal-detection technique. For example, in analysis of samples with the simplest form and a one-dimensional structure (single fibers) or a plane two-dimensional structure (very thin media of biological compounds, thin films, etc.) the OA signal, as a rule, is averaged over the thickness of such samples. The resolution with respect to the depth is determined by the sample thickness.

In analysis of samples with a three-dimensional structure in OAM with indirect detection the longitudinal spatial resolution is determined by the smallest of the two parameters ℓ_α or ℓ_T, ℓ_α being the optical absorption depth ($\ell_\alpha = 1 / \alpha$). Since $\ell_T \propto \omega^{-1/2}$ with $\ell_T \ll \ell_\alpha$, ℓ it is possible to study the integral absorption profile by tunning the modulation frequency ω. Here we should take into account that the relative contribution of separate absorption zones to the resulting OA signal in the gas decreases with depth from the sample surface. The limiting value of "surface" longitudinal resolution is restricted by the upper frequency limit of the microphone in use, f_r. For most of microphones, we have $f_r \leqslant 100$ kHz, which corresponds to $\ell_T \sim 5$ to 10 μm. As the modulation frequency increases, the sensitivity of the OA method with indirect detection drops. Therefore, the choice of sensitivity and longitudinal resolution is based on a compromise.

In OAM with direct detection the longitudinal resolution is determined by the smallest of the three parameters: ℓ_T, ℓ_α, and ℓ_{sp}, ℓ_{sp} being the value of spatial resolution depending on the methods of detection and processing of the OA signal.

The type of OA microscopy under consideration holds promise in analysis of the microstructure of weakly absorbing objects when high sensitivity together with high spatial resolution are required. To identify the type of nonuniformities it is particularly useful to measure the microstructure in various spectral ranges.

10.1.2 Thermally Nonuniform Samples. "Thermal" OA Microscopy

Thermal nonuniformity means here local variations of thermodynamical parameters, and particularly the coefficient of heat diffusion in the sample volume. The effect of this nonuniformity shows up first as nonuniform heating of the sample at different points even with the same absorbed power at these points and, second, as variations in the amplitude and phase of the heat wave propagation from the irradiation zone. These variations are transferred to the acoustic wave that consequently arises and carries away information on the thermal microstructure.

In essence, the OAM works as a thermal microscope in which the process of interaction of the heat wave with local thermal nonuniformities is visualized by recording the acoustic waves (Fig.10.2b). In such a microscope the laser radiation acts as a generator of heat waves in the volume of the radiation-particle interaction. The heat diffusion length ℓ_T determines the maximum depth within which the microstructure of the sample can be visualized. For example, in analysis of condensed media the value of ℓ_T ranges approximately from 1 mm to 1 μm as the modulation frequency varies between 10 Hz and 1 MHz. Fast-response piezoelectric detectors and OAM should be used to record high-frequency acoustic waves. The resolution of the thermal OAM depends on the beam cross-section and the ability of the recording system to detect small relative changes in the amplitude and phase of the OA signal caused by the spatial variations in thermal properties of the sample. Specifically, in OAM with indirect detection the spatial resolution, with $d_L \ll \ell_T$, is no less than ℓ_T. By analogy with the above technique to determine the absorption profile with respect to depth, we can also find the thermal profile of the surface structure of the sample. By analyzing the shape of this profile we can roughly estimate the depth of the subsurface defect in optically opaque samples.

10.1.3 Acoustically Nonuniform Samples. "Acoustic" OA Microscopy

Acoustic nonuniformity here means spatial variations of the parameters responsible for the formation and propagation of acoustic waves (such as the heat
expansion coefficient, elasticity constant, velocity of sound,etc.). The presence of such nonuniformity leads, first, to nonuniform excitation of the
acoustic waves in the volume of the radiation-particle interaction even with
optical and thermal uniformity within this volume and second, to variations
in amplitude and phase of the acoustic wave as the latter propagates from the
zone of irradiation to the detector (Fig.10.2c). When the first process predominates, the resolution depends mainly on the values of ℓ_T and ℓ_{sp}, (Sect.
10.1.2). If the second process predominates, the resolution is determined by
the acoustic wavelength in the sample $\lambda_a = V_s / f$. In most solids and liquids
the sound velocity V_s ranges from $(1 \div 6) \cdot 10^5$ cm/s. With the modulation frequency f = 1 MHz this corresponds to 0.3 - 0.6 cm. To improve the resolution
one should increase the modulation frequency. Its upper limit is restricted
by attenuation of the acoustic waves in the medium and lies at a level of a
few GHz. This frequency limit corresponds to a spatial resolution limit of
several μm. When refining the OAM technique we may widely use acoustic optics,
particularly acoustic lenses, fast-response thin-film piezoelectric detectors,
etc.

   Table 10.1 compares these schemes, which enables the spatial resolution of
OAM in every particular case to be estimated roughly. But in practice, there
are rarely samples in which only one type of nonuniformity is well pronounced.
Therefore, generally the variations of the OA signal in the scanning OAM can
be caused by spatial variations in optical, thermal and acoustic properties of

Table 10.1.   Ultimate spatial resolution of OA microscopes

Type of OAM	Type of detection of OA signal	Resolution	
		Transversal	Longitudinal
Optical	indirect	d_L	$\min(\ell_\alpha, \ell_T)$
	direct	d_L	$\min(\ell_\alpha, \ell_{sp}, \ell_T)$
Thermal	indirect	$d_L + \ell_T$	ℓ_T
	direct	$d_L + \ell_T$	ℓ_T
Acoustic	direct	$d_L + \ell_T$	λ_α

the sample. On the one hand, this makes it rather difficult to obtain differential information on the sample's microstructure from the results of OA
measurements. On the other hand, this materially widens the area of application
of OAM and the class of structures to be analyzed with this technique.

10.1.4 Operation of Laser OA Microscopes

The OA microscope consists of the following principal elements: a laser source,
an optical system for focusing onto a sample; a system of spatial scanning,
and systems or recording and processing the OA signal and a system to visualize
the OA image.

Focusing can be performed using ordinary optical the microscope lenses. To
produce a minimum caustic the laser should operate in a TEM_{00} mode. The beam
can be scanned across the sample in two basic ways: by moving the sample itself with a special movable table, and by spatial scanning the beam with an
optical scanner. The advantage of the first method is high accuracy and reproducibility of the sample position (to 1 μm) and a large field of view (tens
of mm). Its disadvantage is a low time response as well as a potential influence
of attendant vibrations on the ultimate sensitivity of the microscope. Optical
scanners display the best efficiency in fast response. But the field of view
is limited to no more than several mm due to the effect of radiation defocusing
in the process of scanning. Mechanical replacement of a sample is usually performed with step-by-step motors. In scanners the bending of the laser beam can
be carried out with electromagnetic or piezoelectric devices and frame scanning
with a cam-driven mirror.

As a rule, increasing spatial resolution decreases OAS sensitivity because
the relative fraction of absorbed energy responsible for the generation of OA
signal is reduced. So, in practice, the choice of spatial resolution and sensitivity of OAM is based on a compromise. A loss of sensitivity in a high-resolution OAM can be partially compensated for by using higher-power lasers and
by increasing the duration of the measurement. The upper limit on radiation
power in studies of condensed media is determined, as a rule, by the optical
breakdown or heat evaporation. Limits may be imposed by irreversible changes
in the properties which are characteristic, in particular, of biomedical samples.

The relation between the threshold sensitivity and the resolution limit of
an optical OAM has the following form:

$$\alpha_{min} \simeq \frac{1}{I_{thr}(d_L)^2_{min}} \times \begin{array}{ll} (J\alpha)_{min} & \text{CW operation} \\ (E\alpha)_{min} / t_p & \text{pulsed operation} \end{array} \qquad (10.1)$$

where α_{min} is the minimum detectable absorption coefficient, $(d_L)_{min}$ is the minimum possible diameter of the laser spot, I_{thr} is the intensity threshold of damage to the sample, $(J\alpha)_{min}$, $(E\alpha)_{min}$ are the threshold sensitivities of detecting OA signals under continuous and pulsed laser operation, respectively. For example, typical parameter values in (10.1) give $(E\alpha)_{min} = 10^{-9}$ J $\cdot$ cm^{-1}, $t_p = 10^{-8}$s, $(J\alpha)_{min} = 10^{-5}$ W $\cdot$ cm^{-1}, $I_{thr} = 10^7$ W/cm^2 and $d_L \approx 3$ μm for the laser spot, the minimum detectable absorption coefficients for an "optical" OAM under pulsed and CW operation are, respectively, $\alpha_{min}^p = 10^{-1}$ cm^{-1} and $\alpha_{min}^{cw} = 10^{-5}$ cm^{-1}. Besides a high sensitivity, another advantage of CW operation consists in simpler signal processing.

It is possible to compensate the loss of sensitivity in OAM by increasing the storage time of the signal, but this increases the measurement time. For example, 17 min is necessary to measure 1×1 mm^2 of sample with 10 μm resolution and 0.1 s of storage time at a fixed wavelength. Because of the low scanning rate this point-by-point measurement necessitates that one keeps in mind particular details of an OA picture for its subsequent reproduction on the display screen.

It is possible to reproduce both the amplitude and phase images of the object, i.e., the dependence of the amplitude or phase of the OA signal on the beam position, and the brightness image when the beam brightness on the screen is proportional to the amplitude of the OA signal. The second type of image is very useful for quick qualitative estimation of the sample's microstructure.

Of course, OAM needs proper metrological equipment and calibrating techniques. The calibration technique of the OA-cell sensitivity considered in Sect. 6.4 can be applied with small modification to OAM. It is more difficult to refine the techniques for estimating the threshold spatial resolution of OAM particularly longitudinal resolution. This problem can be partially solved by using etalon optical measures, samples with varying thickness as well as by establishing volume measures, for example, as multilayer optical measures.

10.2 Two-Dimensional Optoacoustic Microscopy

It is generally simple to realize, in practice, two-dimensional OAM which is characterized by high transversal resolution and comparatively low longitudinal resolution. It is best used for analysis of plane structures which may be the surface structures of highly absorbing or optically opaque samples or different thin films on optically transparent samples.

10.2.1 Optoacoustic Microscopes with Indirect Detection

In OAS information on absorption in a smaple can be obtained from indirect
detection of the acoustic vibrations in the gas in contact with the sample.
Due to a strong dependence of the OA indirect signal amplitude and phase mainly
on the physical and thermodynamic parameters of the surface layer of the sam-
le, it is advisable to apply this method when studying surface and subsurface
microstructures. This type of OAM can be used to visualize the optical or
thermal structures of a sample since with OAS we may neglect acoustic transport of the energy absorbed in the sample volume in a first approximation.

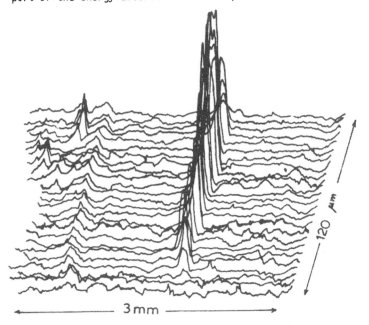

Fig. 10.3. Optoacoustic topogram of a silicon nitride ceramic surface with a
surface defect. Resolution: 30 µm; radiation source is an argon laser; modulation frequency: 600 Hz [10.2]

Figure 10.3 shows a topogram of a silicon nitride ceramic surface with a
surface defect in the form of a crack of 50×100 µm in size. The topogram is
produced with an OAM. The most probable cause of increased OA signal in the
defect region is that the volume of radiation-particle interaction is variable
and this zone is thermally nonuniform. Such a high OAS sensitivity to the surface microgeometry allows it to be used effectively as a microprofilograph.
The potentialities of OAM have been demonstrated by estimating the quality
of such ceramics as Si_3N_4 and SiC widely used as heat-resistant and inert-to-
aggresssive-media materials [10.3]. OAM allows to investigate the microstruc-

272

ture of these types of ceramics and to detect cracks, splinting and foreign microinclusions like Fe. The transversal resolution obtained with a visible laser is about 30 μm, which is comparable with the spot diameter and the length of thermal diffusion in the sample with modulation frequency in use. As already noted, when optical nonuniformity prevails, the resolution must be determined by the dimensions of the beam on the sample surface. This is confirmed by the results of [10.4] where, with the thermal diffusion length from 30 to 50 μm, the transversal resolution equals 5 μm, which is comparable with the dimensions of the laser beam. Then the mechanism of OA image generation is more optical than thermal.

Direct proof of the efficiency of OAM used to visualize subsurface defects is provided by the results of model experiments with metal samples [10.5,6]. In these experiments the thermal nonuniformities near the sample surface were simulated as hollow microvolumes produced by drilling holes parallel to the sample surface [10.5] or by milling a wedge like rectangular groove [10.6]. As the beam was scanned over the surface near such artificial defects the OA signal increased and its phase changed. Qualitatively these effects can be explained by an increase in the relative fraction of absorbed energy which contributed to the formation of an OA signal in gas due to its lower heat removal from the surface into the depth of the sample in the presence of the defects. Formally this is equivalent to heat-wave reflection from the boundary surface of the defect.

In terms of OAS the displacement of the beam on the surface to the zone over the defect corresponds to the changeover from analyzing a thermally thick sample ($\ell_T \ll \ell$, ℓ being the sample thickness) to analyzing a thin sample $\ell_T \gtrsim \ell_d$, ℓ_d being the depth of the defect. The behavior of the OA signal amplitude and phase in the transient zone calculated from a three-dimensional model, is shown in Fig.10.4. It may be seen that the maximum depth at which it is still possible to detect the defects in metals (equivalent to cracks, cavities, etc.) by OAS is equal to ℓ_T. It is somewhat larger when the phase is measured $(1.5-2.0)$ ℓ_T [10.5,7]. Thus, in practice, we can detect thermal nonuniformities in optically opaque media at depths of up to 0.5-1 mm with a modulation frequency up to 10 Hz. The transversal resolution is comparable with ℓ_T and hence grows with increasing modulation frequency since $\ell_T \propto \omega^{-1/2}$. The slope of the frequency dependence of the OA signal from the thermally thick to the thermally thin area of the sample varies from ω^{-1} to $\omega^{-3/2}$. This variation can also be used to identify defects.

The case considered corresponds to the operating conditions of a purely thermal OA microscope. Practically, thermal and optical nonuniformities can

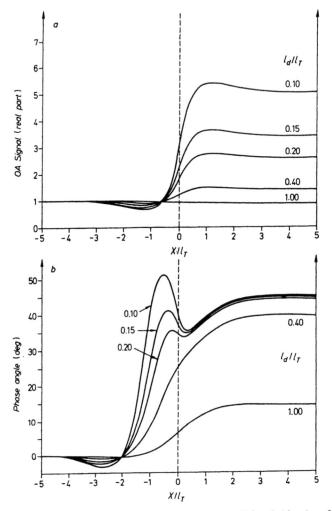

Fig. 10.4 a,b. Amplitude (**a**) and phase (**b**) of OA signal with indirect detection as a function of the position of the laser beam on the surface of a metal sample x/ℓ_T for different depths of a defect ℓ_d/ℓ_T, such as a cavity (marked with figures). The point $x/\ell_T = 0$ corresponds to the boundary between the thermally thick ($x/\ell_T < 0$) and thermally thin ($x/\ell_T > 0$) zones of the sample [10.7]

exist simultaneously, so the development of methods for their identification becomes important. The simplest way to realize these methods is scanning the frequencies of laser radiation and modulation. The change of pattern of the OA image with the laser wavelength may be an indication of the optical image. Vice versa, with the thermal image prevailing the pattern should not depend on the radiation wavelength (this is valid when the coefficient of reflection from the sample surface is constant).

Thus, OAM makes it possible to define rather accurately the location and approximate dimensions of surface and subsurface defects. Similarly with OAM we can also determine the localized absorbing zones on the surface of slightly absorbing crystals. Optoacoustic microscopy may also be applied to other spectral ranges, e.g., to the microwave region [10.8] with a corresponding decrease in spatial resolution.

10.2.2 Optoacoustic Microscope with Direct (Piezoelectric) Detection

The scheme of OAM with direct piezoelectric detection of sound waves is simple enough. It comprises a beam scanner and an optical system focusing the radiation onto the sample with a piezoelectric detector attached to it. Unlike microphone detection, fast-response piezoelectric detectors enable us to improve essentially both the thermal and the acoustic resolution limits of OAM (almost up to fractions of μm).

According to the modulation frequency, there are two types of operating conditions of OAM: at comparatively low modulation frequencies (no higher than 1 MHz) and at very high frequencies (up to 1 GHz). Below we consider the distinctive features of these operations in more detail.

At low-frequency modulation, due to a comparatively large wavelength of the acoustic vibrations (above 0.5 - 1 cm), it is mostly advisable to use such OAM for visualizing the thermal and optical spectrostructure of a sample surface. In this technique one of the possible ways of obtaining separate information on the optical and thermal properties of a sample is to measure simultaneously the amplitude and phase OA images of the object. This can be explained by the fact that the OA signal amplitude depends both on the optical and thermal properties of the object. At the same time, when $\ell_\alpha < \ell_T$, the OA-signal phase depends just on the thermal properties of the object. To support this fact one has only to compare the integrated circuits in Fig.10.5 produced with an ordinary optical microscope (a) and an OA microscope (b,c). The OA images were measured using an Ar laser with $\lambda = 5145$ Å and a modulation frequency 185 kHz. The spatial resolution was 7 μm with a spot diameter of 5 μm and a thermal diffusion length of 7 μm. Comparison of the figures shows that the amplitude and phase OA images are different, the amplitude image copying more faithfully most contrast details of the optical image, for example, the vertical light stripes. The phase OA image characterizes the local thermal properties of the electronic circuit depending, for the most part, on the type and thickness of its individual elements. Thus, OAM provides information on the subsurface thermal structure of a sample even in the case of strong optical contrast, which is important in many practical applications and particularly in semi-

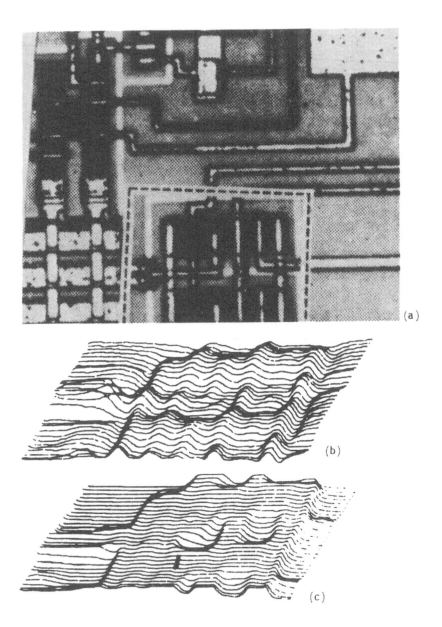

Fig. 10.5 **a-c.** Optical image of an electronic circuit, 250 x 250 μm, produced
with a standard optical microscope (**a**). Amplitude (**b**) and phase (**c**) OA images
of parts of an electronic circuit shown by a dashed line in (**a**) [10.9]

conductor technology. The resolution with respect to depth in OAM with indirect as well as direct detection is in phase measurements twice as good as in amplitude measurements [10.10].

When the OA microscope operates at high frequencies the depth of propagation of heat waves in the sample is small, for example, no longer than 0.1 μm, at frequencies above 100 MHz. At the same time the acoustic wavelengths decrease and become comparable with the wavelength of optical radiation. This enables the OA microscope to be used in analysis of the thermoelastic and mechanical properties of a medium, i.e., in tasks traditionally solved by acoustic microscopy. The only difference between OAM and the ordinary acoustic microscope is that the piezoelectric acoustic source at the input of OAM is replaced by a laser thermooptical source. High-frequency modulation can be accomplished with an electrooptical modulator and also, as shown in Fig. 10.6, under pulsed excitation with a train of several single pulses with width τ_1 and separation τ_2. In [10.11] such an experiment was performed using a Nd:YAG laser with $\lambda = 1.06$ μm, $\tau_1 = 200$ ps, $\tau_2 = 5$ ns, the duration of one train was 200 ns and a repetition frequency of 2.7 kHz. The radiation was focused with a lens onto the sample surface in a spot about 2 μm. The acoustic waves resulting from thermoelastic deformations pass through the sample and enter the immersion liquid (usually H_2O) pass the acoustic lens (radius: 200 μm) and are detected with a piezoelectric film of ZnO which has an upper limit frequency of 800 MHz and a transmission band up 100 - 200 MHz. To produce an image necessary to move the object mechanically along two axes about the lens focus.

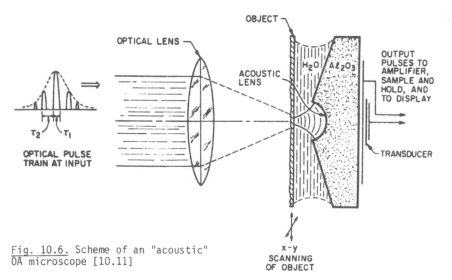

Fig. 10.6. Scheme of an "acoustic" OA microscope [10.11]

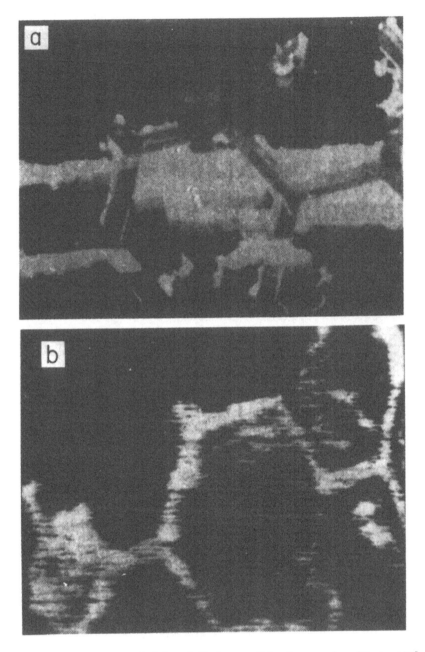

Fig. 10.7 **a,b.** Optical (**a**) and OA images (**b**) of a sample with a metal film of chromium 200-nm thick on a glass substrate [10.11]

Figure 10.7b shows the OA image of a film of chromium on glass produced by measuring the acoustic vibrations at the fourth harmonic (840 MHz) of the modulation frequency. It may be seen that, compared to the optical image (a), in the OA image (b) the chromium hexagonal structure is clearer. To avoid damage of the sample, the laser pulse power was no higher than 1 KW. Thus, the OA microscope under acoustic operation detects local changes in the elasticity of materials.

The resolution of such an acoustic OA microscope is determined mostly by the wavelength of the acoustic vibrations. The ultimate resolution at 0.5 to 1 μm is limited by high sound attenuation in different media at frequencies above 1 GHz. The maximum sensitivity of this OA microscope can be obtained in an analysis of optically opaque samples satisfying the following basic requirements: low reflection of optical radiation from the surface; good acoustic matching with the detector or the immersion liquid and a high efficiency of conversion of absorbed energy to acoustic energy. This can be realized with a high expansion coefficient, low heat capacity and a high modulation frequency. It should be noted that despite a small depth of penetration of heat waves, sometimes the character of an OA image can also be affected by the optical-thermal properties of the sample, for example, variations of the reflection coefficient on the sample surface.

10.3 Three-Dimensional Optoacoustic Microscopy

Three-dimensional microscopy means OAM of the spatial structure of objects with a high resolution in both the transversal and longitudinal directions. The possibility of acquisition of three-dimensional information on an object has partially been touched upon in the study of the profile of sample absorption in depth using OAS by scanning the modulation frequency (Sect. 10.2). A more general problem is how to increase the longitudinal resolution in an OA analysis of weakly absorbing media with direct detection of the OA signal, i.e., how to develop a three-dimensional OA microscope. This problem is in the initial stages of development and below we list just the most general considerations of a feasable solution.

Usually in measuring OA signals the information on the absorbed power along the laser beam is integrated, and the OA image of the object resulting from spatial scanning of the laser beam is, in essence, the two-dimensional projection of a three-dimensional structure. The most general method of acquiring information on the three-dimensional structure is mathematical processing of the measurements of many two-dimensional projections at different angles of the beam axis to the object. This method is based on the well-known fact that it is possible to reconstruct unambiguously any three-dimensional object using an infinite set of its two-dimensional projections [10.12]. This method (tomography) proves good enough in acoustic, optical and X-ray microscopy, and we hope that it has good feasibility in OAM, too. For its correct application in OAM we should have available information on the law of OA signal integration in a longitudinal direction. A preliminary result obtained with two perpendicular scannings of an aluminium piece (with two subsurface holes) was presented in [10.13]. Optothermal deflection detection (the mirage effect) was used in this experiment. A new efficient OA imaging approaches suitable for non-destructive testing. The tomographic-like scanning of the sample allows both a large reduction of the sample illumination and a fast scanning procedure leading to the 3-dimensional localisation of subsurface defects.

Due to a comparatively low velocity of propagation of acoustic waves and rather a wide band of piezoelectric detectors the longitudinal resolution can also be increased (10 to 100 µm) through time gating of the OA signals from separate local zones along the laser beam. Then the problem of time selecting OA signals is substantially facilitated by strongly focusing the laser radiation into the zone of the object under investigation. This is essential due to time focusing of the OA signals from the zones beyond the focal volume and by a decrease in their relative amplitude owing to lower radiation intensity. Under pulsed operation a combination of time and amplitude methods of selecting the signals from the focal volume of the radiation-particle interaction enables us to expect longitudinal resolution up to 0.1 mm. To reduce undesired spatial integration of OA signals on the detector area we must minimize it. In certain cases the spatial resolution can be increased by using acoustic lenses as in acoustic microscopes. Good results may also be obtained from using several detectors at the same time or a detector matrix in combination with correlation methods in processing the data.

We should also mention a possibility of nonlinear OAM that uses different nonlinear effects arising in the interaction of powerful laser radiation with the medium both on the optical and acoustic levels. The simplest version of practically realizing this type of OAM may be an OA microscope scheme with harmonic modulation of focused laser radiation and recording of acoustic signals at the second harmonic from the modulation frequency. A microscopic investigation of the nonlinear properties of a medium increases the efficiency of OA measurements, which is useful, for example, for more reliable identification of foreign inclusions with different nonlinear properties in a sample.

Contrasts in the OA image can be increased, whereby small details are observed against the background of a large constant component of the OA signal by the derivative method [10.14]. This method can be realized through periodic small transverse shifts of the laser spot on the sample surface. Apparently it can also be used in analysis of three-dimensional structure to increase the longitudinal resolution under CW operation. For this purpose the beam caustic should be displaced periodically due to vibrations along the optical axis. Such an operation allows studies of micrononuniformities in depth, with a resolution depending on the caustic length, i.e., on the degree of laser radiation focusing.

10.4 Comparison with Other Methods of Microscopy

Optoacoustic microscopy is still a young field and so its potentialities are only in the initial stage of realization. However, even the results of the very first studies enable us to hope that this method will find wide application and essentially complement the potentialities of such conventional methods of microstructure analysis as optical, acoustic, electronic, and X-ray microscopy. The most general and important characteristics of these methods are presented in Table 10.2, where one can clearly see the distinctive features of OAS itself.

First of all, with its resolution almost the same as that of ordinary optical microscopy, OAM provides much more information since the OA image depends simultaneously on the optical, thermal and elastic properties of the sample. The problem here consits of developing correct and reliable methods to identify the type and character of the structure to be visualized by using, for example, spectral, amplitude, phase, frequency and nonlinear methods of OA signal selection.

The advantage of OAM optical microscopy is that it is possible to analyze the spatial structure of weakly absorbing nonrefractive media without addition and high modulation of laser radiation allows us to investigate the optically opaque media. The latter can be achieved when the OA microscope operates as a thermal microscope.

Like a purely acoustic microscope, the OA microscope with the same resolution and microwave modulation of laser radiation allows us to investigate the elastic properties of optically opaque media. Its specific feature is that we may use a thermooptical acoustic source with the advantage that it is contactless has a variety of types of excitation, a possibility of high localiza-

Table 10.2. Comparison of the basic methods of microscopy

Characteristics	Optical microscopy	Acoustic microscopy	OAM
Resolution: transversal longitudinal	$0.5 - 2$ μm $5 \ - 20$ μm	$5 - 10$ μm (for $\omega/2\pi = 1$ GHz) 100 to 1000 μm	$1 - 5$ μm $0.5 - 3$ μm
Threshold sensiti- vity of absorption	$1 - 10^{-1}$ cm^{-1}		$10^{-1} - 10^{-5}$ cm^{-1}
Type of interaction with medium	Optical interaction (ab- sorption, reflection and scattering)	Acoustic waves	Simultaneous inter- action of optical, heat and acoustic waves with medium
Microstructure analysis			
a) high absorption	yes	yes	yes
b) weak absorption	no	no	yes
c) refractive	yes	no	yes
d) thermal (optically opaque)	no	no	yes
e) elastic or acoustic	no	yes	yes

tion of the radiator volume and it is simple to scan the sample surface. Potentially, the latter property combined with adaptive acoustic lenses enables us, in principle, to hope for a high longitudinal resolution of the acoustic OA microscope. A unique feature of this type of microscope is its ability to visualize the thermoacoustic structure of surface layers with the same resolution as an acoustic microscope but at much lower modulation frequencies (approximately up to 100 - 300 kHz). The high resolution here (3 to 5 µm) is due to resolution its dependence on only the thermal diffusion length ℓ_T rather than on the acoustic wavelength, since at the same frequency $\ell_T \ll \lambda_a$.

Other advantages of OAM over electronic microscopy are that there are no requirements for sample evacuation, and it is possible to analyze thicker samples thermally and optically. A possible way of increasing the resolution here is a hybrid version of electronic-thermal-acoustic microscope based on a standard scanning electronic microscope. The electron beam locally generates heat waves in the surface layer of the sample, and the microstructure is visualized through piezoelectric detection of the attendant acoustic waves in the sample [10.15-17]. In such a microscope, through the electron beams we can obtain much better transverse resolution than in the thermal OA microscope and a better depth resolution compared to the electronic microscope, since heat waves penetrate the sample better than electron beams.

The thermoacoustic detection under electron beams has therefore found important application in thermal-wave imaging at high spatial resolution where micrometer-sized thermal waves are needed. Nonspectroscopic applications of thermal-wave physics, in particular, those involving materials analysis through thermal-wave imaging, and quantitative thin-film thickness measurements were described in [10.17] for the study of semiconductor materials and devices.

The strong illumination of the sample may be avoided by using Fourier or Hadamard spatial coding of the pump beam, as suggested recently in [10.18, 19].

Thermal-wave physics is playing an ever-increasing role in the study of material parameters. There are a multitude of mechanisms by which thermal-waves can be produced, the two most common ones involve the absorption of energy from either an intensity-modulated optical beam or from an intensity modulated electron beam. Several mechanisms are also available for detecting directly or indirectly, the resulting thermal waves. These include gas-microphone OA detection of heat flow from the sample to the surrounding gas optothermal measurement of infrared radiation emitted from the heated sample surface, optical deflection of a beam traversing the periodically heated gaseous or liquid layer just above the sample surface, interferometric detection of the thermoelastic displacements of the surface, optical detection of the thermoelastic deformations, and piezoelectric detection of the thermoacoustic signals generated in the sample (references in [10.17,20]). Experimental results have been shown for depth-resolved thermal-wave analysis in a stereoscopic arrangement with two thermal-wave sources [10.20]

11. Nonspectroscopic Applications of Optoacoustic Spectroscopy

Optoacoustic spectroscopy can also be used to solve nonspectroscopic problems, in particular, to measure and control the parameters of laser radiation, to study the thermal properties of various media, to investigate electron-phonon interaction, etc. In this chapter we consider some interesting areas of such nonspectroscopic applications.

11.1 Control and Stabilization of Laser Parameters

The OA effect has been used for a long time in the, so-called, OA detectors developed to measure weak radiation flows [11.1,2]. The OA Golay detector is the most widely used, in which the radiation flow to be measured is received by a thin absorbing film placed in a nonabsorbing gas. The heat from the film is transferred to the gas, increasing its pressure, so deflecting the reflecting membrane that is an element of the "optical" microphone. The operation of such a microphone is based on fixing the relative displacement of the raster and its autocollimation from the image reflected from the membrane surface. The advantage of such a detector is its spectral nonselectivity, high sensitivity (at the level of thermal threshold) and the large collecting area. In a number of devices a simpler OA (pneumatic) detector with a condenser microphone is useful.

The OA cells described above are quite suited to measure the parameters of laser radiation. Their merits for this task are: 1) low radiation losses (no more than 1-3%), which enables to realize transparent meter the parameters of laser radiation; 2) linearity over a large dynamic range (up to 10^4 - 10^6), which allows measurements in a region of weak flows with sufficient sensitivity and with high powers, in the absence of nonlinear effects in the receiving/recording system; 3) a short path along the optical axis ($\leqslant 1$ - 2 cm) which allows us to introduce OA cells into almost any point of the optical scheme of the instrument without difficulties; 4) a large aperture combined with a slight dependence of the OA signal on the position of the beam across the OA cell. A disadvantage of OA cells when they are used in controlling-measuring operations is their relatively large time constant and hence strong dependence of the signal on the modulation frequency. But in some cases this is not a serious obstacle to their application.

11.1.1 Measuring and Control of Power

The OA signal is proportional to the product of radiation power (energy) and the absorption coefficient in a sample. Therefore, the absorption coefficient being known, OAS allows the radiation power (energy) to be determined and its variation to be controlled. For this purpose we may use, for example, a spectrophone filled with some buffer gas, like N_2 or He, with an admixture of absorbing molecules. Such a spectrophone measures the pulse energy or power under continuous laser operation both with radiation modulation and, in principle, without modulation (by static deflection of the measuring membrane).

A disadvantage of filling the spectrophone chamber with a mixture containing absorbing molecules is a sharp spectral dependence of the power meter as well as its nonlinearity in the region of high intensities due to saturation. When OA signals are formed from radiation absorption in the windows of a spectrophone filled with nonabsorbing gas, this disadvantage does not arise. Absorption in the windows depend slightly on the wavelength, and the effect of saturation in dielectrics is absent up to powers which bring about optical damage. High sensitivity of such a scheme even with weakly-absorbing windows produces a signal-to-noise ratio sufficient for many applications. For example, radiation from a 10-W CO_2 laser in an OA cell with BaF_2 windows and a commercial condenser microphone produces a signal-to-noise ratio of about $10^4 / \sqrt{Hz}$ at the optimum frequency of 150 Hz [11.3,4]. With the windows at the Brewster angle, the total radiation loss in the OA cell does not exceed several percent. Thus, there is a real reason for creating compact transparent power meters not requiring beam splitters.

The OA measurement of absorption in dielectric can be also performed with a piezoelectric detector attached to sample, which is simpler than the scheme above. Besides, the frequency response is constant in a wider range up to several kHz [11.5]. Furthermore, the described scheme to measure radiation power can be simply realized by joining proper acoustic detectors to different optical elements, for example, to windows of the absorption cell, lenses, and the beam splitter.

Liquid OA cells can be utilized to measure the energy parameters. In [11.6] it has been shown that in the visible and near IR ranges high accuracy (about 7%) and good linearity results over a wide dynamic range of energies (up to 6 orders) can be obtained when the OA cell is filled with solutions of organic dyes with a high absorption coefficient (up to 100 cm^{-1}). Under OA signal integration the output signal in such a cell is independent of the beam cross section of the intensity distribution over the cross section, the beam divergence, and the radiation pulse duration (in a range from 15 ns to 10 μs). A simple method for measuring temporal characteristics of short picosecond or nanosecond pulses in the visible and near IR has been described in [11.7]. It relies on the OA detection of two-photon absorption in thallous halides in an otherwise conventional correlation experiment.

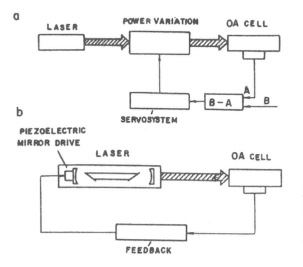

Fig. 11.1 **a,b.** Simplest schemes of stabilization of laser radiation power (**a**) and frequency (**b**) using OA cells

Figure 11.1a presents the simplest system for power stabilization with an OA cell. The signal from microphone A is compared with the given etalon signal B and the error signal B-A through the feedback loop enters the element controlling the level of radiation power. Electrooptical modulators, interferometers, transparent plates at large angles, etc., are usually used as such an element. For example, combining the last element with a spectrophone filled with $NH_3 + N_2$ mixture (diameter: 1cm) suppresses the CO_2-laser power fluctuations in the low-frequency region, with modulation frequency of 200 Hz, by about 10 to 30 times, thus stabilizing the power at a number of lines to a level of 0.3% during a 30-min measurement [11.8].

11.1.2 Frequency Control and Stabilization

The laser frequency can be controlled and identified by the known absorption lines of different gases placed in the spectrophone. When the absorption lines are narrow enough the instability of laser frequency can be estimated from the fluctuations of the OA signal from the spectrophone. For this purpose the influence of power fluctuations must be eliminated, e.g., those due to preliminary power stabilization, by measuring the ratio between the signals from the spectrophone and an additional detector of radiation power. The required narrowing of the absorption line contour can be obtained by reducing the gas pressure.

Optoacoustic stabilization of laser frequency is based on an OA cell used as an element discriminating unambiguously the signal related to the laser frequency [11.9-11]. Then by means of the servosystem this signal stabilizes the frequency, for example, by moving the mirror on the piezoelement (Fig. 11.1b [11.9]). This is essential, especially for effective optical pumping of lasers in the far-IR range [11.12] or for precision magnetospectroscopy [11.13].

A spectrophone can be used not only to measure and stabilize separately the radiation frequency and power, but also their product or the power ratio of two lasers. The scheme given in Fig.9.6 can be used to stabilize the power ratio of lasers. The first spectrophone acts as a zero-order comparison element which through the servoloop controls the power level of one of the lasers. In this case the power of one laser is controlled to that of the second, which is much simpler than to stabilize the power level of each laser separately. This scheme allows the said parameters to be stabilized to 0.5 - 1%.

11.1.3 Search for New Laser Lines

In addition OAS may prove useful in studying active laser media which constitute, as a rule, a complex nonequilibrium mixture of different gases at varying temperatures or pressures. Then the microphone must be joined through a special acoustic channel to the laser tube, that acts as a spectrophone chamber here.

The simplest task is to find potential laser lines ν_p in optical pumping of weakly-absorbing active gaseous media. The technique is reduced to studying the dependence of the OA signal on the probe frequency. The signal maximum denotes effectively pumped transitions. The practical potential of OAS in this respect include a search for laser lines in the far-IR spectral range at optical pumping of the molecules, e.g., CH_3CHF_2 and other molecules with CW CO_2-laser radiation [11.9,12]. Discrete tuning to lines was carried out with a diffraction grating, and continuous tuning within the limits of a Doppler-broadened amplification contour with piezoelectric shift of a cavity mir-

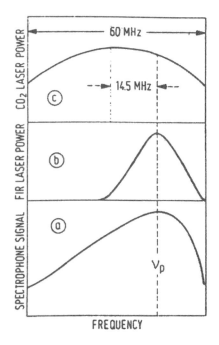

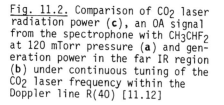

Fig. 11.2. Comparison of CO_2 laser radiation power (**c**), an OA signal from the spectrophone with CH_3CHF_2 at 120 mTorr pressure (**a**) and generation power in the far IR region (**b**) under continuous tuning of the CO_2 laser frequency within the Doppler line R(40) [11.12]

ror (Fig.11.2). In this way two laser lines were found in CH_3CHF_2 in the region of 770 and 663 μm by a nonselective OA Golay detector which detected their power.

It is possible to control the level of pumping power, generation power and to tune, by servosystem, the pumping frequency to the absorption maximum in the active medium using gas OA cells attached to the imput and output windows and directly to the active medium of the optically pumped laser [11.4]. In [11.14] the same technique was used to control the efficiency of optical pumping of CF_4 molecules (at the lasing wavelengths near 16 μm) with a pulsed tunable high-pressure CO_2 laser (Fig.7.3). For this purpose the spectrophone was attached to the cell of a CF_4 laser, cooled down to 120 K.

11.2 Measurement of Thermodynamic Parameters

The dependence of the OA signal on the thermal properties of the medium allows OAS to measure some thermodynamic properties of the medium, including thermal diffusion, sound velocity, etc., and to study phase transitions.

11.2.1 Gaseous Media

In accordance with (2.22-27), in nonresonant spectrophones the dependence of the OA signal on gas heat conductivity K and heat capacity C_V, with part of one of the components in the mixture dominant, has the following forms:

$$P \propto 1 \;/\; \rho C_V \quad , \qquad\qquad\qquad\qquad\qquad\qquad (11.1a)$$

for pulsed and continuous operation, with $\omega\tau_T \gg 1$, and

$$P \propto 1 \;/\; K \quad , \qquad\qquad\qquad\qquad\qquad\qquad\qquad (11.1b)$$

for continuous operation with $\omega\tau_T \ll 1$.

In the simplest case of a binary mixture where one component has a low concentration all the thermodynamic properties affecting the character of the OA signal formation depend on the basic gas component. This enables us to find the heat capacity or heat conductivity (according to the mode of laser operation) of different gases or gas mixtures by comparing the OA signals in these gases and an etalon gas. Nonabsorbing gases can also participate in the measurements. Then we should add a gas in a relatively low concentration (no higher than 0.1 - 1%) with strong absorption at the lines of available lasers. This, for example, may be methane absorbing He-Ne laser radiation with $\lambda = 3.39$ μm. It is also possible to apply OAS to measure the gas heat capacity at constant volume C_V on account of fast heating of these gases by the pulsed radiation, when the role of the walls and gas compression can be neglected.

Information on the influence of various thermodynamic parameters on the formation of OA signals can be obtained from studying the dependence of the OA signal on buffer gas pressure. Figure 11.3 presents such dependences for four gases measured in [11.15] using a nonresonant spectrophone with D = 0.9 cm, L = 6 cm at a modulation frequency of 50 Hz. The behavior of the curves is consistent with the theoretical predictions. The OA signal, for example, is inversely proportional to the gas heat capacity in the high-pressure region with $\omega\tau_T \gg 1$. The presence of a maximum is explained by the fact that the elasticities of the gas and the microphone membrane are equal (Sect.2.3.1). At low pressures the condition $\omega\tau_T \gg 1$ is realized and, according to (2.27), the OA signal is proportional to the gas pressure, and the variation of gas heat conductivity may be important, too.

It is possible to widen the scope of OAS to solve the problems concerned by using resonant spectrophones where additional information on the thermo-dynamic parameters of gases, beside the OA signal amplitude, can also be obtained from the acoustic resonance frequency $\omega_{k,m,n}$ determined from (5.4).

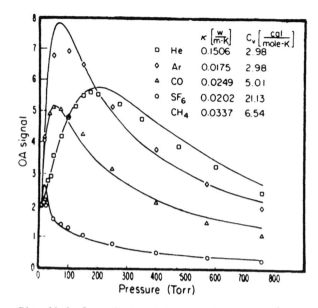

		$\kappa\left[\frac{w}{m\cdot K}\right]$	$C_v\left[\frac{cal}{mole\cdot K}\right]$
□	He	0.1506	2.98
◊	Ar	0.0175	2.98
△	CO	0.0249	5.01
○	SF_6	0.0202	21.13
	CH_4	0.0337	6.54

<u>Fig. 11.3.</u> Dependence of the OA signal amplitude on the pressure of different buffer gases. (——) Calculated values; (□,◊,△,o): experimental results obtained by adding methane to the gases at a pressure of 10 Torr using a He-Ne laser at $\lambda = 3.39$ μm [11.15]

and the system quality Q described by (5.5). The first parameter depends materially on the sound velocity. Q depends on different effects of dissipation of an acoustic wave in a resonant chamber. THOMAS et al. [11.16] systematized the values of these parameters and the signal-to-noise ratio for some gases obtained with a He-Ne laser with $\lambda = 3.39$ μm and a cylindrical resonant spectrophone with $D = L = 10.8$ cm operating at the first radial mode. The measurements were carried out with the absorbing CH_4 molecules added in a low concentration (0.9%) to the gases under study. It was found, in particular, that the maximum signal-to-noise ratios $1.4 \cdot 10^4$ and $1.1 \cdot 10^4$ were obtained in Xe and Kr, respectively. The maximum and minimum values of Q were found for the molecules $SF_6 (Q = 1220)$ and $CO_2 (Q = 250)$, respectively. The frequency of the first resonant longitudinal mode varied from 1522 Hz (for SF_6) to 5074 (for Ne). The calculated and experimentally measured resonant frequencies coincide well, accurately to 1%. The value of the ratio Q_{exp}/Q_{theor}, however, varies from 0.2 to 0.9. The reason for this discrepancy lies in (5.5) for Q_{theor}, which accounts for viscous friction and thermal scattering, but does not take into account other mechanisms of energy dissipation.

These peculiarities make it possible to identify gases by their molecular weight, i.e., to realize rough OA mass spectroscopy through precision measurement of the acoustic resonance frequency, all other parameters being known. In [11.16], for example, it has been shown that this technique enables to differentiate between the $^{12}CO_2$ and $^{13}CO_2$ molecules whose resonant frequencies of the first longitudinal modes equal 3030 and 2994 Hz. Thus, at high Q, OAS allows to estimate isotopic effects in molecules from the resonant-frequency shift and accordingly to identify the isotopic molecules in a mixture.

By measuring the resonant frequency one can estimate the sound velocity V_s and the heat-capacity ratio γ of any mixture, e.g., for $^{13}CO_2$ V_s = 264.4 m/s γ = 1.279 [11.16]. The sound velocity can also be found under pulsed operation from the time in which the sound wave covers the distance between the zone of excitation and the microphone. The accuracy of determining V_s depends on the accuracy of determining the spatial position of the laser beam and the time resolution of the microphone. This technique was used in [11.17] to find the sound velocity in acetylene-air flame with sodium vapor excited by a pulsed-dye laser. It turned out to be $(9.7 \pm 0.5) \cdot 10.4$ cm/s which corresponded to a flame temperature of about $2280 \pm 130°C$.

It should be noted that optothermal spectroscopy (OT) is very useful in determining the heat-diffusion coefficient in gases at different pressures (Chap.4). The measuring technique may take advantage of determining the time of arrival of the heat wave from the zone of excitation to the thermal detector.

A new noncontact OA method for simultaneous measurement of velocity and temperature in a flow to accuracies of 5 cm/s and $0.1°$ C, respectively, was developed in [11.18,19]. It is applicable for pure particulate-free gases and liquids. A pulsed laser generates an acoustic pulse whose propagation in directions along and against the flow are monitored by the deflection of three CW probe beams. This techniques of optical probing of OA pulses provide new opportunities for ultrasonic and spectroscopic measurements in hostile environments (e.g., flames, corrosive fluids, dangerous aerosols, etc.).

11.2.2 Condensed Media

For measuring the thermodynamic parameters in condensed media, OAS with indirect detection of acoustic vibrations in the gas adjacent to the sample is most effective.

Measuring the heat-diffusion coefficient k with OAS is based on the dependence of the OA-signal phase on this parameter. In a first approximation this dependence has the form

$$\Theta = x \sqrt{\frac{\omega}{2k}} \quad , \tag{11.2}$$

where Θ is the OA-signal phase, x is the depth of penetration of heat into the sample. Thus, given k or x, one of these quantities can be determined from (11.2). By this technique, e.g., in [11.20] the value of k for glass was determined to $9.8 \cdot 10^{-3}$ cm^2/s. To first approximation this value is in satisfactory agreement with the tabulated results, like $5 - 7 \cdot 10^{-3}$ cm^2/s. An absorbing layer of black enamel coated on the external side of the window of a glass OA cell served as the source of heat. Such a technique was tried with success in measuring k and the thickness of thin polymer films (from 9 to 30 μm) on a metal substratum [11.21].

At thermal saturation ($\ell_T^S < \ell_S$ and $\ell_T^S \gg \ell_\alpha$) the OA-signal amplitude depends only on the susceptibility coefficient of the sample $B_S = (\rho_S K_S C_S)$ and decreases with increasing frequency as ω. To determine B_S one measures the absolute value of the OA signal and uses (3.22,23), all its constituent parameters known, and compares the OA signals from the sample and the etalon with B_S known. The ratios of these signals for the substance pairs like CdSe/Ge, CdSe/Si and Ge/Si thus obtained in [11.22] are, respectively, 5.54 ± 0.81, 6.86 ± 1.17 and 1.24 ± 0.14. Within an accuracy of 5 to 7% they are in satisfactory agreement with the calculated values.

An important quantity in the annealing processes of an implanted semiconductor by a pulsed radiation, is the thermal conductivity of the damaged region. An experimental method for determination of this quantity for the damaged layer of implanted silicon samples has been reported in [11.23]. The method essentially consists of studying the OA-signal amplitude and phase behavior of implanted silicon in comparison with the crystalline one. Measurements of thermal diffusivities and thermal conductivities of metals, CdS and thin-film by OA technique were reported in [11.24-26].

11.3 Thermal Properties of Solid Surfaces

Furthermore, OAS is extremely sensitive to the state of the surface of optically opaque solids, including semiconductors and metals. Therefore, it can be used to estimate the crystal structure of the surface to study phase transitions, and the process of optical destruction, etc. Generally the expression for the OA signal in indirect detection may be written as

$$P(x, T, t) = K(x, T, t)\, J(t)\, [1 - R(x, T, t)] \quad , \tag{11.3}$$

where J(t) is the radiation power falling onto the sample, $R(x,T,t)$ is the coefficient of reflection from the sample surface depending on the position of the laser beam x, the sample temperature T and the time of irradiation, t. The coefficient $K(x,T,t)$ depends on the optical and geometric properties of the sample. Thermal saturation ($\ell_\alpha \ll \ell_T$), when the OA signal depends neither on optical properties, nor on the absorption coefficient α in particular, is of most interest for analysis of the thermal properties of media. According to (11.3), during beam scanning on the sample surface the variations of the OA signal may also be affected by variations in the reflection coefficient. Therefore, to interpret the measurement results correctly, the reflection co-efficient must be simultaneously controlled by the level of radiation reflec-ted from the sample. Below we consider in more detail possible OA techniques to study the thermal properties of surfaces.

11.3.1 Surface Crystallinity

The thermal and optical properties of substances in crystal and amorphous states are somewhat different. This fact allows approximate diagnosis of these states with OAS, for example, it has been shown [11.27] that as the Ar laser beam shifts from the zone on Si coated with a thin Si film to the clear zone on Si with its crystal structure, the OA signal in gas decreases by about 25%. This results from partial recrystallization of the structure of the Si film as it is deposited on the sample surface. As a consequence it leads to a de-crease in heat conductivity, which causes the OA signal to increase. After the sample was annealed at $1000° \pm 16°C$, within two hours the OA signal in the zone with the film was reduced to the value of the signal in the clear zone. This can be interpreted as restoration of the crystal properties of the Si film.

This technique can also be used to diagnose different defects in the crys-tal structure of semiconductors caused, among other things, by implantation of ions into the surface layer of the sample. Ion implantation produces an amor-phous layer whose thermal and optical properties may differ materially from the properties of a sample in the crystal state. For example, the absorption coefficients in GaAs in the near-IR region in amorphous and crystal states dif-fer by several orders (in the amorphous state it is higher). OAS revealed the distortion of the crystal structure of GaAs caused by implanting Si[+] ions [11.28], using a Nd:YAG CW laser with $\lambda = 1.06$ µm, by scanning the laser spot (diameter about 40 µm) across the sample surface with local zones of ion im-plantation up to 1000 Å in depth. The results are given in Fig.11.4. It may

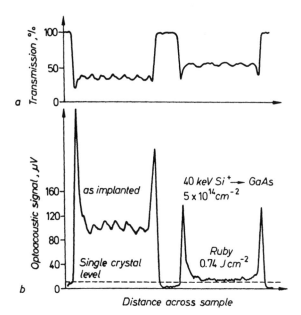

Fig. 11.4. Optical transmission (**a**) and OA signal (**b**) in scanning the beam of a Nd:YAG laser with λ = 1.06 μm on the surface of a GaAs sample before and after irradiation of the layer with implanted Si^+ ions by ruby-laser radiation. The length of the layer in the direction of scanning is about 5 mm [11.28]

be seen that in the region of ion implantation the radiation transmission is smaller and the OA signal is stronger. Increase of the signal in the transient region between the amorphous and crystal structures is attributed in [11.28] to essential changes in the conditions of formation of the OA effect. Figure 11.4 also shows that the action of high-power ruby-laser radiation on the ion layer brings about recrystallization of the surface structure of GaAs, resulting in a decrease of the OA signal.

The advantage of this method of studying defects in crystal structures, compared to such well-known techniques as diffraction of electrons, X-rays or Rutherford scattering, is its relative simplicity, fast response, sufficient spatial resolution and sensitivity. With OAS it is possible, for example, to detect the ion layer with the surface density of ions ranging from 10^{11} to 10^{12} cm^{-2}.

Furthermore, OAS with indirect detection is useful in studying the mechanism of film growth on semiconductors during anodization. This is important for process technology as well as for surface passivation. The efficiency of OAS was demonstrated in [11.29] when observing the growth of Bi_2Te_3 films with a Kr^+ laser with λ = 647 nm and 0.3 W power. WASA et al [11.29] used OA detection of absorption in the film with a piezoelectric detector attached to the sub-stratum. During anodization the OA signal first increased drastically and then slowly changed with small periodic oscillations. These effects are due to two- and three-dimensional growth of the film when, at first, the film grows on the plane to increase the area of some centers and then it begins to grow in thick-ness. It has been shown in [11.29] that the OA signal amplitude for weak ab-sorption is unambiguously related to the film thickness and the value of radi-ation transmission and reflection from the film.

11.3.2 Laser-Light Interaction with Sample

Changes in the surface properties of the sample can be caused also by the ac-tion of the powerful radiation. The interaction of laser radiation with a sample may be followed by an increase in temperature, the destruction and the melting of the sample. To understand these processes, is important for laser technology and for creating optical systems to be employed in high-power lasers. Also OAS enables such processes to be studied, and its distinctive feature is the possibility of simultaneous use of one laser source for simu-lating different effects and one for exciting OA signals to study these effects.

   One of the possible techniques is simultaneous observation of the variations in the OA signal with indirect detection and the reflection coefficient at a constant relatively high level of radiation power. Figure 11.5 shows the typi-cal behavior of such signals when a Ge sample is irradiated by an Ar laser with power of several W [11.30]. It may be seen that after some irradiation

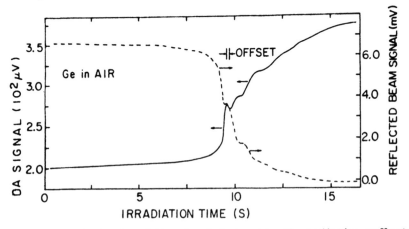

Fig. 11.5. OA signal and the signal induced by the radiation reflected from the surface as a function of time of irradiation, for a Ge sample irradiated by an argon laser of several-watts power. The dimension of the laser spot on the sample is about 10 μm. The modulation frequency is 400 Hz [11.30]

the OA signal increases drastically, caused by the effect of sample destruction. The small negative peak in the OA signal arising during the ninth minute of irradiation corresponds to sample melting in the central zone of the laser beam. A further rather slight increase in the OA signal is connected with the oxidation process and is followed by darkening of the sample surface. Normalizing the OA signal amplitude to the value $1-R$ enables to determine variation in the thermal properties of a sample from (11.3). This technique also makes it possible to define the requirements on radiation power and irradiation duration during which there is no appreciable destruction of the sample surface, yet. In [11.30], for example, the time parameters relating to the processes of thermal etching and oxidation of the sample were determined.

Simultaneous detection of OA and fluorescence signals may increase the reliability of determining optical damages in a sample. For example, the formation of nonradiative centers on the surface of a sample was detected from the increase of the OA signal and at the same time the decrease of the fluorescence signal.

11.3.3 Observation of Phase Transitions

In the region of a phase transition the value of heat capacity C changes considerably. According to (3.22,23), the OA signal with indirect detection depends on C. For optically opaque and thermally thick samples (typical of analysis of metals) the OA signal $P \propto \ell_T^g C^{-1/2} T^{-1}$ and for optically transparent samples $(\ell_\alpha \gg \ell_T)$ $P \propto \ell_T^g C^{-1} T^{-1}$.

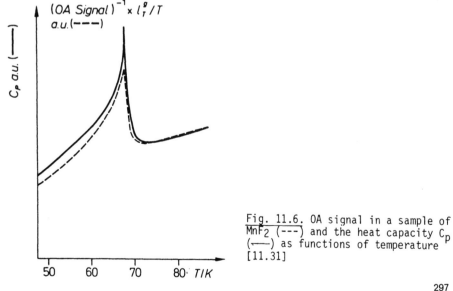

Fig. 11.6. OA signal in a sample of MnF$_2$ (---) and the heat capacity C_p (———) as functions of temperature [11.31]

This dependence enables to investigate different phase transitions in a sample by studying the behavior of the OA signal at varying sample temperature. For correct measurements the OA cell must be calibrated with temperature, using, in particular, a sample with known dependences of its thermal properties on temperature. In [11.31], variations of heat capacity in C_2Cl_3 and MnF at temperatures of 16 and 68 K, respectively, were observed with an Ar laser. These results agree well with the results by other techniques. The temperature calibration of the OA cell was performed in the low-temperature region with a Mn sample. It is shown that the ratio $\ell\frac{g}{T}$ / T and hence the OA cell sensitivity slightly depend on temperature over the range 20 - 300 K. The measurements were carried out in a gas-flowing cryostat, an OA cell with a polipropylene electret microphone built into it. Figure 11.6 presents the results of measurements with MnF corresponding to analysis of an optically transparent sample. It may be seen that the results correlate well with the data available.

12. Conclusion

The main purpose of this monograph is to review the properties and characteristics of OA spectroscopy with laser-radiation sources and the perspectives
of its application in different fields of science and engineering.

The analysis of recent publications yields that the field develops in
several directions. First, the known methods of measurement and the OA-cell
designs undergo continuous improvements. This refers to a widening of the
temperature range for the OA method, the use of various types of modulation
as well as the employment of Hadamar and Fourier transformations, the correlation technique, etc. Second, the development of the OA method is aimed
at studying and using a number of non-completely thermal mechanisms of sound
generation in condensed media. This holds true for such effects as thermal
evaporation, optical breakdown and others. This is explained by the fact that
the efficiency of transformation of light into acoustic energy in the case
of a thermal mechanism varies from 10^{-6} to 0.1% according to the excitation
mode and the type of irradiated media. In the case of optical breakdown it
ranges from 10 to 30%. This makes it possible to design tiny laser-acoustic
transformers which are promising for contactless ultrasonic defectoscopy and
for diagnostics in medicine.

When highly absorbing media are irradiated by powerful ultrashort laser
pulses, they can develop acoustic disturbances with amplitudes of up to hundreds and thousands of atmospheres. It is useful to apply this effect in
medicine.

In recent years we saw the development of methods to acquire information
on composition and properties of the medium by recording the power absorbed
in it using thermal phenomena. Among them there is the opto-thermal method
based on the contact measurement of the sample's temperature (Chap.4); the
opto-radiometric method based on measuring the secondary infrared radiation
[12.1-3]; the opto-refraction methods in which the thermal variations of the
index of refraction are determined from variations in the parameters of the
additional probe beam - its defocusing (the thermal lens technique [12.4-7],

deflection (the deflection method with the use of position - sensitive detectors [12.8-11] or the Moire-band technique [12.12], the phase (interference [12.13,14] and heterodyne methods [12.15,16], diffraction on thermal refraction gratings formed by the intereference of two beams of splitted radiation (the thermal-grating method [12.17-19]; opto-geometrical methods in which the thermal variations in the dimensions of the sample are detected [12.20,21]). Furthermore, the technique can be utilized to determine thermal diffusivity in a variety of thin-film samples with thickness of tens of micrometers [12.22]. The development of these methods and their unification make it possible to widen greatly the limits of applicability.

It can be emphasized with good reason that the OA effect and other calorimetric methods have not exhausted their potentialities yet and the scope of their numerous applications in different fields of science and technology can be widened.

References

Chapter 1

1.1   A.G. Bell: Am. J. Sci. *20*, 305 (1880)
1.2   A.G. Bell: Pholos. Mag. *11*, 510 (1881)
1.3   J. Tyndall: Proc. R. Soc. London *31*, 307 (1881)
1.4   W.G. Roentgen: Philos. Mag. *11*, 308 (1881)
1.5   M.L. Viengerov: Dokl. Akad. Nauk SSSR (Russian) *19*, 687 (1938)
1.6   A.H. Pfund: Science *90*, 326 (1939)
1.7   K.F. Luft: Z. Tech. Phys. *24*, 97 (1943)
1.8   D.W. Hill, T. Powell: *Non-Dispersive Infrared Gas Analyses in Science, Medicine and Industry* (Plenum, New York 1968)
1.9   G. Gorelik: Dokl. Akad. Nauk SSSR (Russian) *54*, 779 (1946)
1.10 M.L. Viengerov, P.V. Slobodskaya: Izv. Akad. Nauk SSSR (Russian), Fiz. *11*, 420 (1947)
1.11 M. Goley: Rev. Sci. Instr. *18*, 347 (1947)
1.12 P.I. Bresler, B.N. Ruzin: Optica i Spectroscopia (Russian) *9*, 22 (1960)
1.13 W.D.Hershberger, E.D. Bush, G.W.Leek: RCA Rev. *7*, 422 (1946)
1.14 Ya.I. Gerlovin: Optica i Spectroscopia (Russian) *7*, 571 (1959)
1.15 W.R. Harshbarger, M.B. Robin: Acc. Chem. Res. *6*, 329 (1973)
1.16 M.L. Viengerov, Ya.I. Gerlovin, N.A. Pankratov: Optica i Spectroscopia (Russian) *1*, 1023 (1956)
1.17 B.I. Stepanov: *Fundamentals of Spectroscopy of Negative Light Flows* (Russian) (Belorussian Univ. Press, Minsk 1961)
1.18 D.O. Gorelik, B.B. Sakharov: *Optoacoustic Effect in Physical-Chemical Measurement* (Russian) (USSR Committee of Standarts, Moscow 1969)
1.19 E.L. Kerr, J.G. Atwood: Appl. Opt. *7*, 915 (1968)
1.20 L.B. Kreuzer: J. Appl. Phys. *42*, 2934 (1971)
1.21 L.B. Kreuzer, C.K.N. Patel: Science *173*, 45 (1971)
1.22 L.B. Kreuzer, N.P. Kenyon, C.K.N. Patel: Science *177*, 349 (1972)
1.23 L.B. Kreuzer: Anal. Chem. *46*, 235A (1974)
1.24 C.K.N. Patel, E.C. Burkhard, C.A. Lambert: Science *184*, 1173 (1974)
1.25 A.S.Gomenyuk, V.P. Zharov, D.D. Ogurok, E.A. Ryabov, O.A. Tumanov, V.O. Shaidurov: Kvantovaya Elektronika (Russian) *1*, 1805 (1974)
1.26 V.B. Anzin, M.V. Glushkov, V.P. Zharov, Yu.V. Kosichkin, V.O. Shaidurov, A.M. Shirokov: Kvantovaya Elektronika (Russian) *2*,1403 (1975)
1.27 A.M. Angus, E.E. Marinero, M.J. Colles: Opt. Commun. *14*, 223 (1975)
1.28 C.F. Dewey Jr., R.D. Kamm, C.E. Hackett: Appl. Phys. Lett. *23*, 633 (1973)
1.29 P.D. Goldan, K. Goto: J. Appl. Phys. *45*, 2058 (1975)
1.30 T.J. Bridges, E.G. Burkhardt: Opt. Commun. *22*, 248 (1977)
1.31 C.K.N. Patel, R.J. Kerl, E.G. Burkhardt: Phys. Rev. Lett. *38*, 1204 (1977)
1.32 M.J. Kavaya, J.S. Margolis, M.S. Shumate: Appl. Opt. *18*, 2801 (1979)
1.33 L.B. Kreuzer: Anal. Chem. *50*, 597A (1978)
1.34 S.P. Belov, A.B. Burenin, L.N. Gershtein, V.V. Korolikhin, A.F. Krupnov: Optica i Spectroscopia (Russian) *35*, 295 (1973)

1.35   A.F. Krupnov, A.V. Burenin: *Molecular Spectroscopy: Modern Research,* Vol. 2 (Academic Press, New York 1976) p. 93
1.36   V.N. Bagratashvili, I.N. Knyazev, V.S. Letokhov, V.V. Lobko: Opt. Commun. *18*, 525 (1976)
1.37   V.N. Bagratashvili, V.P. Zharov, V.V. Lobko: Kvantovaya Elektronika (Russian) *5*, 637 (1978)
1.38   V.P. Zharov, V.S. Letokhov, E.A. Ryabov: Appl. Phys. *12*, 15 (1977)
1.39   V.S. Letokhov: In *Multiphoton Processes,* ed. by J.H. Eberly, P. Lambropoulos (Wiley, New York 1978) p. 331
1.40   A.G. Gomenyuk, V.P. Zharov, V.S. Letokhov, E.A. Raybov: Kvantovaya Elektronika (Russian) *3*, 369 (1976)
1.41   V.S. Letokhov: In *Fundamental and Applied Laser Physics* (Wiley, New York 1973) p. 335
1.42   E.E. Marinero, M. Stuke: Opt. Commun. *30*, 349 (1979)
1.43   J.J. Barrett, M.J. Berry: Appl. Phys. Lett. *34*, 144 (1979)
1.44   A. Rosencwaig: Opt. Commun. *7*, 305 (1973)
1.45   A. Rosencwaig: Science *181*, 657 (1973)
1.46   A. Rosencwaig: Anal. Chem. *47*, 592A (1975)
1.47   A. Rosencwaig: Physics Today *28*, 23 (1975)
1.48   E.L. Kerr: Appl. Opt. *12*, 5250 (1973)
1.49   J.G. Parker: Appl. Opt. *12*, 2974 (1973)
1.50   H.S. Bennet, R.A. Forman: Appl. Opt. *14*, 3031 (1975)
1.51   H.S. Bennet, R.A. Forman: Appl. Opt. *15*, 2405 (1976)
1.52   H.S. Bennet, R.A. Forman: J. Appl. Phys. *48*, 1432 (1977)
1.53   P.E. Nordal, S.O. Kanstad: Opt. Commun. *26*, 367 (1978)
1.54   Y.H. Wong, R.L. Thomas, C.F. Hawkins: Appl. Phys. Lett. *32*, 538 (1978)
1.55   A. Hordvik: J. Opt. Soc. Am. *66*, 1105 (1976)
1.56   A. Hordvik, L. Cholnik: Appl. Opt. *16*, 2919 (1977)
1.57   M.J. Adams, G.F. Kirkbright: Spectrosc. Lett. *9*, 225 (1976)
1.58   E.F. Gross, Ya.Ta. Abolin'sh, A.A. Shyltin: Zh. Techn. Fiz. (Russian) *28*, 832 (1958)
1.59   E.F. Gross, Ya.Ya. Abolin'sh, A.A. Shultin: Zh. Techn. Fiz. (Russian) *28*, 329 (1958)
1.60   A.M. Bonch-Bruevich, T.K. Razumova, I.O. Starobogatov: Pis'ma Zh. Techn. Fiz. (Russian) *1*, 65 (1975)
1.61   C.K.N. Patel, A.C. Tam: Appl. Phys. Lett. *34*, 467 (1979)
1.62   A.M. Bonch-Bruevich, T.K. Razumova, I.O. Starobogatov: Optica i Spectroscopia (Russian) *42*, 82 (1977)
1.63   I.O. Starobogatov: Optica i Spectroscopia (Russian) *42*, 304 (1977)
1.64   T.K. Razumova, I.O. Starobogatov: Optica i Spectroscopia (Russian) *42*, 489 (1977)
1.65   A.C. Tam, C.K.N. Patel, R.J. Kerl: Opt. Lett. *4*, 81 (1979)
1.66   C.K.N. Patel, A.C. Tam: Appl. Phys. Lett. *34*, 760 (1979)
1.67   C.L. Sam, M.L. Shand: Opt. Commun. *31*, 174 (1979)
1.68   C.F. Dewey: Opt. Engin. *13*, 483 (1974)
1.69   L.G. Rosengren: Appl. Opt. *14*, 1960 (1975)
1.70   Y.H. Pao (ed.): *Optoacoustic Spectroscopy and Detection* (Academic Press, New York 1977)
1.71   A. Rosencwaig: *Photoacoustics and Photoacoustic Spectroscopy* (Wiley, New York 1980)
1.72   V.P. Zharov: *Development and Research of Laser Optoacoustic Instruments for Analysis of Gas Media* (Russian), Dissertation, MVTU Moscow (1977)
1.73   M.J. Colles, N.R. Geddes, E. Mehdizadch: *The Optoacoustic Effect,* Contemp. Phys. *20*, 11 (1979)
1.74   H. Walther (ed.): *Laser Spectroscopy of Atoms and Molecules,* Topics Appl. Phys., Vol. 2 (Springer, Berlin, Heidelberg, New York 1976)
1.75   K. Shimoda (ed.): *High-Resolution Laser Spectroscopy,* Topics Appl. Phys. Vol. 13 (Springer, Berlin, Heidelberg, New York 1976)

1.76 V.S. Letokhov, V.P. Chebotayev: *Nonlinear Laser Spectroscopy*, Springer Ser. Opt. Sci., Vol. 4 (Springer, Berlin, Heidelberg, New York 1977)
1.77 V.J. Balyikin, G.I. Bekov, V.S. Letokhov, V.I. Mishin: Uspekhi Fiz. Nauk (Russian) *132*, 293 (1980)
1.78 G.S. Hurst, M.G. Payne, S.O. Kramer, J.P. Young: Red. Mod. Phys. *51*, 767 (1979)
1.79 S.L. Shapiro (ed.): *Ultrashort Light Pulses, Picosecond Techniques and Applications*, Topics Appl. Phys., Vol. 18 (Springer, Berlin, Heidelberg, New York 1976); M.S. Feld, V.S. Letokhov (eds.): *Coherent Nonlinear Optics*, Topics Current Phys., Vol. 21 (Springer, Berlin, Heidelberg, New York 1980) Chap. 6
1.80 E.D. Hinkley (ed.): *Laser Monitoring of the Atmosphere*, Topics Appl. Phys., Vol. 14 (Springer, Berlin, Heidelberg, New York 1976)
1.81 D.K. Killinger, A. Mooradian (eds.): *Optical and Laser Remote Sensing*. Springer Ser. Opt. Sci., Vol. 39 (Springer, Berlin, Heidelberg, New York 1983)
1.82 H. Möenke, L. Möenke: *Einführung in die Laser-Micro-Emissionspectralanalyse* (Akademische Verlagsgesellschaft, Leipzig 1966)
1.83 V.S. Letokhov: *Laser Spektroskopie* (Academic Verlag, 1977); (Vieweg, Braunschweig 1978)
1.84 V.S. Letokhov: Opt. and Laser Technol. Oct., p. 217 (1977); Dec. p. 263 (1977); Febr. p. 15 (1978); June, p. 129 (1978); Oct. p. 247 (1978), Dec. p. 301 (1978); Febr. p. 13 (1979)
1.85 W. Demtröder: *Laser Spectroscopy*, Springer Ser. Chem. Phys., Vol. 5 (Springer, Berlin, Heidelberg, New York 1981)
1.86 L.A. Pakhomyicheva, E.A. Sviridenkov, A.F. Suchkov, L.V. Titova, S.S. Churilov: Pis'ma Zh. Eksp. i Teor. Fiz. (Russian) *12*, 60 (1970)
1.87 N.C. Peterson, M.J. Kurylo, W. Braun, A.M. Bass, R.A. Keller: J. Opt. Soc. Am. *61*, 746 (1971)
1.88 T.E. Deaton, P.A. Depatie, T.W. Walker: Appl. Phys. Lett. *26*, 300 (1975)
1.89 L.G. Rosengren: Infr. Phys. *13*, 173 (1973)
1.90 C. Hurtung, R. Jurgeit: Kvantovaya Elektronika (Russian) *5*, 1825 (1978)
1.91 C. Hurtung, R. Jurgeit: Optika i Spectroscopia (Russian) *46*, 1169 (1979)
1.92 L.M. Dorozhkin, V.P. Zharov, G.N. Makarov, A.A. Puretzky: Pis'ma Zh. Techn. Fiz. (Russian) *6*, 979 (1980)
1.93 A. Hordvik: Appl. Opt. *16*, 2827 (1977)
1.94 P.-E. Nordal, S.O. Kanstad: Physica Scripta. *20*, 659 (1979)
1.95 A.S. Gomenyuk, V.P. Zharov, V.B. Pasetzky: Optika i Spectroscopia (Russian) *51*, 892 (1981)
1.96 S.R.J. Brueck, H. Kildal, L.J. Belanger: Opt. Commun. *34*, 199 (1980)
1.97 J.P. Gordon, R.C.C. Leite, R.S. Moore, S.P.S. Porto, J.R. Whinnery: J. Appl. Phys. *36*, 3 (1965)
1.98 G.W. Flynn: In *Chemical and Biochemical Applications of Lasers*, Vol. 1, ed. by C.B. Moore (Academic, New York 1974) p. 163
1.99 A.C. Boccara, D. Fournier, J. Badoz: Appl. Phys. Lett. *36*, 199 (1980)
1.100 W.B. Jackson, N.M. Amer, A.C. Boccara, D. Fournier: Appl. Opt. *20*, 1333 (1981)
1.101 J. Stone: J. Opt. Soc. Am. *62*, 327 (1972)
1.102 V.M. Gordienko, A.B. Reshilov, V.H. Shmal'gayzen: Vestnik of Moscow State University, Fizika i Astronom. (Russian) *19*, 59 (1978)

Chapter 2

2.1 A.G. Bell: Am. J. Sci. **20**, 305 (1880)
2.2 B.I. Stepanov, O.P. Girin: Zh. Eksp. i Teor. Fiz. (Russian) **20**, 947 (1950)

2.3 R. Kaizer: Canad. J. Phys. **37**, 1499 (1959)
2.4 L.B. Kreuzer: J. Appl. Phys. **42**, 2934 (1971)
2.5 L.G. Rosengren: Inf. Phys. **13**, 109 (1973)
2.6 V.P. Zharov: Development and Research of Laser Optoacoustic Instruments for Analysis of Gas Media, Ph.D. Thesis (Russian), MVTU, Moscow (1977)
2.7 P.G. Kryukov, V.S. Letokhov: Uspekhi Fiz. Nauk (Russian) **99**, 169 (1969)
2.8 V.S. Letokhov, A.A. Makarov: Zh. Eksp. i Teor. Fiz (Russian) **63**, 2064 (1972)
2.9 A.O. Sall: Zh. Techn. Fiz. (Russian) **26**, 157 (1956)
2.10 A.O. Sall: Zh. Techn. Fiz. (Russian) **29**, 330 (1959)
2.11 E.L. Kerr, J.G. Atwood: Appl. Opt. **7**, 915 (1968)
2.12 L.N. Vereshchagina, V.P. Zharov, G.I. Shipov, V.I. Shtepa: Zh. Techn. Fiz. (Russian) **54**, 343 (1984)
2.13 A.V. Burenin: Izv. VUZov, Radiofizika (Russian) **27**, 129 (1974)
2.14 Y.H. Pao (ed.) Optoacoustic Spectroscopy and Detection (Academic Press, New York, 1977)
2.15 R.A. Smith, F.E. Jones, R.P. Chasmar: *The Detection and Measurement of Infrared Radiation* (Clarendon, Oxford 1957)
2.16 V. Tarnov: Techn. Rev. (Bruel & Kjaer Co.) **3**, 3 (1972)
2.17 S.P. Belov, A.B. Burenin, L.N. Gershtein, V.V. Korolikhin, A.F. Krupnov: Optica i Spectroscopia (Russian) **35**, 295 (1973)
2.18 G.M. Sessler: J. Acous. Soc. Amer. **35**, 1354 (1963)
2.19 C.K.N. Patel, R.J. Kerl: Appl. Phys. Letter **30**, 578 (1977)
2.20 M. Gay: La Recherche aerospatiale **113**, 25 (1966)
2.21 A.F. Krupnov, A.V. Burenin: Molecular Spectroscopy: Modern Research, vol. 11 (Academic Press, New York, San Francisco, London (1976) p. 93
2.22 L.J. Gershtein: Izv. VUZov, Radiofizika (Russian) **20**, 223 (1977)
2.23 J.G. Choi, G.J. Diebold: Appl. Opt. **21**, 4087 (1982)
2.24 D.H. Leslie, R.O. Miles, A. Dandridge: J. Physique **44**, Suppl. 6, 537 (1983)

Chapter 3

3.1 E.F. Gross, Ya.Ya. Abolinsh: Zh. Techn. Fiz. (Russian) **28**, 832 (1958)
3.2 C.K.N. Patel, A.C. Tam: Appl. Phys. Lett. **34**, 467 (1979)
3.3 S.R.J. Brueck, H. Kildal, L.J. Belanger: Opt. Commun. **34**, 199 (1980)
3.4 J.P. Gordon, R.C.C. Leite, R.S. Moore, S.P.S. Porto, J.R. Whinnery: J. Appl. Phys. **36**, 3 (1965)
3.5 G.W. Flynn: In: Chemical and Biochemical Applications of Lasers, C.B. Moore (ed.) vol. 1 (Academic Press, New York, 1974) p. 163
3.6 C. Hu, J.R. Whinnery: Appl. Opt. **12**, 72 (1973)
3.7 J. Stone: J. Opt. Soc. Am. **62**, 327 (1972)
3.8 E.L. Moses, C.L. Tang: Opt. Lett. **1**, 115 (1977)
3.9 R.M. White: Report TIS-232, TWT Prod. Sect. General Electric Co., Palo Alto California (1962)
3.10 A.J. De Maria: Proc. IEEE **52**, 96 (1964)
3.11 R.G. Brewer, K.E. Rieckhoff: Phys. Rev. Lett. **13**, 334 (1964)
3.12 L.S. Gourney: J. Acous. Soc. Amer. **40**, 1322 (1966)
3.13 F.V. Bunkin, V.M. Komisarov: Akust. Zh. (Russian) **19**, 305 (1973)
3.14 F.V. Bunkin, M.J. Tribel'skii: Uspekhi Fiz. Nauk (Russian) **130**, 193 (1980)
3.15 A.I. Bozhkov, F.V. Bunkin, A.A. Kolomenskii et al.: Sov. Sci. Rev. A - Phys. Rev. **3**, 459 (Harward, New York 1981)
3.16 L.M. Lyamshev, K.A. Naugol'nyikh: Akust. Zh. (Russian) **27**, 641 (1981)
3.17 A.M. Bonch-Bruevich, T.K. Razumova, I.O. Starobogatov: Pis'ma Zh. Techn. Fiz. (Russian) **1**, 65 (1975)
3.18 A.M. Bonch-Bruevich, T.K. Razumova, I.O. Starobogatov: Optica i Spectroscopia (Russian) **42**, 304 (1977)

3.19 I.O. Starobogatov: Optica i Spectroscopia (Russian) **42**, 304 (1977)
3.20 T.K. Razumova, I.O. Starobogatov: Optica i Spectroscopia (Russian) **42**, 489 (1977)
3.21 A.C. Tam, C.K.N. Patel, R.J. Kerl: Opt. Lett. **4**, 81 (1979)
3.22 C.K.N. Patel, A.C. Tam: Appl. Phys. Lett. **34**, 760 (1979)
3.23 Y. Kohanzadeh, J.R. Whinnery, M.M. Carroll: J. Acous. Soc. Amer. **57**, 67 (1975)
3.24 W. Lahmann, H.J. Ludewig, H. Welling: Anal. Chem. **49**, 549 (1977)
3.25 P. Sladky, R. Danielius, V. Strutkaitis, M. Boudys: Czech. J. Phys. B **27**, 1075 (1977)
3.26 W. Lahmann, H.J. Ludewig: Chem. Phys. Lett. **45**, 177 (1977)
3.27 S. Oda, T. Sawada, H. Kamada: Anal. Chem. **50**, 865 (1978)
3.28 C.K.N. Patel, A.C. Tam: Rev. Mod. Phys. **53**, 517 (1981)
3.29 A.C. Tam, H. Coufal: J. de Physique, **44**, Suppl. 10, Coll. C6, 9 (1983)
3.30 A.M. Bonch-Bruevich, T.K. Razumova, I.O. Starobogatov: Zh. Prikl. Spektroskopii (Russian) **37**, 981 (1982)
3.31 A.M. Bonch-Bruevich, T.K. Razumova, I.O. Starobogatov: Optica i Spectroscopia (Russian) **36**, 692 (1974)
3.32 E.T. Nelson, C.K.N. Patel: Appl. Phys. Lett. **39**, 537 (1981)
3.33 V.S. Gorodetskii, S.V. Egerev, J.B. Osipov, K.A. Naugol'nyikh: Kvantovaya Elektronika (Russian) **5**, 2396 (1978)
3.34 Gu Liu: Appl. Opt. **21**, 955 (1982)
3.35 J.-M. Heritier: Opt. Commun. **44**, 267 (1983)
3.36 A.J. Bozhkov, F.V. Bunkin: Kvantovaya Elektronika (Russian) **2**, 1763 (1975)
3.37 A. Atalar: Appl. Opt. **19**, 3204 (1980)
3.38 T. Kitamori, T. Sawada: Jap. Appl. Phys. **21**, L285 (1982)
3.39 J.-M. Heritier, J.E. Fouquet, A.E. Siegman: Appl. Opt. **21**, 90 (1982)
3.40 A.C. Tam, C.K.N. Patel: Appl. Opt. **18**, 3348 (1979)
3.41 A.C. Tam, C.K.N. Patel: Opt. Lett. **5**, 27 (1980)
3.42 R.M. White: J. Appl. Phys. **34**, 3559 (1963)
3.43 E.F. Carome, N.A. Clark, C.E. Moeller: Appl. Phys. Lett. **4**, 95 (1964)
3.44 N. Kroll: J. Appl. Phys. **36**, 34 (1965)
3.45 S.S. Penner, O.P. Sharma: J. Appl. Phys. **37**, 2304 (1966)
3.46 M.J. Brienza, A.J. DeMaria: Appl. Phys. Lett. **11**, 44 (1967)
3.47 C.M. Percival: J. Appl. Phys. **38**, 5313 (1967)
3.48 R.E. Lee, R.W. White: Appl. Phys. Lett. **12**, 12 (1968)
3.49 J.C. Bushnell, D.J. McClosky: J. Appl. Phys. **39**, 554 (1968)
3.50 A. Hordvik, L. Ckolnik: Appl. Opt. **16**, 2919 (1977)
3.51 A. Hordvik: J. Opt. Soc. Am. **66**, 1105 (1976)
3.52 A. Hordvik, H. Schlossberg: J. Opt. Soc. Am. **65**, 1165 (1975)
3.53 A. Hordvik, H. Schlossberg: Appl. Opt. **16**, 101 (1977)
3.54 W. Jackson, N.M. Amer: J. Appl. Phys. **51**, 3343 (1980)
3.55 A. Hordvik, H. Schlossberg, H. Miller, C. Gallagher: RADC-TR-76-70 (1976)
3.56 M.M. Farrow, R.K. Burnham, M. Auzanneau, S.L. Olsen, N. Purdie, E.M. Eyring: Appl. Opt. **17**, 1093 (1978)
3.57 Y.N. Lokhov, V.S. Mospanov, Y.D. Fiveiski: Kvantovaya Elektronika (Russian) **1**, 67 (1971)
3.58 G.C. Wetsel: Appl. Phys. Lett. **41**, 511 (1982)
3.59 R.J. von Gutfeld, R.L. Melcher: Appl. Phys. Lett. **30**, 257 (1977)
3.60 A.C. Tam, H. Coufal: Appl. Phys. Lett. **42**, 33 (1983)
3.61 M.R. Fisher, D.M. Fasano, N.S. Nogar: Appl. Spectrosc. **36**, 125 (1982)
3.62 H.J. Nussbaumer: *Fast Fourier Transform and Convolution Algorithms*, 2nd ed., Springer Ser. Inform. Sci., Vol. 2 (Springer, Berlin, Heidelberg 1982)
3.63 B. Betz, W. Arnold: J. de Physique **44**, Suppl. 10, Coll. C6, **61** (1983)
3.64 J.B. Callis: J. Res. Nat. Bur. Stand. Sect. A **80**, 413 (1976)

3.65 Y.H. Pao (ed.): Optoacoustic Spectroscopy and Detection (Academic Press, New York, 1977)
3.66 A. Rosencwaig: Photoacoustic and Photoacoustic Spectroscopy (J. Wiley and Sons, New York, Chichester, Brisbane, Toronto, 1980)
3.67 E.L. Kerr: Appl. Opt. **12**, 5250 (1973)
3.68 J.G. Parker: Appl. Opt. **12**, 2974 (1973)
3.69 H.S. Bennett, R.A. Forman: Appl. Opt. **14**, 3031 (1975)
3.70 H.S. Bennett, R.A. Forman: J. Appl. Phys. **48**, 1432 (1977)
3.71 Techn. Digest 4th Intern. Conf. on Photoacoustic, Thermal and Related Sciences, August 4-8, Ecole Polytechnique de Montreal, 1985
3.72 J.F. McClelland, R.N. Knizely: Appl. Opt. **15**, 2658 (1976)
3.73 H.S. Bennett, R.A. Forman: Appl. Opt. **15**, 347 (1976)
3.74 A. Rosencwaig, A. Gersho: J. Appl. Phys. **47**, 64 (1976)
3.75 A. Rosencwaig, A. Gersho: Science **190**, 556 (1975)
3.76 L.C. Aamodt, J.C. Murphy, J.G. Parker: J. Appl. Phys. **48**, 927 (1977)
3.77 H.C. Bennet, R.A. Forman: Appl. Opt. **16**, 2834 (1977)
3.78 H.C. Bennet, R.A. Forman: J. Appl. Phys. **48**, 1217 (1977)
3.79 L.C. Aamodt, J.C. Murphy: J. Appl. Phys. **49**, 3036 (1978)
3.80 A. Rosencwaig: J. Appl. Phys. **49**, 2905 (1978)
3.81 F.A. McDonald, G.C. Wetsel: J. Appl. Phys. **49**, 2323 (1978)
3.82 A. Mandelis, B.S.H. Royce: J. Appl. Phys. **50**, 4330 (1979)
3.83 C.L. Cesar, H. Vargas, J.A. Meyer, L.V.M. Miranda: Phys. Rev. Lett. **42**, 1570 (1979)
3.84 R.S. Quimby, W.M. Yen: Appl. Phys. Lett. **35**, 43 (1979)
3.85 F.A. McDonald: Appl. Opt. **28**, 1363 (1979)
3.86 N.C. Fernelius: J. Appl. Phys. **52**, 650 (1980)
3.87 A. Mandelis, B.S.H. Royce: J. Appl. Phys. **51**, 610 (1980)
3.88 R.L. Thomas, J.J. Pouch, Y.H. Wong, L.D. Favro, P.K. Kuo: J. Appl. Phys. **51**, 1152 (1980)
3.89 F.A. McDonald: Appl. Phys. Lett. **36**, 123 (1980)
3.90 F.A. McDonald: J. Photoacoust. **1**, 171 (1982)
3.91 F.A. McDonald: J. de Physique **44**, Suppl. 10, Coll. C6, 21 (1983)
3.92 P.K. Kuo, L.D. Favro: Appl. Phys. Lett. **40**, 1013 (1983)
3.93 P. Helander: J. Appl. Phys. **54**, 3410 (1983)
3.94 T. Somasundaram, P. Ganguly: J. Physique **44**, Suppl. 10, Coll. C6, 239 (1983)
3.95 J.-P. Monchalin, J.-L. Parpal, L. Bertrand et al.: J. Appl. Phys. **53**, 8525 (1982)
3.96 A. Rosencwaig: Rev. Sci. Instrum. **48**, 1133 (1977)
3.97 D. Cahen, E.J. Lerner, A. Auerbach: Rev. Sci. Instrum. **49**, 1206 (1978)
3.98 D. Bucharme, A. Tessier, R.M. Leblanc: Rev. Sci. Instrum. **50**, 1461 (1979)
3.99 J.F. McClelland, R.N. Knizely: Appl. Opt. **15**, 2967 (1976)
3.100 C.L. Sam, M.L. Shand: Opt. Commun. **31**, 174 (1979)
3.101 A.C. Tam, C.K.N. Patel: Appl. Phys. Lett. **35**, 843 (1979)

Chapter 4

4.1 L.G. Rosengren: Infrar. Phys. **13**, 173 (1973)
4.2 C. Hurtung, R. Jurgeit: Kvantovaya Elektronika (Russian) **5**, 1825 (1978)
4.3 C. Hurtung, R. Jurgeit: Optika i Spectroscopia (Russian) **46**, 1169 (1979)
4.4 L.M. Dorozhkin, V.P. Zharov, G.N. Makarov, A.A. Puretzky: Pis'ma Zh. Techn. (Russian) Fiz. **6**, 979 (1980)
4.5 A. Hordvik: Appl. Opt. **16**, 2827 (1977)
4.6 G.H. Brilmyer, A. Fujishima, K.S.V. Santhanam, A.J. Bard: Anal. Chem. **49**, 2057 (1977)

4.7   Yu. Gershenson, V.B. Rosenshtein, S.L. Umansky: in *Chemistry of Plasma* (Russian), ed. by B.M. Smirnov (Atomizdat, Moscow 1977) Vol.4, p.61
4.8   J.J. Steinfeld, I. Burak, P.G. Sutton, A.V. Novak: J. Chem. Phys. **52**, 5421 (1970)
4.9   T.E. Gough, R.E. Miller, G. Scoles: Appl. Phys. Lett. **30**, 338 (1977)
4.10 R.J. Keyes: *Optical and Infrared Detectors*, Topics Appl. Phys., Vol.19, (Springer, Berlin, Heidelberg 1980)
4.11 R.A. Smith, F.E. Jones, R.P. Chasmar: Detection and Measurement of Infrared Radiation (Osford at the Clarendon Press 1977)
4.12 R. DeWaard, E.M. Wormser: Proc. IRE **47**, 1508 (1959)
4.13 V.K. Novak, N.D. Gavrilova, I.B. Feldman: *Pyroelectric Converters* (Russian) (Sov. Radio, Moscow 1979)
4.14 D.A. Pinnow, T.C. Rich: Appl. Opt. **12**, 984 (1973)
4.15 M. Haas, J.W. Davisson, H.B. Rosenstock, J. Babiskin: Appl. Opt. **14**, 1128 (1975)
4.16 E. Bernal: Appl. Opt. **14**, 314 (1975)
4.17 A. Zaganiaris: Phys. Chem. Glasses **17**, 83 (1976)
4.18 J.A. Harrington, D.A. Gregory, W.F. Otto: Appl. Opt. **15**, 1953 (1976)
4.19 J.A. Harrington, D.A. Gregory: Appl. Opt. **15**, 2075 (1976)
4.20 H.B. Resenstock, M. Haas, D.A. Gregory, J.A. Harrington: Appl. Opt. **16**, 2837 (1977)
4.21 V.G. Artyushenko, E.M. Dianov, E.P. Nikitin: Kvantovaya Elektronika (Russian) **5**, 1065 (1978)
4.22 R.V. Ambartzumian, L.M. Dorozhkin, G.N. Makarov et al: Appl. Phys. **22**, 409 (1980)
4.23 C. Hurtung, R. Jurgeit, H.-H. Ritze: Appl. Phys. **23**, 407 (1980)
4.24 C. Hurtung, R. Jurgeit: Appl. Phys. B, **27**, 39 (1982)
4.25 V.P. Zharov, V.S. Letokhov, S.G. Montanari: J. Physique, **44**, Suppl. 10, C6-573 (1983)
4.26 T. Baumann, F. Dacol, R.L. Melsher: Appl. Phys. Lett., **43**, 71 (1983)
4.27 H. Coufal: Appl. Phys. Lett. **44**, 59 (1984)

Chapter 5

5.1   J.G. Parker: Appl. Opt. **12**, 2974 (1973)
5.2   H.C. Bennet, R.A. Forman: Appl. Opt. **16**, 2834 (1977)
5.3   H.C. Bennet, R.A. Forman: J. Appl. Phys. **48**, 1217 (1977)
5.4   L.G. Rosengren: Appl. Opt. **14**, 1960 (1975)
5.5   G.C. Wetsel: J. Opt. Soc. Am. **70**, 471 (1980)
5.6   S.R.J. Brueck, H. Kildal, L.J. Belanger: Opt. Commun. **34**, 199 (1980)
5.7   F.V. Bunkin, V.M. Komisarov: Akust. Zh. (Russian) **19**, 305 (1973)
5.8   V.P. Zharov: Development and Research of Laser Optoacoustic Instruments for Analysis of Gas. Media, Ph. D. Thesis (Russian) MVTU Moscow, 1977
5.9   L.B. Kreuzer, N.P. Kenyon, C.K.N. Patel: Science **177**, 349 (1972)
5.10 A. Hordvik, L. Ckolnik: Appl. Opt. **16**, 2919 (1977)
5.11 C.W. Bruce, B.Z. Sojka, B.G. Hurd, R.W. Watkins, K.O. White, Z. Derzko: Appl. Opt. **15**, 2970 (1976)
5.12 V.N. Bagratashvili, V.P. Zharov, V.V. Lobko: Kvantovaya Electronika (Russian) **5**, 637 (1978)
5.13 C.F. Dewey: Opt. Engin. **13**, 483 (1974)
5.14 J.F. McClelland, R.N. Knizely: Appl. Opt. **15**, 2967 (1976)
5.15 S.L. Sam, M.L. Shand: Opt. Commun. **31**, 174 (1979)
5.16 A. Nordvik: J. Opt. Soc. Am. **66**, 1105 (1976)
5.17 D.W. Hill, T. Powell: Non-Dispersive Infrared Gas Analyses in Science, Medicine and Industry (Plenum Press, New York, 1968)
5.18 T.E. Deaton, P.A. Depatie, T.W. Walker: Appl. Phys. Lett. **26**, 300 (1975)

5.19 A.G. Gomenyuk, V.P. Zharov, V.S. Letokhov: in *Proc. Soviet-French Symp. Spectroscopic Instruments and Devices for Image Processing* (Acad. Nauk SSSR, Moscow 1977) p.78
5.20 J.G. Peterson, M.E. Thomas, R.J. Nordstrom, E.K. Damon, R.L. Long: Appl. Opt. **18**, 834 (1979)
5.21 Y.H. Pao (ed.): Optoacoustic Spectroscopy and Detection (Academic Press, New York 1977)
5.22 P.D. Goldan, K. Goto: J. Appl. Phys. **45**, 4350 (1974)
5.23 C.F. Dewey Jr., R.D. Kamm, C.E.Hackett: Appl. Phys. Lett. **23**, 633 (1973)
5.24 E. Max, L.G. Rosengren: Opt. Commun. **11**, 4350 (1974)
5.25 W. Schnell, G. Fischer: ZAMP **26**, 133 (1975a)
5.26 R.D. Kamm: J. Appl. Phys. **47**, 3550 (1976)
5.27 R.W. Terhune, J.E. Anderson: Opt. Lett. **1**, 70 (1977)
5.28 K.R. Koch, W. Lahmann: Appl. Phys. Lett. **35**, 289 (1978)
5.29 R. Gerlach, N.M. Amer: Appl. Phys. **23**, 319 (1980)
5.30 R. Gerlach, N.M. Amer: Appl. Phys. Lett. **32**, 228 (1978)
5.31 L.J. Thomas, M.J. Kelly, N.M. Amer: Appl. Phys. Lett. **32**, 736 (1978)
5.32 R. Gerlach, N.M. Amer: Abstr. I Meeting Intern. Photoacoustic Spectroscopy (August 1-3, 1979, Ames, USA) paper WC 4-1
5.33 N. Ioli, P. Violino, M. Meucci: J. Phys. E. **12**, 168 (1979)
5.34 G.W. Bruce, R.G. Pinnick: Appl. Opt. **16**, 1762 (1977)
5.35 E. Nodov: Appl. Opt. **17**, 1110 (1978)
5.36 W.A. McClenny, C.A. Bennet, G.M. Russwurm, R. Richmond: Appl. Opt. **20**, 650 (1981)
5.37 A.S. Gomenyuk, V.P. Zharov, V.B. Pásetzky, Y.O. Shaidurov: Zh. Prikl. Spectroscopii (Russian) **35**, 1112 (1981)
5.38 V.P. Zharov, V.S. Letokhov, S.G. Montanari: Laser Chemistry. **1**, 163 (1983)
5.39 R.S. Quimby, P.M. Selser, W.M. Yen: Appl. Opt. **16**, 2630 (1977)
5.40 R.W. Shaw: Appl. Phys. Lett. **35**, 253 (1979)
5.41 N.C. Fernelius, T.W. Hass: Appl. Opt. **17**, 3348 (1978)
5.42 G. Busse, D. Herboeck: Appl. Opt. **18**, 3959 (1979)
5.43 V.S. Letokhov: Laser Spectroscopy, Acad. Verl. Berlin, Heidelberg, New York, 1977
5.44 V.P. Zharov: Optoacoustic Method in Laser Spectroscopy, in New Methods of Spectroscopy (Nauka, Novosibirsk, 1982) p. 126
5.45 M.M. Farrow, R.K. Burnham, M. Auzanneau, S.L. Olsen, N. Purdie, E.M. Eyring: Appl. Opt. **17**, 1093 (1978)
5.46 V.Yu. Baranov, E.P. Velikhov, Yu.R. Kolomiiskii, V.S. Letokhov, V.G. Niz'ev, V.D. Pis'mennyi, E.A. Ryabov: Kvantovaya Electronika (Russian) **6**, 1062 (1979)
5.47 V.P. Zharov, V.S. Letokhov, E.A. Ryabov: Appl. Phys. Lett. **12**, 15 (1977)
5.48 T.F. Deutsch: Opt. Lett. **1**, 25 (1977)
5.49 Yu.A. Gorokhov, S.V. Efimovskii, I.N. Knyazev, V.V. Lobko: Pis'ma Zh. Techn. Fiz. (Russian) **4**, 1481 (1978)
5.50 J.C. Murphy, L.C. Aamodt: J. Appl. Phys. **48**, 3502 (1977)
5.51 C. Pichon, M. Le Liboux, D. Fournier, A.C. Boccara: Appl. Phys. Lett. **35**, 435 (1979)
5.52 J.E. Allen Jr., W.R. Anderson, D.R. Crosley: Opt. Lett. **1**, 118 (1977)
5.53 C.H. Townes, A.L. Schawlow: *Microwave Spectroscopy* (McGraw-Hill, New York 1955)
5.54 M.J. Kavaya, J.S. Margolis, M.S. Shumate: Appl. Opt. **18**, 280 (1979)
5.55 A.F. Krupnov, A.A. Mel'nikov, V.A. Skvortsov: Izv. VUZov, Radiofizika (Russian) **23**, 874 (1980)
5.56 D.O. Gorelik, B.B. Sakharov: *Optoacoustic Effect in Physical-Chemical Measurement* (Russian) (USSR Committee of Standarts, Moscow, 1969)
5.57 A.Di. Lieto, P. Minguzzi, M. Tonelli: Appl. Phys. B, **27**, 1 (1982)
5.58 A.F. Krupnov, A.A. Mel'nikov, V.A. Skvortsov: Optica i Spectroscopia (Russian) **46**, 1012 (1979)

5.59 C.K.N. Patel, R.J. Kerl, E.G. Burkhardt: Phys. Rev. Lett. **38**, 1204 (1977)
5.60 C.K.N. Patel, R.J. Kerl: Opt. Commun. **24**, 294 (1978)
5.61 W. Schnell, G. Fischer: Opt. Lett. **2**, 67 (1978)
5.62 R.A. Grane: Appl. Opt. **17**, 2097 (1978)
5.63 E. Kritchman, S. Shtrikman, M. Slatkine: J. Opt. Soc. Am. **68**, 1257 (1978)
5.64 P.E. Nordal, S.O. Kanstad: Opt. Commun. **26**, 367 (1978)
5.65 M.J. Adams, A.A. King, G.F. Kirkbright: Analyst. **101**, 73 (1976)
5.66 M.J. Adams, B.C. Beadle, A.A. King, G.F. Kirkbright: Analyst. **101**, 553 (1976)
5.67 M.J. Adams, G.F. Kirkbright: Analyst. **102**, 281 (1977)
5.68 M.J. Adams, B.C. Beadle, G.F. Kirkbright: Analyst. **102**, 569 (1977)
5.69 R.G. Gray, V.A. Fishman, A.J. Bard: Anal. Chem. **49**, 697 (1977)
5.70 A. Rosencwaig, A.P. Ginsberg, J.W. Koepke: Inorg. Chem. **15**, 2540 (1976)
5.71 W.C. Ferrell Jr., Y. Haven: J. Appl. Phys. **48**, 3984 (1977)
5.72 E.M. Monahan Jr., A.W. Nolle: J. Appl. Phys. **48**, 3519 (1977)
5.73 E.E. Marinero, M. Stuke: Rev. Sci. Instrum. **50**, 241 (1979)
5.74 A.S. Gomenyuk, V.P. Zharov, D.D. Ogurok, E.A. Ryabov, O.A. Tumanov, V.O. Shaidurov: Kvantovaya Electronika (Russian) **1**, 1805 (1974)
5.75 C.F. Dewey Jr., J.H. Flint: see ref. 5.32 paper WC 3-1
5.76 R.F. Adamowicz, K.P. Koo: Appl. Opt. **18**, 2938 (1979)
5.77 J.-M. Heritier: Opt. Commun. **44**, 267 (1983)
5.78 Yu.R. Kolomiisky, V.P. Zharov: J. Photoacoust. **1**, 49 (1982)
5.79 P. Costa Ribeiro, M. Labrunie, J.P. von der Weid, O.G. Symko: J. Appl. Phys. **53**, 8378 (1982)
5.80 P.S. Bechthold: J. Photoacoust. **1**, 87 (1982)
5.81 C.N. Ironside, R.G. Denning: J. Phys. E. **15**, 142 (1982)
5.82 P. Helander: J. Photoacoust. **1**, 251 (1982)
5.83 P. Helander, L. Lundström: J. Appl. Phys. **54**, 5069 (1983)
5.84 Y. Sugitani, A. Uejima, K. Kato: J. Photoacoust. **1**, 217 (1982)
5.85 G.F. Kirkbright, R.M. Miller, A. Rzadkiewicz: J. Physique **44**, Suppl. 10, C6-243 (1983)

Chapter 6

6.1 R.A. Smith, F.E. Jones, R.P. Chasmar: *The Detection and Measurement of Infrared Radiation*. (Oxford at the Clarendon Press 1957)
6.2 A.O. Sall: Optica i Spectroscopia (Russian) **6**, 219 (1959)
6.3 V.P. Zharov: *Development and Research of Laser Optoacoustic Instruments for Analysis of Gas Media* (Russian) Dissertation, MVTU Moscow (1977)
6.4 A.F. Krupnov, A.V. Burenin: *Molecular Spectroscopy: Modern Research*, Vol. 2 (Academic, New York, 1976) p. 93
6.5 L.J. Gershtein: Izv. VUZov, Radiofizika (Russian) **20**, 223 (1977)
6.6 T.E. Deaton, P.A. Depatic, T.W. Walker: Appl. Phys. Lett. **26**, 300 (1975)
6.7 V.P. Novikov, M.A. Novikov: Izv. Akad. Nauk. SSSR, Fiz. (Russian) **49**, 751 (1985)
6.8 A. Hordvik: Appl. Opt. **16**, 2827 (1977)
6.9 H.C. Bennet, R.A. Forman: Appl. Opt. **16**, 2834 (1977)
6.9 A. Hordvik: J. Opt. Soc. Am. **66**, 1105 (1976)
6.10 A. Hordvik, L. Cholnik: Appl. Opt. **16**, 2919 (1977)
6.11 E.L. Kerr: Appl. Opt. **12**, 5250 (1973)
6.12 J.G. Parker: Appl. Opt. **12**, 2974 (1973
6.13 A. Atalar: Appl. Opt. **19**, 3204 (1980)
6.14 C.K.N. Patel, A.C. Tam: Appl. Phys. Lett. **34**, 467 (1979)
6.15 A.C. Tam, C.K.N. Patel, R.J. Kerl: Opt. Lett. **4**, 81 (1979)
6.16 W. Lahmann, H.J. Ludewig: Chem. Phys. Lett **45**, 177 (1977)
6.17 W. Jackson, N.M. Amer: J. Appl. Phys. **51**, 3343 (1980)

6.18 D.O. Gorelik, B.B. Sakharov: Optoacoustic Effect in Physical-Chemical Measurement (Russian) (USSR Committee of Standarts, Moscow 1969)
6.19 D.K. Kollerov: *Metological Foundations of Gas-Analitical Measurements* (Russian) (USSR Committee of Standarts, Moscow 1971)
6.20 A.O. Sall: *Infrared Gas-Analitical Measurements* (Russian) (USSR Committee of Standarts, Moscow 1971)
6.21 J.D. Winefordner (ed.): *Trace Analysis Spectroscopic Methods for Elements* (Wiley, New York 1976)
6.22 S.B. Brodersen: J. Opt. Soc. Am. **43**, 877 (1953)
6.23 C. Young, R.E. Chapman: J. Quant. Spectrosc. Radiat. Transfer. **14**, 679 (1974)
6.24 A.S. Gomenyuk, V.P. Zharov, D.D. Ogurok, E.A. Ryabov, O.A. Tumanov, V.O. Shaidurov: Kvantovaya Elektronika (Russian) **1**, 1805 (1974)
6.25 E.A. Ryabov: Kvantovaya Elektronika (Russian) **2**, 138 (1975)
6.26 M.S. Shumate, R.F. Menzies, J.S. Margolis, L.G. Rosengren: Appl. Opt. **15**, 2480 (1976)
6.27 G.C. Wetsel Jr., F.A. McDonald: Appl. Phys. Lett. **30**, 252 (1977)
6.28 J.G. Roark, R.A. Palmer, J.S. Hatchison: Chem. Phys. Lett. **60**, 112 (1978)
6.29 Y.C. Teng, B.S.H. Royce: J. Opt. Soc. Am. **70**, 557 (1980)
6.30 A.B. Antipov, Yu.N. Ponomarev: Zh. Prikl. Spectroscopii (Russian) **30**, 362 (1979)
6.31 A.B. Antipov, Yu.N. Ponomarev: Kvantovaya Elektronika (Russian) **1**, 1345 (1974)
6.32 E.C. Wente: Phys. Rev. **19**, 333 (1922)
6.33 H.D. Arnold, J.B. Crandall: Phys. Rev. **10**, 22 (1917)
6.34 A.N. Brodnikovsky, V.P. Zharov, N.I. Koroteev: Kvantovaya Electronika (Russian) **12**, 2422 (1985)
6.35 L.M. Dorozhkin, V.P. Zharov, G.N. Makarov, A.A. Puretzky: Pis'ma Zh. Techn. (Russian) Fiz. **6**, 979 (1980)
6.36 J.C. Murphy, L.C. Aamodt: Appl. Phys. Lett. **31**, 728 (1977)
6.37 M. Slatkine: Appl. Opt. **20**, 2880 (1981)
6.38 A.S. Gomenyuk, V.P. Zharov, V.B. P'asetzky, V.O. Shaydurov: Zh. Prikl. Spektroskopii (Russian) **39**, 1029 (1983)

Chapter 7

7.1   V.P. Zharov, S.G. Montanari: J. Photoacoust. **1**, 355 (1983)
7.2   A.F. Krupnov, A.V. Burenin: *Molecular Spectroscopy: Modern Research*, Vol. 2 (Academic Press, New York 1976), p.93
7.3   V.J. Alekhnovich, A.S. Gomenyuk, V.P. Zharov, O.N. Kompanetz, V.B.P. P'asetzky, V.O. Shaydurov: Abst. XXI Coll. Spectr. Inter. (Cambridge, 1979) p.224
7.4   V.N. Bagratashvili, I.N. Knyazev, V.S. Letokhov, V.V. Lobko: Kvantovaya Elektronika (Russian) **3**, 1011 (1976)
7.5   H. Walther (ed.): *Laser Spectroscopy of Atoms and Molecules*, Topics Appl. Phys., Vol. 2 (Springer, Berlin, Heidelberg, New York, 1976)
7.6   K. Shimoda (ed.): *High-Resolution Laser Spectroscopy*, Topics Appl. Phys. Vol. 13 (Springer, Berlin, Heidelberg, New York, 1976)
7.7   Yu.A. Gorokhov, S.V. Efimovskii, I.N. Knyazev, V.V. Lobko: *Pis'ma Zh. Techn. Fiz.* (Russian) **4**, 1481 (1978)
7.8   A.B. Antipov, V.A. Sapozhnikova: Zh. Prikl. Spectroscopii (Russian) **28**, 636 (1978)
7.9   A.B. Antipov, V.A. Kapitonov, V.P. Lopasov: Zh. Prikl. Spectroscopii (Russian) **30**, 1043 (1979)
7.10 R.N. Dixon, D.A. Haner, C.R. Webster: Chem. Phys. **22**, 199 (1977)
7.11 E.E. Marinero, A.M. Angus, M.J. Colles: Opt. Commun. **14**, 226 (1975)

7.12 G. Stella, Y. Gelfand, H. Smith: Chem. Phys. Lett. **39**, 146 (1976)
7.13 R.G. Bray, M.J. Berry: J. Chem. Phys. **71**, 4909 (1979)
7.14 B.G. Ageev, A.B. Antipov, A.A. Pomeshchenko, Yu.N. Ponomarev: Optica
      i Spectroscopia (Russian) **40**, 600 (1976)
7.15 L.B. Kreuzer, C.K.N. Patel: Science **173**, 45 (1971)
7.16 V.P. Zharov: *Development and Research of Laser Optoacoustic Instruments
      for Analysis of Gas Media*, Ph. D. Thesis (Russian), MVTU, Moscow (1977)
7.17 V.P. Zharov, Yu.R. Kolomiisky, V.V. Lobko: *Abstr. Conf. on Molecular
      Spectroscopy High Resolution* (Izd. Nauka, Novosibirsk, 1979) p.149
7.18 K.R. German, W.S. Cornall: J. Opt. Soc. Am. **71**, 1452 (1981)
7.19 K. Walzer, M. Tacke, G. Busse: Infr. Phys. **19**. 175 (1979)
7.20 K. Walzer, M. Tacke, G. Busse: J. Chem. Phys. **73**, 3095 (1980)
7.21 D.M. Cox, A. Gnauck: J. Mol. Spectrosc. **81**, 207 (1980)
7.22 P. Anderson, V. Persson: Appl. Opt. **23**, 192 (1984)
7.23 G.L. Loper, G.R. Sasaki, M.A. Stamps: Appl. Opt. **21**, 1648 (1982)
7.24 R.T. Menzies, M.S. Shumate: Appl. Opt. **15**, 2025 (1976)
7.25 M.S. Shumate, R.F. Menzies, J.S. Margolis, L.G. Rosengren: Appl. Opt.
      15, 2480 (1976)
7.26 T.E. Deaton, P.A. Depatie, T.W. Walker: Appl. Phys. Lett. **26**, 300 (1975)
7.27 C.W. Bruce, H.O. White: J. Opt. Soc. Am. **66**, 1088 (1976)
7.28 K.O. Whate, W.R. Watkins, C.W. Bruce, R.E. Meredith, F.G. Smith: Appl.
      Opt. **17**, 2711 (1978)
7.29 A.S. Gomenyuk, V.P. Zharov, D.D. Ogurok et al.: Kvantovaya Elektronika
      (Russian) **1**, 1805 (1974)
7.30 J.G. Peterson, M.E. Thomas, R.J. Nordstrom et al.: Appl. Opt. **18**, 834
      (1979)
7.31 G.L. Loper, M.A. O'Neil, J.A. Gelbwachs: Appl. Opt. **22**, 3701 (1983)
7.32 J. Hinderling, M.W. Sigrist, F.K. Kneubühl: J. Physique **44**, Suppl. 10,
      C6-559 (1983)
7.33 C.W. Bruce, A.V. Jelinek: Appl. Opt. **21**, 4101 (1982)
7.34 G. Krueger: Appl. Opt. **21**, 2841 (1982)
7.35 C.W. Bruce, B.Z. Sojka, B.G. Hurd et al.: Appl. Opt. **15**, 2970 (1976)
7.36 J.S. Ryan, M.H. Hubert, R.A. Grane: Appl. Opt. **22**, 711 (1983)
7.37 T.J. Bridges, E.G. Burkhardt: Opt. Commun. **22**, 248 (1977)
7.38 A. Kaldor, W.B. Olson, A.G. Maki: Science **176**, 508 (1972)
7.39 P. Minguzzi, S. Profeti, M. Tonelli, A.Di Lieto: Opt. Commun. **42**, 237
      (1982)
7.40 A.F. Krupnov, A.A. Mel'nikov, V.A. Skvortsov: Optica i Spectroscopia
      (Russian) **46**, 1012 (1979)
7.41 A. Di Lieto, P. Minguzzi, M. Tonelli: Appl. Phys. **B27**, 1 (1982)
7.42 P. Minguzzi, M. Tonelli, A. Carrozzi: J. Mol. Spectr. **96**,294 (1982)
7.43 M. Tonelli, P. Minguzzi, A. Di Lieto: J. Physique **44**, Suppl. 10, 553
      (1983)
7.44 M.J. Kavaya, J.S. Margolis, M.S. Shumate: Appl. Opt. **18**, 2801 (1979)
7.45 A.F. Krupnov, A.A. Mel'nikov, V.A. Skvortsov: Izv. VUZov, Radiofizika
      (Russian) **23**, 874 (1980)
7.46 V.S. Letokhov, V.P. Chebotayev: *Nonlinear Laser Spectroscopy*, Springer
      Ser. Opt. Sci., Vol. 4 (Springer, Berlin, Heidelberg, New York, 1977)
7.47 M.S. Sorem, A.L. Schawlow: Opt. Commun. **5**, 148 (1972)
7.48 E.E. Marinero, M. Stuke: Opt. Commun. **30**, 349 (1979)
7.49 A. Di Lieto, P. Minguzzi, M. Tonelli: Opt. Commün. **31**, 25 (1979)
7.50 M. Inguscio, A. Moretti, F. Strumia: Opt. Commun. **30**, 355 (1979)
7.51 J.N. Dahiya, K. Iqbal, H.G. Kraft et al.: Infr. Phys. **22**, 77 (1982)
7.52 C.Hurtung, R. Jurgeit, H.H. Ritze: Appl. Phys. **23**, 407 (1980)
7.53 A.F. Krupnov: Zh. Eksp. i Teor. Fiz. (Russian) **69**, 1981 (1975)
7.54 V.P. Zharov, V.S. Letokhov, E.A. Ryabov: Appl. Phys. **12**, 15 (1977)
7.55 V.S. Letokhov: In: *Multiphoton Processes*, J.H. Eberly, P. Lambropoulos
      (eds.) (Wiley, New York, 1978) p.331

7.56 V.S. Letokhov, A.A. Makarov: Uspekhi Fiz. Nauk (Russian) **134**, 45(1981)
7.57 E.A. Ryabov: Kvantovaya Elektronika (Russian) **2**, 138 (1975)
7.58 P.G. Kryukov, V.S. Letokhov: Uspekhi Fiz. Nauk (Russian) **99**, 169 (1969)
7.59 V.P. Lopasov, S.B. Ponomaryova, Yu.N. Ponomaryov, B.A. Tikhomirov:
      Kvantovaya Elektronika (Russian) **7**, 2582 (1980)
7.60 A.B. Antipov, V.E. Zuev, V.P. Lopasov, Yu.N. Ponomarev: Appl. Opt. **18**,
      3014 (1979)
7.61 V.S. Letokhov, C.B. Moore: In: *Chem. and Biochem. Appl. of Lasers*, Vol. 3,
      C.B. Moore (ed.) (Academic Press, London, 1977) p.1
7.62 R.V. Ambartzumian, V.S. Letokhov: In: *Chem. and Biochem. Appl. of Lasers*,
      Vol. 3, C.B. Moore (ed.) (Academic Press, London, 1977) p.167
7.63 V.Yu. Baranov, E.P. Velikhov, Yu.R. Kolomiisky et al.: Kvantovaya Elec-
      tronika (Russian) **6**, 1062 (1979)
7.64 J.L. Lyman, K.M. Leary: J. Chem. Phys. **69**, 1858 (1978)
7.65 V.N. Bagratashvili, I.N. Knyazev, V.S. Letokhov, V.V. Lobko: Opt. Commun.
      18, 525 (1976)
7.66 D.M. Cox: Opt. Commun. **24**, 336 (1978)
7.67 N. Presser, J.R. Barker, R.J. Gordon: J. Chem. Phys. **78**, 2163 (1983)
7.68 S.S. Alimpiev, N.V. Karlov, A.M. Prokhorov, B.G. Sartakov, E.M. Khokhlov:
      Kvantovaya Elektronika (Russian) **6**, 2597 (1979)
7.69 J.G. Black, E. Yablonovitch, N. Bloembergen: Phys. Rev. Lett. **38**, 1131
      (1977)
7.70 J.G. Black, P. Kolodner, M.J. Schultz, E. Yablonovitch, N. Bloembergen:
      Phys. Rev. **A19**, 404 (1979)
7.71 A.S. Akhmanov, V.N. Bagratashvili, V.Yu. Baranov, Yu.R. Kolomisky, V.S.
      Letokhov, V.P. Pismennyi, E.A. Ryabov: Opt. Commun. **23**, 357 (1977)
7.72 T.F. Deutsch: Opt. Lett. **1**, 25 (1977)
7.73 S.S. Alimpiev, N.V. Karlov, B.G. Sartakov, E.M. Khokhlov: Opt. Commun.
      26, 25 (1978)
7.74 T. Fukumi: Opt. Commun. **30**, 351 (1979)
7.75 I.N. Knyazev, N.P. Kuzmina, V.S. Letokhov, V.V. Lobko, A.A. Sarkisian:
      Appl. Phys. **22**, 429 (1980)
7.76 N.V. Akulin, S.S. Alimpiev, N.V. Karlov, A.M. Prokhorov, B.G. Sartakov,
      E.M. Khokhlov: 10th Intern. Quant. Electr. Confer., (Atlanta, 1978)
      paper Q8
7.77 G.P. Quigley: Opt. Lett. **4**, 84 (1979)
7.78 L.M. Dorozhkin, V.P. Zharov, G.N. Makarov, A.A. Puretzky: Pis'ma Zh.
      Techn. Fiz. (Russian) **6**, 979 (1980)
7.79 R.V. Ambartzumian, L.M. Dorozhkin, G.N. Makarov, A.A. Puretzky, B.A.
      Chayanov: Appl. Phys. **22**, 409 (1980)
7.80 N.V. Chekalin, V.S. Dolzhikov, Yu.R. Kolomisky, V.S. Letokhov, E.A.
      Ryabov: Phys. Lett. **59A**, 243 (1976)
7.81 R.V. Ambartzumian, V.M. Apatin, A.V. Evseev, N.P. Furzikov: Kvantovaya
      Elektronika (Russian) **7**, 1998 (1980)
7.82 N.V. Chekalin, V.S. Doljikov, Yu.R. Kolomisky, V.S. Letokhov, V.N.
      Lokhman, E.A. Ryabov: Appl. Phys. **13**, 311 (1977)
7.83 S.Yu. Nechaev, Yu.N. Ponomarev: Kvantovaya Elektronika (Russian) **2**,
      1400 (1975)
7.84 J.J. Barrett, J.M. Berry: Appl. Phys. Lett. **34**, 144 (1979)
7.85 D.R. Siebert, G.A. West, J.J. Barrett: Appl. Opt. **19**, 53 (1980)
7.86 G.A. West, D.R. Siebert, J.J. Barrett: J. Appl. Phys. **51**, 2823 (1980)
7.87 G.A. West, J.J. Barrett: Opt. Lett. **4**, 395 (1980)
7.88 S.A. Akhmanov, N.J. Koroteev: Uspekhi Fiz. Nauk (Russian) **123**, 405 (1977)
7.89 A. Yariv: *Quantum Electronics* (Wiley, New York, 1975)
7.90 G. Gorelik: Dokl. Akad. Nauk SSSR (Russian) **54**, 779 (1946)
7.91 M.L. Viengerov, P.V. Slobodskaya: Izv. Akad. Nauk SSSR (Russian) Fiz. **11**,
      420 (1947)
7.92 P.V. Slobodskaya: Izv. Akad. Nauk SSSR, Fiz. (Russian) **12**, 656 (1978)

7.93 P.V. Slobodskaya, E.S. Gasilevich: Optica i Spectroscopia (Russian) **7**, 97 (1959)
7.94 R. Tripodi: J. Chem. Phys. **52**, 3298 (1970)
7.95 P.V. Slobodskaya, E.S. Gasilevich: Optica i Spectroscopia (Russian) **8**, 678 (1960)
7.96 T.L. Cottrell, I.M. Macfarlane, A.W. Read, A.H. Young: Trans. Faradey Soc. **62**, 2655 (1966)
7.97 R. Tripodi, W.G. Vincenti: J. Chem. Phys. **55**, 2207 (1971)
7.98 T.F. Hunter, D. Rumbles, M.G. Stock: J. Chem. Soc. (Far. Trans. II) **70**, 1010 (1974)
7.99 T. Aoki, M. Katayma: Jap. J. Appl. Phys. **10**, 1303 (1971)
7.100 J.G. Parker, D.N. Ritke: J. Chem. Phys. **59**, 3713 (1973)
7.101 N.M. Lawandy: Infr. Phys. **20**, 131 (1980)
7.102 B.F. Gordiets, V.Ya. Panchenko: Pis'ma Zh. Techn. Fiz. (Russian) **4**, 1396 (1978)
7.103 E. Rohlfing, H. Rabitz, J. Gelfand, R. Miles: J. Chem. Phys. **81**, 820 (1984)
7.104 R. Tripodi, W.G. Vincenti: J. Chem. Phys. **55**, 2207 (1971)
7.105 P.V. Slobodskaya, Yu.N. Obraztsov, L.K. Sukhareva, N.V. Shlyakhtenko: Optika i Spectroscopia (Russian) **38**, 66 (1975)
7.106 E.N. Ritkyin', A.P. Burtsev, P.V. Slobodskaya: Zh. Prikl. Spectroscopii (Russian) **24**, 347 (1976)
7.107 M.Y. Perrin: Chem. Phys. Lett. **94**, 434 (1983)
7.108 T.E. Hunter: J. Chem. Soc. Amer. **1967**A, 1804 (1974)
7.109 A.B. Antipov, V.A. Kapitonov, Yu.N. Ponomarev: Optica i Spectroscopia (Russian) **49**, 53 (1981)
7.110 V.P. Zharov, S.G. Montanari: Optika i Spectroscopia (Russian) **51**, 124 (1981)
7.111 K. Frank, P. Hess: Chem. Phys. Lett. **68**, 540 (1979)
7.112 Ya.I. Gerlovin: Optica i Spectroscopia (Russian) **7**, 571 (1959)
7.113 W.R. Harshbarger, M.B. Robin: Acc. Chem. Res. **6**, 329 (1973)
7.114 P.I. Bresler, B.N. Ruzin: Optika i Spectroskopia (Russian) **9**, 22 (1960)
7.115 Y.H. Pao (ed.): *Optoacoustic Spectroscopy and Detection* (Academic Press, New York, 1977)
7.116 R.B. Hall, A. Kaldor: J. Chem. Phys. **70**, 4027 (1979)
7.117 A.M. Angus, E.E. Marinero, M.J. Colles: Opt. Commun. **14**, 223 (1975)
7.118 C.K.N. Patel: Science **202**, 167 (1978)
7.119 C.K.N. Patel: J. Opt. Soc. Am. **66**, 1079 (1976)
7.120 R.J.H. Voorhoeve, C.K.N. Patel, L.E. Trimble, R.J. Kerl: Science **200**, 157 (1978)
7.121 R.E. Richton, L.A. Farrow: Abstr. I Meeting Intern. on Photoacoustic Spectroscopy (August 1-3, 1979, Ames, USA) paper ThC6-1
7.122 E.E. Marinero, M. Stuke: Rev. Sci. Instrum. **50**, 241 (1979)
7.123 M.J. Colles, E.E. Marinero: *Laser in Chemistry*, M. West (ed.) (Elsevier, Amsterdam, 1977)
7.124 C.K.N. Patel, R.J. Kerl, E.G. Burkhardt: Phys. Rev. Lett. **38**, 1204 (1977)
7.125 V.B. Anzin, M.V. Glushkov, V.P. Zharov et al.: Kvantovaya Elektronika (Russian) **2**, 1403 (1975)
7.126 V.E. Zuev, A.B. Antipov, V.A. Sapozhnikova, K. Fox: J. Chem. Phys. **68**, 1315 (1978)
7.127 E.D. Hinkley: Opto-electronics **4**, 69 (1972)
7.128 C. Hurtung, R. Jurgeit: Optika i Spectroscopia (Russian) **46**, 1169 (1979)
7.129 P. Klein, P. Hess: Acoustica **33**, 198 (1975)
7.130 M. Huetz-Auvert, F. Lepoutre, G. Louis: J. Physique **38**, 283 (1977)
7.131 J.T. Yardley, M.N. Fertig, C.B. Moore: J. Chem. Phys. **52**, 1450 (1970)

Chapter 8

8.1   L. Pauling: *The Nature of the Chemical Bond*, 3rd ed. (Cornell U. Press, Ithaca NY 1960)
8.2   C.K.N. Patel, A.C. Tam: Appl. Phys. Lett. **34**, 467 (1979)
8.3   C.K.N. Patel, A.C. Tam, R.J. Kerl: J. Chem. Phys. **71**, 1470 (1979)
8.4   A.C. Tam, C.K.N. Patel: Appl. Opt. **18**, 3348 (1979)
8.5   M.O. Bulanin: J. Mol. Structure **19**, 59 (1973)
8.6   S.R.J. Brueck, H. Kildal, L.J. Belanger: Opt. Commun. **34**, 199 (1980)
8.7   A.R. Kukudzhanov, V.I. Aleknovich, A.S. Gomenyuk, Yu.A. Kudriavtsev: Pis'ma Zh. Techn. Fiz. (Russian) **7**, 102 (1981)
8.8   E.T. Nelson, C.K.N. Patel: Nature **286**, 368 (1980)
8.9   E.T. Nelson, C.K.N. Patel: Appl. Phys. Lett. **39**, 537 (1981)
8.10 A. Hordvik: Appl. Opt. **16**, 2827 (1977)
8.11 A. Hordvik, L. Cholnik: Appl. Opt. **16**, 2919 (1977)
8.12 A. Hordvik, H. Schlossberg: Appl. Opt. **16**, 101 (1977)
8.13 J.A. Burt, K.J. Ebeling, D.E. Ethimiades: Opt. Commun. **32**, 59 (1980)
8.14 I. Masahide, S. Hiroyoshi: Opt. Commun. **44**, 229 (1983)
8.15 V.P. Novikov, M.A. Novikov: Pis'ma Zh. Techn. Fiz. (Russian), **8**, 6 (1982)
8.16 D. Ghardon, J. Huard: Appl. Phys. Lett. **41**, 341 (1982)
8.17 E.H. Bohan, A.W. Molle: J. Appl. Phys. **48**, 3519 (1977)
8.18 P. Helander, L. Lundström, D. Söderderg: Abstr. I Meeting Intern. on Photoacoustic Spectroscopy (August 1-3, 1979, Ames, USA) paper WC2-1
8.19 P. Helander: J. Appl. Phys. **54**, 3410 (1983)
8.20 A.C. Tam, C.K.N. Patel: Appl. Phys. Lett. **35**, 843 (1979)
8.21 A.M. Bonch-Bruevich, T.K. Razumova, I.O. Starobogatov: Optica i Spectroscopia (Russian) **42**, 82 (1977)
8.22 A.M. Bonch-Bruevich, T.K. Razumova, I.O. Starobogatov: Zh. Prikl. Spectroscopii (Russian) **37**, 981 (1982)
8.23 Y. Bae, J.J. Song, Y.B. Kim: J. Appl. Phys. **53**, 615 (1982)
8.24 P. Horn, A. Schmid, P. Bräunlich: IEEE J. QE-**19**, 1169 (1983)
8.25 C.K.N. Patel, A.C. Tam: Appl. Phys. Lett. **34**, 760 (1979)
8.26 Y.H. Pao (ed.): *Optoacoustic Spectroscopy and Detection* (Academic Press, New York, 1977)
8.27 A. Rosencwaig: *Photoacoustics and Photoacoustic Spectroscopy* (Wiley, New York, 1980)
8.28 J.F. McClelland: Anal. Chem. **55**, 89A (1983)
8.29 P.E. Nordal, S.O. Kanstad: Opt. Commun. **22**, 185 (1977)
8.30 S.O. Kanstad, P.E. Nordal: Infr. Phys. **19**, 413 (1979)
8.31 P.E. Nordal, S.O. Kanstad: Opt. Commun. **26**, 367 (1978)
8.32 M.A. Afromowitz, P.S. Yen, S.S. Yee: J. Appl. Phys. **48**, 209 (1977)
8.33 N.C. Fernelius: J. Appl. Phys. **51**, 650 (1980)
8.34 E.L. Kerr: Appl. Opt. **12**, 5250 (1973)
8.35 J.G. Parker: Appl. Opt. **12**, 2974 (1973)
8.36 J.M. McDavid, K.L. Lee, S.S. Lee, M.A. Afromowitz: J. Appl. Phys. **49**, 6112 (1978)
8.37 N.C. Fernelius: J. Appl. Phys. **51**, 1756 (1980)
8.38 N.C. Fernelius: Appl. Opt. **21**, 481 (1982)
8.39 J.C. Murphy, L.C. Aamodt: Appl. Phys. Lett. **31**, 728 (1977)
8.40 H.S. Bennet, R.A. Forman: J. Appl. Phys. **48**, 1432 (1977)
8.41 A. Rosencwaig, J.B. Willis: J. Appl. Phys. **51**, 4361 (1980)
8.42 S.A. Francis, A.H. Ellison: J. Opt. Soc. Am. **49**, 131 (1959)
8.43 T. Inagaki, K. Kagami, E.T. Arakawa: Appl. Opt. **21**, 949 (1982)
8.44 T. Inagaki, M. Motosuga, K. Yamamori: Phys. Rev. B **28**, 1740 (1983)
8.45 P.R. Muessig, G.J. Diebold: J. Appl. Phys. **54**, 4251 (1983)
8.46 V.P. Novikov, M.A. Novikov: Abstr. III, Intern. Topical Meeting on Photoacoustic and Photothermal Spectroscopy (Paris 1983) Rep.4.9/1

8.47 L. Velluz, M. Legran, M. Grosican: *Optical Circular Dichroism* (Verlag Chemie, Weinheim 1965)
8.48 R.A. Palmer, J.C. Roark, J.C. Robinson, J.L. Howell: Abstr. I Meeting Intern. on Photoacoustic Spectroscopy (August 1-3, 1979, Ames, USA) paper ThA3-1
8.49 M.A. Novikov, V.P. Novikov: Pis'ma Zh. Techn. Fiz. (Russian) **6**, 141 (1980)
8.50 J.D. Saxe, T.R. Faulkner, F.S. Richardson: Chem. Phys. Lett. **68**, 71 (1979)
8.51 J.D. Caxe, T.R. Faulkner, F.S. Richardson: J. Appl. Phys. **50**, 8204 (1979)
8.52 D. Fournier, A.C. Boccara, J. Badoz: Appl. Phys. Lett. **32**, 640 (1978)
8.53 S. Malkin, D. Cahen: Photochem. Photobiol. **91**, 131 (1978)
8.54 M.G. Rockley: Chem. Phys. Lett. **50**, 427 (1977)
8.55 M.N. Alentsev: Zh. Eksp. i. Teor. Fiz. (Russian) **21**, 133 (1951)
8.56 W. Lahmann, H.J. Ludewig: Chem. Phys. Lett. **45**, 177 (1977)
8.57 I.O. Starobogatov: Optica i Spectroscopia (Russian) **42**, 304 (1977)
8.58 M.G. Rockley, M. Waugh: Chem. Phys. Lett. **54**, 597 (1978)
8.59 A. Rosencwaig, M.J. Weber, R.A. Saroyan: Abstr. I Meeting Intern. on Photoacoustic Spectroscopy (August 1-3, 1979, Ames, USA) paper FA5-1
8.60 J.C. Murphy, L.C. Aamodt: J. Appl. Phys. **48**, 3502 (1977)
8.61 R.S. Quimby, W.M. Ven: Opt. Lett. **3**, 181 (1978)
8.62 S. Malkin, N. Lasser-Ross, D. Cahen: Abstr. I Meeting Intern. on Photoacoustic Spectroscopy (August 1-3, 1979, Ames, USA) paper WB10P-1
8.63 A.C. Tam: Appl. Phys. Lett. **37**, 978 (1980)
8.64 D. Cahen: Appl. Phys. Lett. **33**, 810 (1978)
8.65 J.-M. Heritier, A.E. Siegman: IEEE J. QE-**19**, 1551 (1983)
8.66 M. Bernstein, L.J. Rothberg, K.S. Peters: Chem. Phys. Lett. **91**, 315 (1982)

Chapter 9

9.1 E.D. Hinkley (ed.): *Laser Monitoring of the Atmosphere*, Topics Appl. Phys. Vol. 14 (Springer, Berlin, Heidelberg, New York, 1976)
9.2 J. Melngailis: IEEE Trans. GE-10, 2 (1972)
9.3 V.P. Zharov, A.S. Gomenyuk, V.B. P'aset'zki, E.G. Tokhtuev: In proc. Intern. Symp. on Metrological Assurance of Measurements for Environmental Controle, Leningrad (IMEKO, VNIIM 1981) p.80
9.4 V.P. Zharov, A.S. Gomenyuk, V.B. P'iasetzky, E.G. Tokhtuev: Zh. Prikl. Spektroskopii (Russian) **39**, 1029 (1983)
9.5 C.K.N. Patel: Science **202**, 167 (1978)
9.6 K.R. Koch, W. Lehmann: Appl. Phys. Lett. **35**, 289 (1978)
9.7 N. Joli, P. Violino, M. Meucci: J. Phys. E.: Sci. Instrum. **12**, 168 (1979)
9.8 P.C. Claspy, C. Ha, Y.H. Pao: Appl. Opt. **16**, 2972 (1977)
9.9 A.M. Angus, E.E. Marinero, M.J. Colles: Opt. Commun. **14**, 223 (1975)
9.10 A.S. Gomenyuk, V.P. Zharov, D.D. Ogurok et al.: Kvantovaya Elektronika (Russian) **1**, 1805 (1974)
9.11 L.B. Kreuzer: J. Appl. Phys. **42**, 2934 (1971)
9.12 C.F. Dewey Jr., R.D. Kamm, C.E. Hackett: Appl. Phys. Lett. **23**, 633 (1973)
9.13 P.D. Goldan, K. Goto: J. Appl. Phys. **45**, 2058 (1975)
9.14 R.D. Kamm: J. Appl. Phys. **47**, 3550 (1976)
9.15 R. Gerlach, N.M. Amer: Appl. Phys. **23**, 319 (1980)
9.16 L.B. Kreuzer, C.K.N. Patel: Science **173**, 45 (1971)
9.17 C.K.N. Patel, E.C. Burkhard, C.A. Lambert: Science **184**, 1173 (1974)

9.18 C.K.N. Patel, R.J. Kerl: Appl. Phys. Lett. **30**, 578 (1977)
9.19 L.B. Kreuzer, N.P. Kenyon, C.K.N. Patel: Science **177**, 349 (1972)
9.20 R.A. Grane: Appl. Opt. **17**, 2097 (1978)
9.21 P.C. Claspy, Y.H. Pao, S. Kwong, E. Nodov: Appl. Opt. **15**, 1506 (1976)
9.22 E. Nodov: Appl. Opt. **17**, 1110 (1978)
9.23 E. Max, L.-G. Rosengren: Opt. Commun. **11**, 4350 (1974)
9.24 E. Kritshman, S. Shtrikman, M. Slatkine: J. Opt. Soc. Amer. **68**, 1257, (1978)
9.25 V.P. Zharov: *Development and Research of Laser Optoacoustic Instruments for Analysis of Gas Media*, Ph. D. Thesis (Russian) (MVTU, Moscow, 1977)
9.26 C.K.N. Patel: Opt. Quant. Electr. **8**, 145 (1976)
9.27 A. Kaldor, W.B. Olson, A.G. Maki: Science **176**, 508 (1972)
9.28 Y.H. Pao (ed.): *Optoacoustic Spectroscopy and Detection* (Academic Press, New York, 1977)
9.29 V.I. Shtepa, L.N. Vereshchagina, R.R. Osmanov, F.N. Putilin, V.P. Zharov: J. Photoacoustics **1**, 181 (1982)
9.30 D.R. Siebert, G.A. West, J.J. Barrett: Appl. Opt. **19**, 53 (1980)
9.31 W. Lahmann, H.J. Ludewig, H. Welling: Anal. Chem. **49**, 549 (1977)
9.32 S. Oda, T. Sawada, H. Kamada: Anal. Chem. **50**. 865 (1978)
9.33 S. Oda, T. Sawada, H. Kamada: Bunseki Kagaku **27**, N 51 (1978)
9.34 A.C. Tam, C.K.N. Patel, R.J. Kerl: Opt. Lett. **4**, 81 (1978)
9.35 Y. Kohanzadeh, J.R. Whinnery, M.M. Carroll: J. Acous. Soc. Amer. **57**, 67 (1975)
9.36 E. Volgtman, A, Jurgensen, J. Winefordner: Anal. Chem. **53**, 1442 (1981)
9.37 W. Schrepp, R. Stumpe, J.I. Kim, H. Walther: Appl. Phys. B**32**, 207 (1983)
9.38 S.R.J. Brueck, H. Kildal, L.J. Belanger: Opt. Commun. **34**, 199 (1980)
9.39 A.R. Kukudzhanov, V.I. Alekhnovich, A.S. Gomenyuk, Yu.A. Kudriavtsev: Pis'ma Zh. Techn. Fiz. (Russian) **7**, 102 (1981)
9.40 G.C. Wetsel Jr., F.A. McDonald: Abstr. I Meeting Intern. on Photoacoustic Spectroscopy (August 1-3, 1979, Ames, USA) paper ThAS-I
9.41 U. Madvaliev: *Photoacoustic Spectroscopy of Solids*, Ph. D. Thesis, Moscow, State University (1979)
9.42 P.E. Nordal, S.O. Kanstad: Opt. Commun. **22**, 185 (1977)
9.43 A. Rosencwaig, S.S. Hall: Anal. Chem. **47**, 549 (1975)
9.44 S.L. Castedum, G.F. Kirkbright: Abstr. I Meeting Intern. on Photoacoustic Spectroscopy (August 1-3, 1979, Ames, USA) paper FB3-1
9.45 B. Lehmann, M. Wahlen, R. Zumbrunn, H. Oechger: Appl. Phys. **13**, 157 (1977)
9.46 W. Schnell, G. Fisher: Opt. Lett. **2**, 67 (1978)
9.47 H. Trautmann, K.W. Rothe, J. Wanner, H. Walther: Appl. Phys. **24**, 49 (1981)
9.48 A.S. Gomenyuk, V.P. Zharov, V.S. Letokhov, E.A. Ryabov: Kvantovaya Elektronika (Russian) **3**, 369 (1976)
9.49 B.M. Golubitskyi, M.V. Tantachev: Izv. Akad. Nauk SSSR Fiz. Atmos. and Okean (Russian) **12**, 442 (1976)
9.50 R.W. Terhune, J.E. Anderson: Opt. Lett. **1**, 70 (1977)
9.51 G.W. Bruce, R.G. Pinnick: Appl. Opt. **16**, 1762 (1977)
9.52 D.M. Roesslen, F.R. Faxvog: J. Opt. Soc. Amer. **69**, 1699 (1979)
9.53 S.A. Schleusener, J.D. Lindberg, K.O. White: Appl. Opt. **14**, 2364 (1975)
9.54 S.A. Schleusener, J.P. Lindberg, K.O. White, R.L. Johnson: Appl. Opt. **15**, 2546 (1976)
9.55 J.R. Mouchalin, J.M. Gague, J.L. Parpal, L. Bertrand: Appl. Phys. Lett. **35**, 360 (1979)
9.56 S. Oda, T. Sawada, T. Moriguchi, H. Kamada: Anal. Chem. **52**, 650 (1980)
9.57 M. Nomura, S. Oda, T. Sawada: J. Photoacoust. **1**, 121 (1982)
9.58 V.P. Zharov, S.G. Montanari: Zh. Prikl. Spectroscopii (Russian) **41**, 401 (1984)
9.59 I.N. Knyazev, N.P. Kuzmina, V.S. Letokhov, V.V. Lobko, A.A. Sarkisian: Appl. Phys. **22**, 429 (1980)

9.60 V.P. Zharov, S.G. Montanari: Optika i Spectroscopia **51**, 124 (1981)
9.61 P. Perlmutter, S. Shtrikman, M. Slatkine: Appl. Opt. **18**, 2267 (1979)
9.62 L.B. Kreuzer: Anal. Chem. **50**, 597A (1978)
9.63 V.P. Zharov, V.S. Letokhov, S.G. Montanari: Laser Chem. **1**, 163 (1983)
9.64 V.P. Zharov, V.S. Letokhov, S.G. Montanari, L.M. Tumanova: Dokl. Akad. Nauk, SSSR (Russian) **269**, 1079 (1983)
9.65 V.P. Zharov, S.G. Montanari: J. Photoacoust. **1**, 355 (1982-83)
9.66 V.P. Zharov, V.S. Letokhov, S.G. Montanari: L.M. Tumanova: Abstr. III Intern. Topical Meeting on Photoacoustic and Photothermal Spectroscopy (Paris 1983) Report 7.14/1
9.67 S. Oda, T. Sawada: Anal. Chem. **53**, 471 (1981)
9.68 E. Voigtman, A. Jurgensen, J.D. Winefordner: Anal. Chem. **53**, 1921 (1981)
9.69 E. Voigtman, A. Jurgensen, J.D. Winefordner: J. Liquid Chromatog. **6**, 1275 (1983)
9.70 A.C. Gomeniouk, V.P. Jarov, S.G. Montanari et al: Tzavaux du III Symposium Franco-Sovietique en Instrumentation, Aussoig, 21-25 octobre, 1984, p. 157-161

Chapter 10

10.1 A. Rosencwaig: J. Appl. Phys. **51**, 2210 (1980)
10.2 Y.H. Wong, R.L. Thomas, C.F. Hawkins: Appl. Phys. Lett. **32**, 538 (1978)
10.3 J.J. Pouch, R.L. Thomas, Y.H. Wong, J. Schuldies, J. Srivivasan: J. Opt. Soc. Am. **70**, 562 (1980)
10.4 M. Luukala, A. Penttinen: Electr. Lett. **15**, 326 (1979)
10.5 G. Busse: Appl. Phys. Lett. **35**, 759 (1979)
10.6 Y.H. Wong, R.L. Thomas, J.J. Pouch: Appl. Phys. Lett. **35**, 368 (1979)
10.7 R.L. Thomas, J.J. Pouch, Y.H. Wong, L.D. Favro, P.K. Kuo: J. Appl. Phys. **51**, 1152 (1980)
10.8 S. Perkowitz, G. Busse: Opt. Lett. **5**, 228 (1980)
10.9 A. Rosencwaig, G. Busse: Appl. Phys. Lett. **36**, 725 (1980)
10.10 G. Busse, A. Rosencwaig: Appl. Phys. Lett. **36**, 815 (1980)
10.11 H.K. Wickramasinghe, R.C. Bray, V. Jipson, C.F. Quate, J.R. Salcedo: Appl. Phys. Lett. **33**, 923 (1978)
10.12 R.K. Mueller, M. Kaveh, G. Wade: Proc. IEEE **67**, 567 (1979)
      G.T. Herman (ed.): *Image Reconstruction from Projections*, Topics Appl. Phys., Vol. 32 (Springer, Berlin, Heidelberg 1979)
10.13 D. Fournier, F. Lepoutre, A.C. Boccara: J. Physique **44**, Suppl. 10, C6, 479 (1983)
10.14 G. Busse: Opt. and Laser Techn. **12**, 149 (1980)
10.15 E. Brandis, A. Rosencwaig: Appl. Phys. Lett. **37**, 98 (1980)
10.16 A. Rosencwaig: Science **218**, 223 (1982)
10.17 A. Rosencwaig: J. Physique **44**, Suppl. 10, C6, 437 (1983)
10.18 H. Coufal, V. Moller, S. Schneider: Appl. Opt. **21**, 117 (1982)
10.19 H. Coufal, V. Moller, S. Schneider: Appl. Opt. **21**, 2339 (1982)
10.20 G. Busse: J. Physique **44**, Suppl. 10, C6, 427 (1983)

Chapter 11

11.1 M. Goley: Rev. Sci. Instrum. **18**, 347 (1947)
11.2 R.A. Smith, F.E. Jones, R.P. Chasmar: *The Detection and Measurement of Infrared Radiation* (Clarendon, Oxford, 1957)
11.3 G. Busse, K.N. Berget, D. Rogalski: Opt. Commun. **28**, 341 (1979)
11.4 G. Busse, H. Schutz: Infr. Phys. **19**, 313 (1979)
11.5 M.M. Farrow, R.K. Burnham, M. Auzanneau, S.L. Olsen, N. Purdie, E.M. Eyring: Appl. Opt. **17**, 1093 (1978)

11.6 A.P. Onokhov, T.K. Razumova, I.O. Starobogatov: Optiko-mekhan. Prom. (Russian) **1**, 37 (1980)
11.7 A. Schmid, P. Horn, P. Bräunlich: Appl. Phys. Lett. **43**, 151 (1983)
11.8 V.P. Zharov: *Optoacoustic Method in Laser Spectroscopy*, in New Methods of Spectroscopy (Russian) (Nauka, Novosibirsk, 1982) p.126
11.9 G. Busse, E. Basel, A. Pfaller: Appl. Phys. **12**, 387 (1977)
11.10 D. Souihac, A. Gundjian: Appl. Opt. **21**, 1478 (1982)
11.11 M.J. Kavaya, R.T. Menzies, V.P. Oppenheim: IEEE J. QE-19, 1234 (1983)
11.12 G. Busse, R. Thurmaier: Appl. Phys. Lett. **31**, 194 (1977)
11.13 G. Busse, R. Thurmainer, A. Hadni: Infr. Phys. **19**, 125 (1979)
11.14 Yu.A. Gorokhov, S.V. Efimovskii, I.N. Knyazev, V.V. Lobko: Pis'ma Zh. Techn. Fiz. (Russian) **4**, 1481 (1978)
11.15 D.R. Wake, N.M. Amer: Appl. Phys. Lett. **34**, 379 (1979)
11.16 L.J. Thomas, M.J. Kelly, N.M. Amer: Appl. Phys. Lett. **32**, 736 (1978)
11.17 J.E. Allen Jr., W.R. Anderson, D.R. Crosley: Opt. Lett. **1**, 118 (1977)
11.18 W. Zapka, A.C. Tam: Appl. Phys. Lett. **40**, 1015 (1982)
11.19 A.C. Tam, W. Zapka, H. Coufal, B. Sullivan: J. Physique **44**, Suppl. 10, C6, 203 (1983)
11.20 M.J. Adams, G.F. Kirkbright: Spectr. Lett. **9**, 225 (1976)
11.21 G.F. Kirkbright, H.J. Adams: ESN-Europ. Spectrosc. News N14, 22 (1977)
11.22 V.E. Lyamov, V. Madvaliev, R. Shikhlinskay: Akust. Zh. (Russian) **25**, 427 (1979)
11.23 T. Papa, F. Scuderi: Opt. Commun. **41**, 431 (1982)
11.24 G. Rousset, F. Lepoutre, L. Bertrand: J. Appl. Phys. **54**, 2383 (1983)
11.25 C.L. Cesar, H. Vargas, J. Mendes Filho: Appl. Phys. Lett. **43**, 555 (1983)
11.26 R.T. Swimm: Appl. Phys. Lett. **42**, 955 (1983)
11.27 J.F. McClelland, R.N. Knisely: Appl. Phys. Lett. **35**, 585 (1979)
11.28 R.A. McFarlane, L.D. Hess: Appl. Phys. Lett. **36**, 137 (1980)
11.29 K. Wasa, K. Isuboushi, N. Mikoshiba: Jap. J. Appl. Phys. **19**, 1475 (1980)
11.30 J.F. McClelland, R.N. Knisely: Appl. Phys. Lett. **35**, 121 (1979)
11.31 C. Pichon, M. Le Liboux, D. Fournier, A.C. Boccara: Appl. Phys. Lett. **35**, 435 (1979)

Chapter 12

12.1 P.E. Nordal, S.O. Kanstad: Phys. Scr. **20**, 659 (1979)
12.2 P.E. Nordal, S.O. Kanstad: Appl. Phys. Lett. **38**, 386 (1981)
12.3 W.P. Leung, A.C. Tam: Opt. Lett. **9**, 93 (1984)
12.4 R.C.C. Leite, R.S. Moore, J.R. Whinnery: Appl. Phys. Lett. **5**, 141 (1964)
12.5 J.R. Whinnery: Acc. Chem. Res. **7**, 225 (1974)
12.6 J.M. Harris, N.J. Dovichi: Anal. Chem. **52**, 695A (1980)
12.7 N.J. Dovichi, T.G. Nolan, W.A. Weimer: Anal. Chem. **56**, 1700 (1984)
12.8 A.C. Boccara, D. Fournier, J. Badoz: Appl. Phys. Lett. **36**, 130 (1980)
12.9 W.B. Jackson, N.M. Amer, A.C. Boccara, D. Fournier: Appl. Opt. **20**, 1333 (1981)
12.10 S.W. Kizirnis, R.J. Brecha, B.N. Ganguly et al.: Appl. Opt. **23**, 3873 (1984)
12.11 K.V. Reddy: Rev. Sci. Instrum. **54**, 422 (1983)
12.12 I. Glatt, Z. Karny, O. Kafri: Appl. Opt. **23**, 274 (1984)
12.13 J. Stone: J. Opt. Soc. Am. **62**, 317 (1972)
12.14 D.A. Cremers, R.A. Keller: Appl. Opt. **21**, 1654 (1982)
12.15 C.C. Davis: Appl. Phys. Lett. **36**, 515 (1980)
12.16 C.C. Davis, S.J. Petuchowski: Appl. Opt. **20**, 2539 (1981)
12.17 M.J. Pelletier, H.R. Thorsheim, J.M. Harris: Anal. Chem. **54**, 239 (1982)
12.18 R.J.D. Miller, R. Casalegno, K.A. Nelson, M.D. Fayer: Chem. Phys. **72**, 371 (1982)

12.19 H.J. Eichler, P. Günter, D.W. Pohl: *Laser-Induced Dynamic Gratings*, Springer Ser. Opt. Sci., Vol.50 (Springer, Berlin, Heidelberg 1986)
12.20 J.B. Callis: J. Res. NBS **80**A, 413 (1976)
12.21 N.M. Amer: J. Phys. **44**, C6-10, 185 (1983)
12.22 H. Coufal, P. Hefferle: Appl. Phys. A**38**, 213 (1985)

Main Notation

A,B   types of molecules

A_m membrane area

A_i cross section of laser beam

A_d area of detector

C   density of particles (atoms, molecules)

C_m microphone capacity

c   heat capacity

C_p heat capacity at constant pressure

C_v heat capacity at constant volume

d_L diameter of laser beam

d_m thickness of membrane

d   distance between the membrane and the fixed electrode

e   electron charge

e_{31}^p piezoelectric constant

E   internal energy of particle, energy of radiation

E_{abs} absorbed energy in a sample

E_J energy of j-th quantum state

E_L, E_T fractions of absorbed energy transformed to luminescence and heat, respectively

E_{st}, E_p energies of Stokes and pumping waves, respectively

ΔE_{ph} change internal energy (by a mole) corrected with product formation photochemical reactions

$<E>$ average energy absorbed per molecule

F   factor of selectivity

f   modulation frequency

Δf detection bandwidth

f_L laser pulsation rate

g_j degeneracy of j-th quantum state

g_s gain of spontaneous Raman scattering

g_{CD} dissymmetry factor

H_s informative capacity of spectrometer

H_{ga} informative capacity of gas analyzer

h   Planck's constant

h(t)   response of detection sample

i_{sc} current in a shorted circuit

I   intensity of laser radiation

I_s saturation intensity of quantum transition

I_{st} intensity of probe wave at the Stokes frequency

J   radiation power

J_{abs} absorbed power in the sample

j   rotational quantum number

k_i thermal diffusion rate of material i (where i can take the subscripts s, g, and b for solids, gases and substrata respectively)

$k*$ coefficient of gas thermal diffusion of a definite pressure

K   thermal conductivity of material

K_B Boltzmann's constant

L   sample length

ℓ thickness or length of sample

ℓ_{sp} value of spatial resolution

ℓ_α optical absorption depth

ℓ_T	length of thermal diffusion: in gas (ℓ_T^g), sample (ℓ_T^S), substratum (ℓ_T^b) and membrane (ℓ_T^m)	R_{mic}	acoustic resistance of microphone
ℓ_g	length of gas column	R	reflection coefficient
ℓ_T^W	length of thermal diffusion in window or in wall	r	radius
		r_c	spectrophone chamber radius
M_{mem}	mass of the microphone membrane	r_L	radius of laser beam
M_{min}	minimum detectable mass of molecules	S	coefficient of selectivity
		s	integral intensity of line
M_A	mass concentration of aerosols	T	temperature
		T_0	initial temperature
M	gas molecular weight	T^*	membrane tension
m	number of azimuthal mode	$\mathcal{T}$	transmission coefficient
N_v	density of molecules in v-th vibrational state	t_p	duration of laser pulse
		U_{oc}	voltage in an open-circuit
N	number of vines the beam passes through the cell	U_p	polarizing voltage
		U_N	voltage noise
N	number of detectable components	$\bar{u}$	displacement (in a medium)
N_v^o	equilibrium population of the vibrational state V without laser excitation	V	volume
		V_m	volume microphone
		v	index of vibrational level
N_A	Avogadro's number	$\mathcal{V}$	velocity
N_0	density of molecules in normal condition	V_s	sound velocity (in a medium)
		V_L	light velocity
n	index of radial modes	V_T	thermal velocity of molecules
$n_{v,j}$	molecular density in the vibrational-rotational state (V,j)	W	rate of excitation
		w_0	waist of the laser beam (at the level $1/e^2$
n^*	ratio Y_{mem}/Y_p	x,y,z	coordinates
$\tilde{n}$	refractive index	Z_v^{rot}	statistical sum of rotational states
$<n>$	average number of absorbed quanta per molecule		
P	total gas pressure	$Z(\omega)$	acoustic resistance of microphone
P_s	mean value of spontaneous polarization	α	absorption coefficient per unit length
$\mathcal{P}$	amplitude of OA signal	α_E	extinction coefficient
Q	quality factor	α_v	value of volume (bulk) absorption
Q	laser radiation energy flux absorbed per unit time and converted into heat	α_{st}	absorption coefficient at the Stokes frequency
q_{vj}	fraction of molecules in the vibrational state V at the rotational sublevel j	α_s	value of surface absorption
		α_{sc}	scattering coefficient
R	resistance	α_b	background absorption factor

β_v	coefficient of thermal volume expansion	$\Delta\lambda^{or}, \Delta\nu$	spectral resolution
β_L	coefficient of thermal linear expansion	μ	depth of frequency modulation
Γ	half-width of spectral line (at half-height)	ν	radiation frequency
$\gamma = \frac{C_p}{C_v}$	ratio between gas heat capacity at constant pressure C_p and volume C_v	ν_L	laser frequency, luminescence frequency
		ν_0	center of absorption line
γ_B	broadening parameter	ν_p	pumping frequency
γ_s	ratio of surface absorption factors in two windows	ν_{st}	Stokes frequency
		ξ	ratio r_{eff}/r_L
δ	depth of modulation	ρ	density of sample
δ_d	damping decrement	σ	absorption cross-section determined by total populations of vibrational levels
δ_i	relative error of signal		
ε_A	accommodation coefficient	σ_c	electric conductivity
$\varepsilon_{rr}, \varepsilon_{\theta\theta}$	radial and azimuthal elastic tension, respectively	σ_j	cross section of the transition between rotational-vibrational states
ε_d	dielectric constants		
η	quantum yield	σ_p	Poisson's coefficient
η_L	quantum yield luminescence	$\frac{d\sigma}{d\Omega}$	differential cross section of Raman scattering
η_r	quantum yield of radiative processes	τ_{rad}	time of radiative relaxation
η_{nr}	quantum yield of nonradiative processes	τ_T	time of thermal relaxation
		τ_{rot}	time of rotational relaxation
η_{pa}	quantum yield of photoactive processes (photoionization, photodissociation, etc.)	τ_{vib}	time of vibrational relaxation
		τ_{V-T}	time of vibrational-translational relaxation
η_{ph}	quantum yields of photochemical reactions	τ_W	relaxation time of molecules on the cell walls
η_{pv}	quantum yield of photovoltaic effect	τ_m	time constant of microphone
		τ_d	time constant of acoustic detector
$\varkappa_s$	surface absorption factor (determined from the ratio of the absorbed power in the surface layer to the total power)	Φ	energy flux of laser pulse
		Φ_s	saturation flux for absorbing transition
		Φ^*	deformation potential
$\varkappa_{ij}$	absorption (concentration) factor	φ	phase difference
$\varkappa_M$	mass coefficient of absorption	ψ	average energy of translational motion of molecules per unit volume
Λ	free path of molecules		
λ	wavelength of laser light	Y_{mem}	elasticity of membrane
λ_a	wavelength of sound	Y_p, Y_b	elasticity of gas in the pre-membrane and back-membrane volumes
λ_T	wavelength of thermal wave		
Δ	fractional population difference density between two transition levels	Y_{res}	resultant elasticity
		Ω	frequency detuning

ω	modulation frequency
ω_{mic}	resonant frequency of the microphone
$\omega_{k,m,n}$	resonant frequency of OA cell

Subject Index

Absorption 7,19,21,138,171-178

Acoustic impedance 31,52

Acoustic pressure 24-29,46-51,64-67,96-99

Acoustic velocity 45,48-51,96

Aerosol 246-251

Aluminium 216,217

Ammonia 155-157,234-236

Analytical applications 229-264

Atom detection 237

Background signals, correction for 84-107

Balloon-borne measurements 233

Beam

    electron 238

    laser 42,117

    molecular 76

Benzene 197,198,207,208,224

Boundary conditions 25,48-49,63

Boxcar (gate integration technique) 119,197,198

Buffer gas 139,140

Cadmium 238

Calibration 137-145

Calorimetry 9-15

Catalysis 193-196

Cells 17,30,42,51-63,67-71,96-119

C_2H_4 234-236

Chemical reactions 190-196

Chromatographic applications 240,258-264

CO 159,162,170,234,235,261

CO_2 142,150-163,167-169,172,175-178,184,189,234-237,261

Collision 12,13,75-78,182-190

Comparison methods 9-11,72,73,215,281-284

Cross section (absorption) 7,21,22,174,205

Correlation technique 121

Damping of acoustic modes (Q factor) 96-99

Deflection beam 15,299,300

Depth-profiling 210,211,267

Design consideration for optoacoustic systems 115-121,148-153,229-232,265-271

    see also Cells

Detection limits 126-131,155,215,234,235,238-241,244

Diaphragms (in optoacoustics) 90,97,106,166

Dichroism, studies in 218-222

Differential cells 59,93-95,103,244

Dissociation 127,175,176

Doppler-free spectroscopy 163-166

Dye lasers (applications) 181,197-200,204-208, 238,239

Electrostriction 47,87,88

Energy transfer in samples 7,8,12, 13,19-25,74,75,191,192,222

Etalons 137-140,199,230-232,256-258,271

Ethanol 262

Excited states, studies of 166-171

Explosives, detection of 236

Filters, electronic 150

Fluorescence 7,8,13,222-226

Fourier transform 149,283

Frequency
  modulation 27-28,92,150,239,244, 253
  stabilization 287-288

Gas, analysis 16-44,148-196

Gaussian laser beam 42

Grating diffraction 151,152

Hadamard transform 283

Harmonic modulation 27,28,50,63

Heat conduction 25,57,63

Heat source 48

Helmholtz resonance 97,102-104

High-resolution spectroscopy 148-166

History, optoacoustic 2-5

H_2O, studies in 155,156,159-161,184

Impurity detection 229-264

Incoherent sources 16-18,192,222

Intensity of excitation 7,19,24

Interferometer 149

Intracavity absorption 9

Isotope analysis 241-245

Laser characteristics 6-7,125,148-161,234,235,238

Lifetimes 11-13,21,182-190

Life profile 28,156

Liquids, study of 45-56,197-201,205-208,224-226,237-239

Low temperatures 107-109,176,200,201, 239

Magnetic fields 110,111,162

Metals, study of 215-218

Microphone
  design of 29-32,51,52,69,70
  electronics for 32-39
  noise 33,128-131

Microscopy 265-284

Minimum detectable concentration
  in gases 234-235
  in liquids 238

Mode acoustic 96-102

Molecular spectroscopy, study of 154-182

Multicomponent mixtures, analysis of 251-264

Multipass arrangement for excitation 116-118

Multiphoton absorption 173-176,205-207

Multiplex excitation 149,150,263,283

NH_3, detection of 234,235

NO, study of 155,162,170,171,194,233-236

NO_2, study of 155,191,192,234-236

Noise 33,128-131,135

Nondispersive infrared systems 16-17

Nonlinear effects 126,139,171-182, 205-208,280

Nonlinear spectroscopy 163-165

O_2, study of 184-185

Optical systems 116-119

Optoacoustic 2

Optoacoustic cells *see* Cells

Phase angle measurements 186

Phase transition studies 297,298

Photoacoustic (optoacoustic with indirect detection) 61-70,210-218

Photoactive media 222-228

Photochemical studies 190-193,226, 227

Photolysis 196

Photophone 2

Physical description of opto-acoustic 18-19,45-48,61-64

Piezoelectric detection 51-60

Polarization modulation 218-222

Pollution detection 223-237

Powder, study of 203-205

Preamplifiers 32-39

Propagation of laser beams *see* Gaussian laser beam

Q, quality factor *see* Damping of acoustic modes

Quantum yields 222-228

Quenching 226

Radiative lifetime 11-13

Raman spectroscopy 179-182,207,208

Rate equations 19-24

Refraction methods 14,15,299,300

Relaxation time *see* Lifetimes

Resolution

 in spectroscopy 125,151-166

 in microscopy 266-279

Sample chamber design

 for gas samples 30-32,42

 for liquid samples 52,53,55

 for solid samples 61,67-69

Saturation, optical 22-24,171-173

Scattering of radiation within sample cell 87

Selectivity 251-264

Sensitivity 122-125

 see also Detection limits

Signal-to-noise ratio 39,40

Solids

 optoacoustic technique 56-71

 study of 201-205

Stark spectroscopy 162,163

Surface loss 99

Surface studies 208-218,240,241

Temperature distribution 63

Thermal

 conductivity 25

 diffusivity 25

 lens effect 15

 time 25

 wave 266-269

Thin layer chromatography 240

Three-dimensional microscopy 279-281

Trace analysis 233-264

Tunable laser sources 154-162

Two-dimensional microscopy 271-279

Uranium hexafluoride 158-160

Vapor detection and identification 236

Velocity, sound 48-51,99

Vibrational transitions 19-24

Viscosity loss 99

Wavelength modulation *see* Frequency, modulation

Window materials 85

Zeeman effect *see* Magnetic fields

Springer Series in Optical Sciences

Editorial Board: J.M. Enoch D.L. MacAdam A.L. Schawlow K. Shimoda T. Tamir

1   **Solid-State Laser Engineering**
    By W. Koechner

2   **Table of Laser Lines in Gases and Vapors**
    3rd Edition
    By R. Beck, W. Englisch, and K. Gürs

3   **Tunable Lasers and Applications**
    Editors: A. Mooradian, T. Jaeger, and
    P. Stokseth

4   **Nonlinear Laser Spectroscopy** 2nd Edition
    By V. S. Letokhov and V. P. Chebotayev

5   **Optics and Lasers** Including Fibers and
    Optical Waveguides 3rd Edition
    By M. Young

6   **Photoelectron Statistics**
    With Applications to Spectroscopy and
    Optical Communication By B. Saleh

7   **Laser Spectroscopy III**
    Editors: J. L. Hall and J. L. Carlsten

8   **Frontiers in Visual Science**
    Editors: S. J. Cool and E. J. Smith III

9   **High-Power Lasers and Applications**
    2nd Printing
    Editors: K.-L. Kompa and H. Walther

10 **Detection of Optical and Infrared Radiation**
    2nd Printing   By R. H. Kingston

11 **Matrix Theory of Photoelasticity**
    By P. S. Theocaris and E. E. Gdoutos

12 **The Monte Carlo Method in Atmospheric Optics**
    By G. I. Marchuk, G. A. Mikhailov,
    M. A. Nazaraliev, R. A. Darbinian, B. A. Kargin,
    and B. S. Elepov

13 **Physiological Optics**
    By Y. Le Grand and S. G. El Hage

14 **Laser Crystals** Physics and Properties
    By A. A. Kaminskii

15 **X-Ray Spectroscopy** By B. K. Agarwal

16 **Holographic Interferometry**
    From the Scope of Deformation Analysis of
    Opaque Bodies
    By W. Schumann and M. Dubas

17 **Nonlinear Optics of Free Atoms and Molecules**
    By D. C. Hanna, M. A. Yuratich, D. Cotter

18 **Holography in Medicine and Biology**
    Editor: G. von Bally

19 **Color Theory and Its Application in Art and
    Design** 2nd Edition By G. A. Agoston

20 **Interferometry by Holography**
    By Yu. I. Ostrovsky, M. M. Butusov,
    G. V. Ostrovskaya

21 **Laser Spectroscopy IV**
    Editors: H. Walther, K. W. Rothe

22 **Lasers in Photomedicine and Photobiology**
    Editors: R. Pratesi and C. A. Sacchi

23 **Vertebrate Photoreceptor Optics**
    Editors: J. M. Enoch and F. L. Tobey, Jr.

24 **Optical Fiber Systems and Their Components**
    An Introduction By A. B. Sharma,
    S. J. Halme, and M. M. Butusov

25 **High Peak Power Nd : Glass Laser Systems**
    By D. C. Brown

26 **Lasers and Applications**
    Editors: W. O. N. Guimaraes, C. T. Lin,
    and A. Mooradian

27 **Color Measurement** Theme and Variations
    2nd Edition By D. L. MacAdam

28 **Modular Optical Design**
    By O. N. Stavroudis

29 **Inverse Problems of Lidar Sensing of the
    Atmosphere** By V. E. Zuev and I. E. Naats

30 **Laser Spectroscopy V**
    Editors: A. R. W. McKellar, T. Oka, and
    B. P. Stoicheff

31 **Optics in Biomedical Sciences**
    Editors: G. von Bally and P. Greguss

32 **Fiber-Optic Rotation Sensors**
    and Related Technologies
    Editors: S. Ezekiel and H. J. Arditty

33 **Integrated Optics: Theory and Technology**
    2nd Edition By R. G. Hunsperger 2nd Printing

34 **The High-Power Iodine Laser**
    By G. Brederlow, E. Fill, and K. J. Witte

35 **Engineering Optics** By K. Iizuka

36 **Transmission Electron Microscopy** Physics of
    Image Formation and Microanalysis
    By L. Reimer

37 **Opto-Acoustic Molecular Spectroscopy**
    By V. S. Letokhov and V. P. Zharov

38 **Photon Correlation Techniques**
    Editor: E. O. Schulz-DuBois

39 **Optical and Laser Remote Sensing**
    Editors: D. K. Killinger and A. Mooradian

40 **Laser Spectroscopy VI**
    Editors: H. P. Weber and W. Lüthy

41 **Advances in Diagnostic Visual Optics**
    Editors: G. M. Breinin and I. M. Siegel